Python 深度学习算法实践(影印版)
Hands-On Deep Learning
Algorithms with Python

Sudharsan Ravichandiran 著

图书在版编目(CIP)数据

Python 深度学习算法实践：影印版：英文/(印)苏达桑·拉维尚迪兰著. —南京：东南大学出版社，2020.8

书名原文：Hands-On Deep Learning Algorithms with Python

ISBN 978‑7‑5641‑8969‑3

Ⅰ.①P… Ⅱ.①苏… Ⅲ.①软件工具-程序设计-英文 ②机器学习-算法-英文 Ⅳ.①TP311.561 ②TP181

中国版本图书馆 CIP 数据核字(2020)第 116739 号

图字：10‑2020‑157 号

© 2019 by PACKT Publishing Ltd.

Reprint of the English Edition, jointly published by O'Reilly Media, Inc. and Southeast University Press, 2020. Authorized reprint of the original English edition, 2020 PACKT Publishing Ltd, the owner of all rights to publish and sell the same.

All rights reserved including the rights of reproduction in whole or in part in any form.

英文原版由 PACKT Publishing Ltd 出版 2019。

英文影印版由东南大学出版社出版 2020。此影印版的出版和销售得到出版权和销售权的所有者 —— PACKT Publishing Ltd 的许可。

版权所有，未得书面许可，本书的任何部分和全部不得以任何形式重制。

Python 深度学习算法实践(影印版)

出版发行：东南大学出版社
地　　址：南京四牌楼 2 号　　邮编：210096
出 版 人：江建中
网　　址：http://www.seupress.com
电子邮件：press@seupress.com
印　　刷：常州市武进第三印刷有限公司
开　　本：787 毫米×980 毫米　　16 开本
印　　张：32
字　　数：627 千字
版　　次：2020 年 8 月第 1 版
印　　次：2020 年 8 月第 1 次印刷
书　　号：ISBN 978‑7‑5641‑8969‑3
定　　价：119.00 元

本社图书若有印装质量问题，请直接与营销部联系。电话(传真)：025‑83791830

To my adorable mom, Kasthuri, and to my beloved dad, Ravichandiran.

- Sudharsan Ravichandiran

Packt.com

Subscribe to our online digital library for full access to over 7,000 books and videos, as well as industry leading tools to help you plan your personal development and advance your career. For more information, please visit our website.

Why subscribe?

- Spend less time learning and more time coding with practical eBooks and Videos from over 4,000 industry professionals

- Improve your learning with Skill Plans built especially for you

- Get a free eBook or video every month

- Fully searchable for easy access to vital information

- Copy and paste, print, and bookmark content

Did you know that Packt offers eBook versions of every book published, with PDF and ePub files available? You can upgrade to the eBook version at www.packt.com and as a print book customer, you are entitled to a discount on the eBook copy. Get in touch with us at customercare@packtpub.com for more details.

At www.packt.com, you can also read a collection of free technical articles, sign up for a range of free newsletters, and receive exclusive discounts and offers on Packt books and eBooks.

Contributors

About the author

Sudharsan Ravichandiran is a data scientist, researcher, artificial intelligence enthusiast, and YouTuber (search for Sudharsan reinforcement learning). He completed his bachelor's in information technology at Anna University. His area of research focuses on practical implementations of deep learning and reinforcement learning, which includes natural language processing and computer vision. He is an open source contributor and loves answering questions on Stack Overflow. He also authored a best-seller, *Hands-On Reinforcement Learning with Python*, published by Packt Publishing.

I would like to thank my most amazing parents and my brother, Karthikeyan, for inspiring and motivating me. I am forever grateful to my Sœur who always has my back. I can't thank enough my editors, Unnati and Naveen for their hard work and dedication. Without their support, it would have been impossible to complete this book.

About the reviewers

Sujit S Ahirrao is a computer vision and machine learning researcher and software developer who's mostly experienced in image processing and deep learning. He graduated in electronics and telecommunication from University of Pune. He made his way into the field of artificial intelligence through start-ups and has been a part of in-house R&D teams at well-established firms. He pursues his interest in contributing to education, healthcare, and scientific research communities with his growing skills and experience.

Bharath Kumar Varma currently works as a lead data scientist at an Indian tech start-up called MTW Labs, with clients in India and North America. His primary areas of interest are deep learning, NLP, and computer vision. He is a seasoned architect focusing on machine learning projects, vision and text analytics solutions, and is an active member of the start-up ecosystem. He holds a M.Tech degree from IIT Hyderabad, with a specialization in data science, and is certified in various other technological and banking-related certifications. Aside from his work, he actively participates in teaching and mentoring data science enthusiasts and contributes to the community by networking and working with fellow enthusiasts in groups.

Doug Ortiz is an experienced enterprise cloud, big data, data analytics, and solutions architect who has architected, designed, developed, re-engineered, and integrated enterprise solutions. His other areas of expertise include Amazon Web Services, Azure, Google Cloud, business intelligence, Hadoop, Spark, NoSQL databases, and SharePoint. He is also the founder of Illustris, LLC.

Packt is searching for authors like you

If you're interested in becoming an author for Packt, please visit authors.packtpub.com and apply today. We have worked with thousands of developers and tech professionals, just like you, to help them share their insight with the global tech community. You can make a general application, apply for a specific hot topic that we are recruiting an author for, or submit your own idea.

Table of Contents

Section 2: Fundamental Deep Learning Algorithms

Preface

Deep learning is one of the most popular domains in the **artificial intelligence** (**AI**) space, which allows you to develop multi-layered models of varying complexities. This book introduces you to popular deep learning algorithms—from basic to advanced—and shows you how to implement them from scratch using TensorFlow. Throughout the book, you'll gain insights into each algorithm, the mathematical principles behind it, and how to implement them in the best possible manner.

The book starts by explaining how you can build your own neural network, followed by introducing you to TensorFlow; the powerful Python-based library for machine learning and deep learning. Next, you will get up to speed with gradient descent variants, such as NAG, AMSGrad, AdaDelta, Adam, Nadam, and more. The book will then provide you with insights into the working of **Recurrent Neural Networks** (**RNNs**) and **Long Short-Term Memory** (**LSTM**) and how to generate song lyrics with RNN. Next, you will master the math for convolutional and Capsule networks, widely used for image recognition tasks. Towards the concluding chapters, you will learn how machines understand the semantics of words and documents using CBOW, skip-gram, and PV-DM. Then you will explore various GANs such as InfoGAN and LSGAN and also autoencoders such as contractive autoencoders, VAE, and so on.

By the end of this book, you will be equipped with the skills needed to implement deep learning in your own projects.

Who this book is for

If you are a machine learning engineer, data scientist, AI developer, or anyone who wants to focus on neural networks and deep learning, this book is for you. Those who are completely new to deep learning, but have some experience in machine learning and Python programming, will also find this book helpful.

What this book covers

Chapter 1, *Introduction to Deep Learning*, explains the fundamentals of deep learning and helps us to understand what artificial neural networks are and how they learn. We will also learn to build our first artificial neural network from scratch.

Chapter 2, *Getting to Know TensorFlow*, helps us to understand one of the most powerful and popular deep learning libraries called TensorFlow. You will understand several important functionalities of TensorFlow and how to build neural networks using TensorFlow to perform handwritten digits classification.

Chapter 3, *Gradient Descent and Its Variants*, provides an in-depth understanding of gradient descent algorithm. We will explore several variants of gradient descent algorithm such as SGD, Adagrad, ADAM, Adadelta, Nadam, and many more and learn how to implement them from scratch.

Chapter 4, *Generating Song Lyrics Using RNN*, describes how an RNN is used to model sequential datasets and how it remembers the previous input. We will begin by getting a basic understanding of RNN then we will deep dive into its math. Next, we will learn how to implement RNN in TensorFlow for generating song lyrics.

Chapter 5, *Improvements to the RNN*, begins by exploring LSTM and how exactly LSTM overcomes the shortcomings of RNN. Later, we will learn about GRU cell and how bidirectional RNN and deep RNN work. At the end of the chapter, we will learn how to perform language translation using seq2seq model.

Chapter 6, *Demystifying Convolutional Networks*, helps us to master how convolutional neural networks work. We will explore how forward and backpropagation of CNNs work mathematically. We will also learn about various architectures of CNN and Capsule networks and implement them in TensorFlow.

Chapter 7, *Learning Text Representations*, covers the state-of-the-art text representation learning algorithm known as word2vec. We will explore how different types of word2vec models such as CBOW and skip-gram work mathematically. We will also learn how to visualize the word embeddings using TensorBoard. Later we will learn about doc2vec, skip-thoughts and quick-thoughts models for learning the sentence representations.

Chapter 8, *Generating Images Using GANs*, helps us to understand one of the most popular generative algorithms called GAN. We will learn how to implement GAN in TensorFlow to generate images. We will also explore different types of GANs such as LSGAN and WGAN.

Chapter 9, *Learning More about GANs*, uncovers various interesting different types of GANs. First, we will learn about CGAN, which conditions the generator and discriminator. Then we see how to implement InfoGAN in TensorFlow. Moving on, we will learn to convert photos to paintings using CycleGAN and how to convert text descriptions to photos using StackGANs.

Chapter 10, *Reconstructing Inputs Using Autoencoders*, describes how autoencoders learn to reconstruct the input. We will explore and learn to implement different types of autoencoders such as convolutional autoencoders, sparse autoencoders, contractive autoencoders, variational autoencoders, and more in TensorFlow.

Chapter 11, *Exploring Few-Shot Learning Algorithms*, describes how to build models to learn from a few data points. We will learn what is few-shot learning and explore popular few-shot learning algorithms such as siamese, prototypical, relation, and matching networks.

To get the most out of this book

Those who are completely new to deep learning, but who have some experience in machine learning and Python programming, will find this book helpful.

Download the example code files

You can download the example code files for this book from your account at www.packt.com. If you purchased this book elsewhere, you can visit www.packt.com/support and register to have the files emailed directly to you.

You can download the code files by following these steps:

1. Log in or register at www.packt.com.
2. Select the **SUPPORT** tab.
3. Click on **Code Downloads & Errata**.
4. Enter the name of the book in the **Search** box and follow the onscreen instructions.

Once the file is downloaded, please make sure that you unzip or extract the folder using the latest version of:

- WinRAR/7-Zip for Windows
- Zipeg/iZip/UnRarX for Mac
- 7-Zip/PeaZip for Linux

The code bundle for the book is also hosted on GitHub at https://github.com/PacktPublishing/Hands-On-Deep-Learning-Algorithms-with-Python. In case there's an update to the code, it will be updated on the existing GitHub repository.

We also have other code bundles from our rich catalog of books and videos available at https://github.com/PacktPublishing/. Check them out!

Download the color images

We also provide a PDF file that has color images of the screenshots/diagrams used in this book. You can download it here: http://www.packtpub.com/sites/default/files/downloads/9781789344158_ColorImages.pdf.

Conventions used

There are a number of text conventions used throughout this book.

CodeInText: Indicates code words in text, database table names, folder names, filenames, file extensions, pathnames, dummy URLs, user input, and Twitter handles. Here is an example: "Compute J_plus and J_minus."

A block of code is set as follows:

```
J_plus = forward_prop(x, weights_plus)
J_minus = forward_prop(x, weights_minus)
```

Any command-line input or output is written as follows:

```
tensorboard --logdir=graphs --port=8000
```

Bold: Indicates a new term, an important word, or words that you see on screen. For example, words in menus or dialog boxes appear in the text like this. Here is an example: "Any layer between the input layer and the output layer is called a **hidden layer**."

Warnings or important notes appear like this.

Tips and tricks appear like this.

Get in touch

Feedback from our readers is always welcome.

General feedback: If you have questions about any aspect of this book, mention the book title in the subject of your message and email us at customercare@packtpub.com.

Errata: Although we have taken every care to ensure the accuracy of our content, mistakes do happen. If you have found a mistake in this book, we would be grateful if you would report this to us. Please visit www.packt.com/submit-errata, selecting your book, clicking on the Errata Submission Form link, and entering the details.

Piracy: If you come across any illegal copies of our works in any form on the internet, we would be grateful if you would provide us with the location address or website name. Please contact us at copyright@packt.com with a link to the material.

If you are interested in becoming an author: If there is a topic that you have expertise in, and you are interested in either writing or contributing to a book, please visit authors.packtpub.com.

Reviews

Please leave a review. Once you have read and used this book, why not leave a review on the site that you purchased it from? Potential readers can then see and use your unbiased opinion to make purchase decisions, we at Packt can understand what you think about our products, and our authors can see your feedback on their book. Thank you!

For more information about Packt, please visit packt.com.

Section 1: Getting Started with Deep Learning

1

In this section, we will get ourselves familiarized with deep learning and will understand the fundamental deep learning concepts. We will also learn about powerful deep learning framework called TensorFlow, and set TensorFlow up for all of our future deep learning tasks.

The following chapters are included in this section:

- Chapter 1, *Introduction to Deep Learning*
- Chapter 2, *Getting to Know TensorFlow*

Introduction to Deep Learning

1

Deep learning is a subset of machine learning inspired by the neural networks in the human brain. It has been around for a decade, but the reason it is so popular right now is due to the computational advancements and availability of the huge volume of data. With a huge volume of data, deep learning algorithms outperform classic machine learning. It has already been transfiguring and extensively used in several interdisciplinary scientific fields such as computer vision, **natural language processing** (**NLP**), speech recognition, and many others.

In this chapter, we will learn about the following topics:

- Fundamental concepts of deep learning
- Biological and artificial neurons
- Artificial neural network and its layers
- Activation functions
- Forward and backward propagation in ANN
- Gradient checking algorithm
- Building an artificial neural network from scratch

What is deep learning?

Deep learning is just a modern name for artificial neural networks with many layers. What is *deep* in deep learning though? It is basically due to the structure of the **artificial neural network** (**ANN**). ANN consists of some *n* number of layers to perform any computation. We can build an ANN with several layers where each layer is responsible for learning the intricate patterns in the data. Due to the computational advancements, we can build a network even with 100s or 1000s of layers deep. Since the ANN uses deep layers to perform learning we call it as deep learning and when ANN uses deep layers to learn we call it as a deep network. We have learned that deep learning is a subset of machine learning. How does deep learning differ from machine learning? What makes deep learning so special and popular?

The success of machine learning lies in the right set of features. Feature engineering plays a crucial role in machine learning. If we handcraft the right set of features to predict a certain outcome, then the machine learning algorithms can perform well, but finding and engineering the right set of features is not an easy task.

With deep learning, we don't have to handcraft such features. Since deep ANNs employ several layers, it learns the complex intrinsic features and multi-level abstract representation of data by itself. Let's explore this a bit with an analogy.

Let's suppose we want to perform an image classification task. Say, we are learning to recognize whether an image contains a dog or not. With machine learning, we need to handcraft features that help the model to understand whether the image contains a dog. We send these handcrafted features as inputs to machine learning algorithms which then learn a mapping between the features and the label (dog). But extracting features from an image is a tedious task. With deep learning, we just need to feed in a bunch of images to the deep neural networks, and it will automatically act as a feature extractor by learning the right set of features. As we have learned, ANN uses multiple layers; in the first layer, it will learn the basic features of the image that characterize the dog, say, the body structure of the dog, and, in the succeeding layers, it will learn the complex features. Once it learns the right set of features, it will look for the presence of such features in the image. If those features are present then it says that the given image contains a dog. Thus, unlike machine learning, with DL, we don't have to manually engineer the features, instead, the network will itself learns the correct set of features required for the task.

Due to this interesting aspect of deep learning, it is substantially used in unstructured datasets where extracting features are difficult, such as speech recognition, text classification, and many more. When we have a fair amount of huge datasets, deep learning algorithms are good at extracting features and mapping the extracted features to their labels. Having said that, deep learning is not just throwing a bunch of data points to a deep network and getting results. It's not that simple either. We would have numerous hyperparameters that act as a tuning knob to obtain better results which we will explore in the upcoming sections.

Although deep learning performs better than conventional machine learning models, it is not recommended to use DL for smaller datasets. When we don't have enough data points or the data is very simple, then the deep learning algorithms can easily overfit to the training dataset and fail to generalize well on the unseen dataset. Thus, we should apply deep learning only when we have a significant amount of data points.

The applications of deep learning are numerous and almost everywhere. Some of the interesting applications include automatically generating captions to the image, adding sound to the silent movies, converting black-and-white images to colored images, generating text, and many more. Google's language translate, Netflix, Amazon, and Spotify's recommendations engines, and self-driving cars are some of the applications powered by deep learning. There is no doubt that deep learning is a disruptive technology and has achieved tremendous technological advancement in the past few years.

In this book, we will learn from the basic deep learning algorithms as to the state of the algorithms by building some of the interesting applications of deep learning from scratch, which includes image recognition, generating song lyrics, predicting bitcoin prices, generating realistic artificial images, converting photographs to paintings, and many more. Excited already? Let's get started!

Biological and artificial neurons

Before going ahead, first, we will explore what are neurons and how neurons in our brain actually work, and then we will learn about artificial neurons.

A **neuron** can be defined as the basic computational unit of the human brain. Neurons are the fundamental units of our brain and nervous system. Our brain encompasses approximately 100 billion neurons. Each and every neuron is connected to one another through a structure called a **synapse**, which is accountable for receiving input from the external environment, sensory organs for sending motor instructions to our muscles, and for performing other activities.

A neuron can also receive inputs from the other neurons through a branchlike structure called a **dendrite**. These inputs are strengthened or weakened; that is, they are weighted according to their importance and then they are summed together in the cell body called the **soma**. From the cell body, these summed inputs are processed and move through the **axons** and are sent to the other neurons.

The basic single biological neuron is shown in the following diagram:

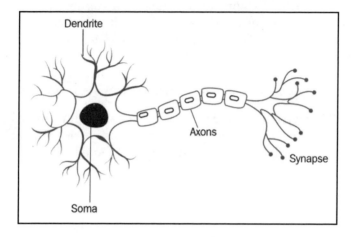

Now, let's see how artificial neurons work. Let's suppose we have three inputs x_1, x_2, and x_3, to predict output y. These inputs are multiplied by weights w_1, w_2, and w_3 and are summed together as follows:

$$x_1 . w_1 + x_2 . w_2 + x_3 . w_3$$

But why are we multiplying these inputs by weights? Because all of the inputs are not equally important in calculating the output y. Let's say that x_2 is more important in calculating the output compared to the other two inputs. Then, we assign a higher value to w_2 than the other two weights. So, upon multiplying weights with inputs, x_2 will have a higher value than the other two inputs. In simple terms, weights are used for strengthening the inputs. After multiplying inputs with the weights, we sum them together and we add a value called bias, b:

$$z = (x_1 . w_1 + x_2 . w_2 + x_3 . w_3) + b$$

If you look at the preceding equation closely, it may look familiar? Doesn't z look like the equation of linear regression? Isn't it just the equation of a straight line? We know that the equation of a straight line is given as:

$$z = mx + b$$

Here m is the weights (coefficients), x is the input, and b is the bias (intercept).

Well, yes. Then, what is the difference between neurons and linear regression? In neurons, we introduce non-linearity to the result, z, by applying a function $f(\cdot)$ called the **activation** or **transfer function**. Thus, our output becomes:

$$y = f(z)$$

A single artificial neuron is shown in the following diagram:

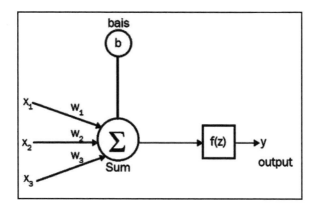

So, a neuron takes the input, x, multiples it by weights, w, and adds bias, b, forms z, and then we apply the activation function on z and get the output, y.

ANN and its layers

While neurons are really cool, we cannot just use a single neuron to perform complex tasks. This is the reason our brain has billions of neurons, stacked in layers, forming a network. Similarly, artificial neurons are arranged in layers. Each and every layer will be connected in such a way that information is passed from one layer to another.

A typical ANN consists of the following layers:

- Input layer
- Hidden layer
- Output layer

Each layer has a collection of neurons, and the neurons in one layer interact with all the neurons in the other layers. However, neurons in the same layer will not interact with one another. This is simply because neurons from the adjacent layers have connections or edges between them; however, neurons in the same layer do not have any connections. We use the term **nodes** or **units** to represent the neurons in the artificial neural network.

A typical ANN is shown in the following diagram:

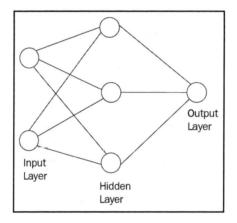

Input layer

The **input layer** is where we feed input to the network. The number of neurons in the input layer is the number of inputs we feed to the network. Each input will have some influence on predicting the output. However, no computation is performed in the input layer; it is just used for passing information from the outside world to the network.

Hidden layer

Any layer between the input layer and the output layer is called a **hidden layer**. It processes the input received from the input layer. The hidden layer is responsible for deriving complex relationships between input and output. That is, the hidden layer identifies the pattern in the dataset. It is majorly responsible for learning the data representation and for extracting the features.

There can be any number of hidden layers; however, we have to choose a number of hidden layers according to our use case. For a very simple problem, we can just use one hidden layer, but while performing complex tasks such as image recognition, we use many hidden layers, where each layer is responsible for extracting important features. The network is called a **deep neural network** when we have many hidden layers.

Output layer

After processing the input, the hidden layer sends its result to the output layer. As the name suggests, the output layer emits the output. The number of neurons in the output layer is based on the type of problem we want our network to solve.

If it is a binary classification, then the number of neurons in the output layer is one that tells us which class the input belongs to. If it is a multi-class classification say, with five classes, and if we want to get the probability of each class as an output, then the number of neurons in the output layer is five, each emitting the probability. If it is a regression problem, then we have one neuron in the output layer.

Exploring activation functions

An **activation function**, also known as a **transfer function**, plays a vital role in neural networks. It is used to introduce non-linearity in neural networks. As we learned before, we apply the activation function to the input, which is multiplied by weights and added to the bias, that is, $f(z)$, where $z = (input * weights) + bias$ and $f(\cdot)$ is the activation function. If we do not apply the activation function, then a neuron simply resembles the linear regression. The aim of the activation function is to introduce a non-linear transformation to learn the complex underlying patterns in the data.

Now let's look at some of the interesting commonly used activation functions.

The sigmoid function

The **sigmoid function** is one of the most commonly used activation functions. It scales the value between 0 and 1. The sigmoid function can be defined as follows:

$$f(x) = \frac{1}{1 + e^{-x}}$$

It is an S-shaped curve shown as follows:

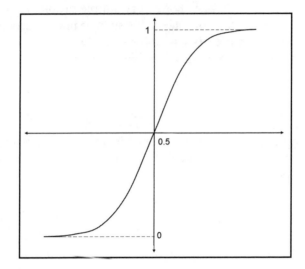

It is differentiable, meaning that we can find the slope of the curve at any two points. It is **monotonic**, which implies it is either entirely non-increasing or non-decreasing. The sigmoid function is also known as a **logistic** function. As we know that probability lies between 0 and 1 and since the sigmoid function squashes the value between 0 and 1, it is used for predicting the probability of output.

The sigmoid function can be defined in Python as follows:

```
def sigmoid(x):

    return 1/ (1+np.exp(-x))
```

The tanh function

A **hyperbolic tangent (tanh)** function outputs the value between -1 to +1 and is expressed as follows:

$$f(x) = \frac{1 - e^{-2x}}{1 + e^{-2x}}$$

It also resembles the S-shaped curve. Unlike a sigmoid function which is centered on 0.5, the tanh function is 0 centered, as shown in the following diagram:

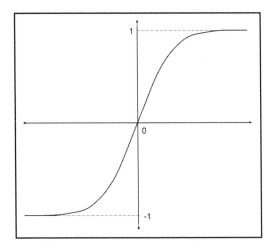

Similar to the sigmoid function, it is also a differentiable and monotonic function. The tanh function is implemented as follows:

```
def tanh(x):
    numerator = 1-np.exp(-2*x)
    denominator = 1+np.exp(-2*x)

    return numerator/denominator
```

The Rectified Linear Unit function

The **Rectified Linear Unit (ReLU)** function is another one of the most commonly used activation functions. It outputs a value from o to infinity. It is basically a **piecewise** function and can be expressed as follows:

$$f(x) = \begin{cases} 0 & for\ x < 0 \\ x & for\ x \geq 0 \end{cases}$$

That is, $f(x)$ returns zero when the value of x is less than zero and $f(x)$ returns x when the value of x is greater than or equal to zero. It can also be expressed as follows:

$$f(x) = max(0, x)$$

The ReLU function is shown in the following figure:

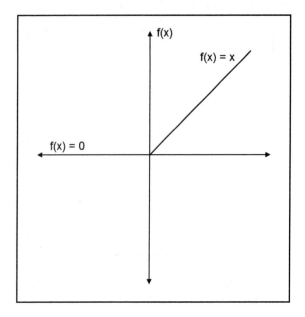

As we can see in the preceding diagram, when we feed any negative input to the ReLU function, it converts it to zero. The snag for being zero for all negative values is a problem called **dying ReLU**, and a neuron is said to be dead if it always outputs zero. A ReLU function can be implemented as follows:

```
def ReLU(x):
    if x<0:
        return 0
    else:
        return x
```

The leaky ReLU function

Leaky ReLU is a variant of the ReLU function that solves the dying ReLU problem. Instead of converting every negative input to zero, it has a small slope for a negative value as shown:

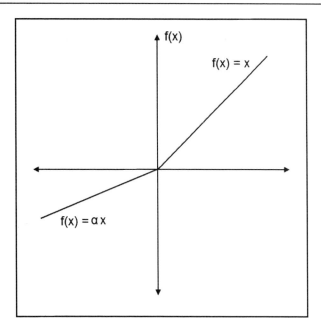

Leaky ReLU can be expressed as follows:

$$f(x) = \begin{cases} \alpha x & for\ x < 0 \\ x & for\ x \geq 0 \end{cases}$$

The value of α is typically set to 0.01. The leaky ReLU function is implemented as follows:

```
def leakyReLU(x,alpha=0.01):
    if x<0:
        return (alpha*x)
    else:
        return x
```

Instead of setting some default values to α, we can send them as a parameter to a neural network and make the network learn the optimal value of α. Such an activation function can be termed as a **Parametric ReLU** function. We can also set the value of α to some random value and it is called as **Randomized ReLU** function.

The Exponential linear unit function

Exponential linear unit (**ELU**), like Leaky ReLU, has a small slope for negative values. But instead of having a straight line, it has a log curve, as shown in the following diagram:

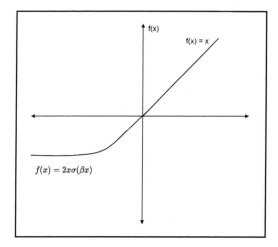

It can be expressed as follows:

$$f(x) = \begin{cases} \alpha(e^x - 1) & for\ x < 0 \\ x & for\ x \geq 0 \end{cases}$$

The `ELU` function is implemented in Python as follows:

```
def ELU(x,alpha=0.01):
    if x<0:
        return ((alpha*(np.exp(x)-1)))
    else:
        return x
```

The Swish function

The **Swish** function is a recently introduced activation function by Google. Unlike other activation functions, which are monotonic, Swish is a non-monotonic function, which means it is neither always non-increasing nor non-decreasing. It provides better performance than ReLU. It is simple and can be expressed as follows:

$$f(x) = x\sigma(x)$$

Here, $\sigma(x)$ is the sigmoid function. The Swish function is shown in the following diagram:

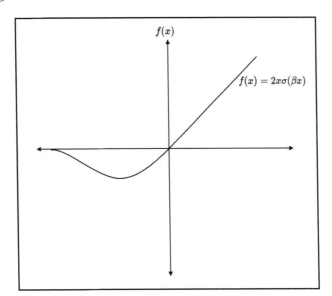

We can also reparametrize the Swish function and express it as follows:

$$f(x) = 2x\sigma(\beta x)$$

When the value of β is 0, then we get the identity function $f(x) = x$.

It becomes a linear function and, when the value of β tends to infinity, then $f(x)$ becomes $2max(0, x)$, which is basically the ReLU function multiplied by some constant value. So, the value of β acts as a good interpolation between a linear and a nonlinear function. The swish function can be implemented as shown:

```
def swish(x,beta):
    return 2*x*sigmoid(beta*x)
```

The softmax function

The **softmax function** is basically the generalization of the sigmoid function. It is usually applied to the final layer of the network and while performing multi-class classification tasks. It gives the probabilities of each class for being output and thus, the sum of softmax values will always equal 1.

It can be represented as follows:

$$f(x_i) = \frac{e^{x_i}}{\sum_j e^{x_j}}$$

As shown in the following diagram, the softmax function converts their inputs to probabilities:

$$\begin{bmatrix} 0.5 \\ 1.3 \\ 1.1 \end{bmatrix} \rightarrow \boxed{\text{Softmax}} \rightarrow \begin{bmatrix} 0.198 \\ 1.440 \\ 1.360 \end{bmatrix}$$

The `softmax` function can be implemented in Python as follows:

```
def softmax(x):
    return np.exp(x) / np.exp(x).sum(axis=0)
```

Forward propagation in ANN

In this section, we will see how an ANN learns where neurons are stacked up in layers. The number of layers in a network is equal to the number of hidden layers plus the number of output layers. We don't take the input layer into account when calculating the number of layers in a network. Consider a two-layer neural network with one input layer, x, one hidden layer, h, and one output layer, y, as shown in the following diagram:

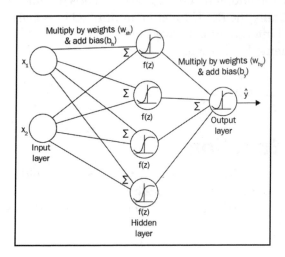

Let's consider we have two inputs, x_1 and x_2, and we have to predict the output, $\hat{y}$. Since we have two inputs, the number of neurons in the input layer will be two. We set the number of neurons in the hidden layer to four, and, the number of neurons in the output layer to one. Now, the inputs will be multiplied by weights, and then we add bias and propagate the resultant value to the hidden layer where the activation function will be applied.

Before that, we need to initialize the weight matrix. In the real world, we don't know which input is more important than the other so that we can weight them and compute the output. Therefore, we will randomly initialize weights and the bias value. The weight and the bias value between the input to the hidden layer are represented by W_{xh} and b_h, respectively. What about the dimensions of the weight matrix? The dimensions of the weight matrix must be *number of neurons in the current layer x number of neurons in the next layer*. Why is that?

Because it is a basic matrix multiplication rule. To multiply any two matrices, AB, the number of columns in matrix A must be equal to the number of rows in matrix B. So, the dimension of the weight matrix, W_{xh}, should be *number of neurons in the input layer x number of neurons in the hidden layer*, that is, 2 x 4:

$$z_1 = XW_{xh} + b_h$$

The preceding equation represents, $z_1 = \text{input} \times \text{weights} + \text{bias}$. Now, this is passed to the hidden layer. In the hidden layer, we apply an activation function to z_1. Let's use the sigmoid σ activation function. Then, we can write:

$$a_1 = \sigma(z_1)$$

After applying the activation function, we again multiply result a_1 by a new weight matrix and add a new bias value that is flowing between the hidden layer and the output layer. We can denote this weight matrix and bias as W_{hy} and b_y, respectively. The dimension of the weight matrix, W_{hy}, will be the *number of neurons in the hidden layer x the number of neurons in the output layer*. Since we have four neurons in the hidden layer and one neuron in the output layer, the W_{hy} matrix dimension will be 4 x 1. So, we multiply a_1 by the weight matrix, W_{hy}, and add bias, b_y, and pass the result z_2 to the next layer, which is the output layer:

$$z_2 = a_1 W_{hy} + b_y$$

Now, in the output layer, we apply a sigmoid function to z_2, which will result an output value:

$$\hat{y} = \sigma(z_2)$$

This whole process from the input layer to the output layer is known as **forward propagation**. Thus, in order to predict the output value, inputs are propagated from the input layer to the output layer. During this propagation, they are multiplied by their respective weights on each layer and an activation function is applied on top of them. The complete forward propagation steps are given as follows:

$$z_1 = XW_{xh} + b_h$$

$$a_1 = \sigma(z_1)$$

$$z_2 = a_1 W_{hy} + b_y$$

$$\hat{y} = \sigma(z_2)$$

The preceding forward propagation steps can be implemented in Python as follows:

```python
def forward_prop(X):
    z1 = np.dot(X,Wxh) + bh
    a1 = sigmoid(z1)
    z2 = np.dot(a1,Why) + by
    y_hat = sigmoid(z2)
    return y_hat
```

Forward propagation is cool, isn't it? But how do we know whether the output generated by the neural network is correct? We define a new function called the **cost function** (J), also known as the **loss function** (L), which tells us how well our neural network is performing. There are many different cost functions. We will use the mean squared error as a cost function, which can be defined as the mean of the squared difference between the actual output and the predicted output:

$$J = \frac{1}{n} \sum_{i=1}^{n} (y_i - \hat{y}_i)^2$$

Here, n is the number of training samples, y is actual output, and $\hat{y}$ is the predicted output.

Okay, so we learned that a cost function is used for assessing our neural network; that is, it tells us how good our neural network is at predicting the output. But the question is where is our network actually learning? In forward propagation, the network is just trying to predict the output. But how does it learn to predict the correct output? In the next section, we will examine this.

How does ANN learn?

If the cost or loss is very high, then it means that our network is not predicting the correct output. So, our objective is to minimize the cost function so that our neural network predictions will be better. How can we minimize the cost function? That is, how can we minimize the loss/cost? We learned that the neural network makes predictions using forward propagation. So, if we can change some values in the forward propagation, we can predict the correct output and minimize the loss. But what values can we change in the forward propagation? Obviously, we can't change input and output. We are now left with weights and bias values. Remember that we just initialized weight matrices randomly. Since the weights are random, they are not going to be perfect. Now, we will update these weight matrices (W_{xh} and W_{hy}) in such a way that our neural network gives a correct output. How do we update these weight matrices? Here comes a new technique called **gradient descent**.

With gradient descent, the neural network learns the optimal values of the randomly initialized weight matrices. With the optimal values of weights, our network can predict the correct output and minimize the loss.

Now, we will explore how the optimal values of weights are learned using gradient descent. Gradient descent is one of the most commonly used optimization algorithms. It is used for minimizing the cost function, which allows us to minimize the error and obtain the lowest possible error value. But how does gradient descent find the optimal weights? Let's begin with an analogy.

Imagine we are on top of a hill, as shown in the following diagram, and we want to reach the lowest point on the hill. There could be many regions that look like the lowest points on the hill, but we have to reach the lowest point that is actually the lowest of all.

That is, we should not be stuck at a point believing it is the lowest point when the global lowest point exists:

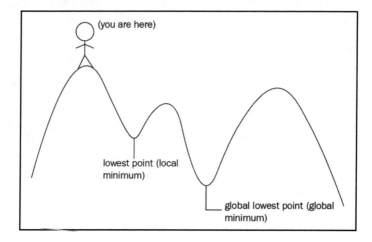

Similarly, we can represent our cost function as follows. It is a plot of cost against weights. Our objective is to minimize the cost function. That is, we have to reach the lowest point where the cost is the minimum. The solid dark point in the following diagram shows the randomly initialized weights. If we move this point downward, then we can reach the point where the cost is the minimum:

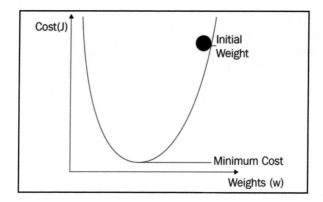

But how can we move this point (initial weight) downward? How can we descend and reach the lowest point? Gradients are used for moving from one point to another. So, we can move this point (initial weight) by calculating a gradient of the cost function with respect to that point (initial weights), which is $\frac{\partial J}{\partial w}$.

Gradients are the derivatives that are actually the slope of a tangent line as illustrated in the following diagram. So, by calculating the gradient, we descend (move downward) and reach the lowest point where the cost is the minimum. Gradient descent is a first-order optimization algorithm, which means we only take into account the first derivative when performing the updates:

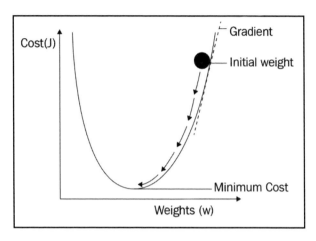

Thus, with gradient descent, we move our weights to a position where the cost is minimum. But, still, how do we update the weights?

As a result of forward propagation, we are in the output layer. We will now **backpropagate** the network from the output layer to the input layer and calculate the gradient of the cost function with respect to all the weights between the output and the input layer so that we can minimize the error. After calculating gradients, we update our old weights using the weight update rule:

$$W = W - \alpha \frac{\partial J}{\partial W}$$

This implies *weights = weights -α * gradients*.

What is α? It is called the **learning rate**. As shown in the following diagram, if the learning rate is small, then we take a small step downward and our gradient descent can be slow.

If the learning rate is large, then we take a large step and our gradient descent will be fast, but we might fail to reach the global minimum and become stuck at a local minimum. So, the learning rate should be chosen optimally:

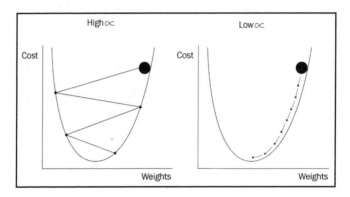

This whole process of backpropagating the network from the output layer to the input layer and updating the weights of the network using gradient descent to minimize the loss is called **backpropagation**. Now that we have a basic understanding of backpropagation, we will strengthen our understanding by learning about this in detail, step-by-step. We are going to look at some interesting math, so put on your calculus hats and follow the steps.

So, we have two weights, one W_{xh}, input to hidden layer weights, and the other W_{hy}, hidden to output layer weights. We need to find the optimal values for these two weights that will give us the fewest errors. So, we need to calculate the derivative of the cost function J with respect to these weights. Since we are backpropagating, that is, going from the output layer to the input layer, our first weight will be W_{hy}. So, now we need to calculate the derivative of J with respect to W_{hy}. How do we calculate the derivative? First, let's recall our cost function, J:

$$J = \frac{1}{n} \sum_{i=1}^{n} (y_i - \hat{y}_i)^2$$

We cannot calculate the derivative directly from the preceding equation since there is no

W_{hy} term. So, instead of calculating the derivative directly, we calculate the partial derivative. Let's recall our forward propagation equation:

$$\hat{y} = \sigma(z_2)$$

$$z2 = a_1 W_{hy} + b_y$$

First, we will calculate a partial derivative with respect to $\hat{y}$, and then from $\hat{y}$ we will calculate the partial derivative with respect to z_2. From z_2, we can directly calculate our derivative W_{hy}. It is basically the chain rule. So, the derivative of J with respect to W_{hy} becomes as follows:

$$\frac{\partial J}{\partial W_{hy}} = \frac{\partial J}{\partial \hat{y}} \cdot \frac{\partial \hat{y}}{\partial z_2} \cdot \frac{dz_2}{dW_{hy}} \tag{1}$$

Now, we will compute each of the terms in the preceding equation:

$$\frac{\partial J}{\partial \hat{y}} = (y - \hat{y})$$

$$\frac{\partial \hat{y}}{\partial z_2} = \sigma'(z_2)$$

Here, σ' is the derivative of our sigmoid activation function. We know that the sigmoid function is $\sigma(z) = \dfrac{1}{1+e^{-z}}$, so the derivative of the sigmoid function would be $\sigma'(z) = \dfrac{e^{-z}}{(1+e^{-z})^2}$:

$$\frac{d_{z_2}}{dW_{hy}} = a_1$$

Thus, substituting all the preceding terms in equation *(1)* we can write:

$$\boxed{\frac{\partial J}{\partial W_{hy}} = (y - \hat{y}).\sigma'(z_2).a_1} \tag{2}$$

Now we need to compute a derivative of J with respect to our next weight, W_{xh}.

Similarly, we cannot calculate the derivative of W_{xh} directly from J as we don't have any W_{xh} terms in J. So, we need to use the chain rule. Let's recall the forward propagation steps again:

$$\hat{y} = \sigma(z_2)$$

$$z2 = a_1 W_{hy} + b_y$$

$$a_1 = \sigma(z_1)$$

$$z_1 = X W_{xh} + b$$

Now, according to the chain rule, the derivative of J with respect to W_{xh} is given as:

$$\frac{\partial J}{\partial W_{xh}} = \frac{\partial J}{\partial \hat{y}} \cdot \frac{\partial \hat{y}}{\partial z_2} \cdot \frac{\partial z_2}{\partial a_1} \cdot \frac{\partial a_1}{\partial z_1} \cdot \frac{dz_1}{dW_{xh}} \tag{3}$$

We have already seen how to compute the first terms in the preceding equation; now, we will see how to compute the rest of the terms:

$$\frac{\partial z_2}{\partial a_1} = W_{hy}$$

$$\frac{\partial a_1}{\partial z_1} = \sigma'(z_1)$$

$$\frac{dz_1}{dw_{xh}} = X$$

Thus, substituting all the preceding terms in equation *(3)* we can write:

$$\boxed{\frac{\partial J}{\partial w_{xh}} = (y - \hat{y}) . \sigma'(z_2) . W_{hy} . \sigma'(z_1) . x} \tag{4}$$

After we have computed gradients for both weights, W_{hy} and W_{xh}, we will update our initial weights according to the weight update rule:

$$W_{hy} = W_{hy} - \alpha \frac{\partial J}{\partial W_{hy}} \tag{5}$$

$$W_{xh} = W_{xh} - \alpha \frac{\partial J}{\partial W_{xh}} \tag{6}$$

That's it! This is how we update the weights of a network and minimize the loss. If you don't understand gradient descent yet, no worries! In Chapter 3, *Gradient Descent and Its Variants*, we will go into the basics and learn gradient descent and several variants of gradient descent in more detail. Now, let's see how to implement the backpropagation algorithm in Python.

In both the equations *(2)* and *(4)*, we have the term $(y - \hat{y}) \cdot \sigma'(z_2)$, so instead of computing them again and again, we just call them delta2:

```
delta2 = np.multiply(-(y-yHat),sigmoidPrime(z2))
```

Now, we compute the gradient with respect to W_{hy}. Refer to equation *(2)*:

```
dJ_dWhy = np.dot(a1.T,delta2)
```

We compute the gradient with respect to W_{xh}. Refer to equation *(4)*:

```
delta1 = np.dot(delta2,Why.T)*sigmoidPrime(z1)

dJ_dWxh = np.dot(X.T,delta1)
```

We will update the weights according to our weight update rule equation *(5)* and *(6)* as follows:

```
Wxh = Wxh - alpha * dJ_dWhy
Why = Why - alpha * dJ_dWxh
```

The complete code for the backpropagation is given as follows:

```
def backword_prop(y_hat, z1, a1, z2):
    delta2 = np.multiply(-(y-y_hat),sigmoid_derivative(z2))
    dJ_dWhy = np.dot(a1.T, delta2)

    delta1 = np.dot(delta2,Why.T)*sigmoid_derivative(z1)
    dJ_dWxh = np.dot(X.T, delta1)

    Wxh = Wxh - alpha * dJ_dWhy
```

```
Why = Why - alpha * dJ_dWxh


    return Wxh,Why
```

Before moving on, let's familiarize ourselves with some of the frequently used terminologies in neural networks:

- **Forward pass**: Forward pass implies forward propagating from the input layer to the output layer.
- **Backward pass**: Backward pass implies backpropagating from the output layer to the input layer.
- **Epoch**: The epoch specifies the number of times the neural network sees our whole training data. So, we can say one epoch is equal to one forward pass and one backward pass for all training samples.
- **Batch size**: The batch size specifies the number of training samples we use in one forward pass and one backward pass.
- **Number of iterations**: The number of iterations implies the number of passes where *one pass = one forward pass + one backward pass*.

Say that we have 12,000 training samples and that our batch size is 6,000. It will take us two iterations to complete one epoch. That is, in the first iteration, we pass the first 6,000 samples and perform a forward pass and a backward pass; in the second iteration, we pass the next 6,000 samples and perform a forward pass and a backward pass. After two iterations, our neural network will see the whole 12,000 training samples, which makes it one epoch.

Debugging gradient descent with gradient checking

We just learned how gradient descent works and how to code the gradient descent algorithm from scratch for a simple two-layer network. But implementing gradient descent for complex neural networks is not a simple task. Apart from implementing, debugging a gradient descent for complex neural network architecture is again a tedious task. Surprisingly, even with some buggy gradient descent implementations, the network will learn something. However, apparently, it will not perform well compared to the bug-free implementation of gradient descent.

If the model does not give us any errors and learns something even with buggy implementations of the gradient descent algorithm, how can we evaluate and ensure that our implementation is correct? That is why we use the gradient checking algorithm. It will help us to validate our implementation of gradient descent by numerically checking the derivative.

Gradient checking is basically used for debugging the gradient descent algorithm and to validate that we have a correct implementation.

Okay. So, how does gradient checking works? In gradient checking, we basically compare the numerical and analytical gradients. Wait! What are numerical and analytical gradients?

Analytical gradient implies the gradients we calculated through backpropagation. **Numerical gradients** are the numerical approximation to the gradients. Let's explore this with an example. Assume we have a simple square function, $f(x) = x^2$.

The analytical gradient for the preceding function is computed using the power rule as follows:

$$f'(x) = 2x \tag{7}$$

Now, let's see how to approximate the gradient numerically. Instead of using the power rule to calculate gradients, we calculate gradients using a definition of the gradients. We know that the gradient or slope of a function basically gives us the steepness of the function.

Thus, the gradient or slope of a function is defined as follows:

$$\text{Slope} = \frac{\text{change in y}}{\text{change in x}}$$

A gradient of a function can be given as follows:

$$f'(x) \approx \lim_{\epsilon \to 0} \frac{f(x + \epsilon) - f(x - \epsilon)}{2\epsilon} \tag{8}$$

We use the preceding equation and approximate the gradients numerically. It implies that we are calculating the slope of the function manually, instead of using power rule as shown in the following diagram:

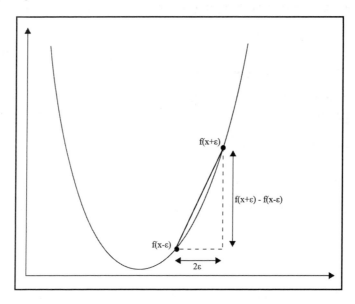

Computing gradients through power rule *(7)* and approximating the gradients numerically *(8)* essentially gives us the same value. Let's see how *(7)* and *(8)* give the same value in Python.

Define the square function:

```
def f(x):
    return x**2
```

Define the epsilon and input value:

```
epsilon = 1e-2
x=3
```

Compute the analytical gradient:

```
analytical_gradient = 2*x

print analytical_gradient
```

```
6
```

Compute the numerical gradient:

```
numerical_gradient = (f(x+epsilon) - f(x-epsilon)) / (2*epsilon)

print numerical_gradient

6.000000000012662
```

As you may have noticed, computing numerical and analytical gradients of the square function essentially gave us the same value, which is 6 when $x = 3$.

While backpropagating the network, we compute the analytical gradients to minimize the cost function. Now, we need to make sure that our computed analytical gradients are correct. So, let's validate that we approximate the numerical gradients of the cost function.

The gradients of J with respect to W can be numerically approximated as follows:

$$J'(W) = \lim_{\epsilon \to 0} \frac{J(W + \epsilon) - J(W - \epsilon)}{2\epsilon} \tag{9}$$

It is represented as follows:

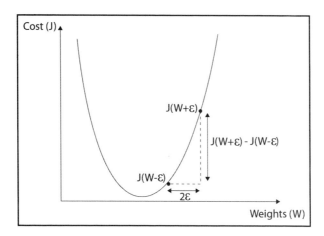

We check whether the analytical gradients and approximated numerical gradients are the same; if not, then there is an error in our analytical gradient computation. We don't want to check whether the numerical and analytical gradients are exactly the same; since we are only approximating the numerical gradients, we check the difference between the analytical and numerical gradients as an error. If the difference is less than or equal to a very small number say, *1e-7*, then our implementation is fine. If the difference is greater than *1e-7*, then our implementation is wrong.

Instead of calculating the error directly as the difference between the numerical gradient and the analytical gradient, we calculate the relative error. It can be defined as the ratio of differences to the ratio of an absolute value of the gradients:

$$relative\ error = \frac{||analytical\ grad - numerical\ grad||_2}{||analytical\ grad||_2 + ||numerical\ grad||_2} \qquad (10)$$

When the value of relative error is less than or equal to a small threshold value, say, *1e-7*, then our implementation is fine. If the relative error is greater than *1e-7*, then our implementation is wrong. Now let's see how to implement gradient checking algorithm in Python step-by-step.

First, we calculate the weights. Refer equation *(9)*:

```
weights_plus = weights + epsilon
weights_minus = weights - epsilon
```

Compute `J_plus` and `J_minus`. Refer equation *(9)*:

```
J_plus = forward_prop(x, weights_plus)
J_minus = forward_prop(x, weights_minus)
```

Now, we can compute the numerical gradient as given in *(9)* as follows:

```
numerical_grad = (J_plus - J_minus) / (2 * epsilon)
```

Analytical gradients can be obtained through backpropagation:

```
analytical_grad = backword_prop(x, weights)
```

Compute the relative error as given in equation *(10)* as follows:

```
numerator = np.linalg.norm(analytical_grad - numerical_grad)
denominator = np.linalg.norm(analytical_grad) +
np.linalg.norm(numerical_grad)
relative_error = numerator / denominator
```

If the relative error is less than a small threshold value, say `1e-7`, then our gradient descent implementation is correct; otherwise, it is wrong:

```
if relative_error < 1e-7:
        print ("The gradient is correct!")
else:
        print ("The gradient is wrong!")
```

Thus, with the help of gradient checking, we make sure that our gradient descent algorithm is bug-free.

Putting it all together

Putting all the concepts we have learned so far together, we will see how to build a neural network from scratch. We will understand how the neural network learns to perform the XOR gate operation. The XOR gate returns 1 only when exactly only one of its inputs is 1, else it returns 0 as shown in the following table:

Input(x)		Output(y)
x_1	x_2	y
0	0	0
0	1	1
1	0	1
1	1	0

Building a neural network from scratch

To perform the XOR gate operation, we build a simple two-layer neural network, as shown in the following diagram. As you can see, we have an input layer with two nodes: a hidden layer with five nodes and an output layer comprising one node:

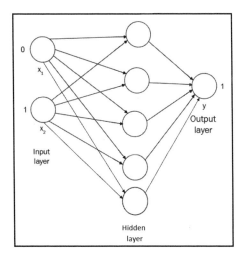

We will understand step-by-step how a neural network learns the XOR logic:

1. First, import the libraries:

```
import numpy as np
import matplotlib.pyplot as plt
%matplotlib inline
```

2. Prepare the data as shown in the preceding XOR table:

```
X = np.array([ [0, 1], [1, 0], [1, 1],[0, 0] ])
y = p.array([ [1], [1], [0], [0]])
```

3. Define the number of nodes in each layer:

```
num_input = 2
num_hidden = 5
num_output = 1
```

4. Initialize the weights and bias randomly. First, we initialize the input to hidden layer weights:

```
Wxh = np.random.randn(num_input,num_hidden)
bh = np.zeros((1,num_hidden))
```

5. Now, we initialize the hidden to output layer weights:

```
Why = np.random.randn (num_hidden,num_output)
by = np.zeros((1,num_output))
```

6. Define the sigmoid activation function:

```
def sigmoid(z):
    return 1 / (1+np.exp(-z))
```

7. Define the derivative of the sigmoid function:

```
def sigmoid_derivative(z):
    return np.exp(-z)/((1+np.exp(-z))**2)
```

8. Define the forward propagation:

```
def forward_prop(X,Wxh,Why):
    z1 = np.dot(X,Wxh) + bh
    a1 = sigmoid(z1)
    z2 = np.dot(a1,Why) + by
    y_hat = sigmoid(z2)
    return z1,a1,z2,y_hat
```

9. Define the backward propagation:

```
def backword_prop(y_hat, z1, a1, z2):
    delta2 = np.multiply(-(y-y_hat),sigmoid_derivative(z2))
    dJ_dWhy = np.dot(a1.T, delta2)
    delta1 = np.dot(delta2,Why.T)*sigmoid_derivative(z1)
    dJ_dWxh = np.dot(X.T, delta1)

    return dJ_dWxh, dJ_dWhy
```

10. Define the cost function:

```
def cost_function(y, y_hat):
    J = 0.5*sum((y-y_hat)**2)
    return J
```

11. Set the learning rate and the number of training iterations:

```
alpha = 0.01
num_iterations = 5000
```

12. Now, let's start training the network with the following code:

```
cost =[]

for i in range(num_iterations):
    z1,a1,z2,y_hat = forward_prop(X,Wxh,Why)
    dJ_dWxh, dJ_dWhy = backword_prop(y_hat, z1, a1, z2)

    #update weights
    Wxh = Wxh -alpha * dJ_dWxh
    Why = Why -alpha * dJ_dWhy

    #compute cost
    c = cost_function(y, y_hat)

    cost.append(c)
```

13. Plot the cost function:

```
plt.grid()
plt.plot(range(num_iteratins),cost)

plt.title('Cost Function')
plt.xlabel('Training Iterations')
plt.ylabel('Cost')
```

As you can observe in the following plot, the loss decreases over the training iterations:

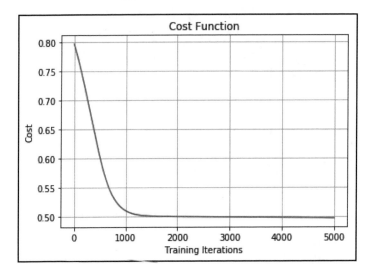

Thus, in this chapter, we got an overall understanding of artificial neural network and how they learn.

Summary

We started off the chapter by understanding what deep learning is and how it differs from machine learning. Later, we learned how biological and artificial neurons work, and then we explored what is input, hidden, and output layer in the ANN, and also several types of activation functions.

Going ahead, we learned what forward propagation is and how ANN uses forward propagation to predict the output. After this, we learned how ANN uses backpropagation for learning and optimizing. We learned an optimization algorithm called gradient descent that helps the neural network to minimize the loss and make correct predictions. We also learned about gradient checking, a technique that is used to evaluate the gradient descent. At the end of the chapter, we implemented a neural network from scratch to perform the XOR gate operation.

In the next chapter, we will learn about one of the most powerful and popularly used deep learning libraries called **TensorFlow**.

Questions

Let's evaluate our newly acquired knowledge by answering the following questions:

1. How does deep learning differ from machine learning?
2. What does the word *deep* mean in deep learning?
3. Why do we use the activation function?
4. Explain dying ReLU problem.
5. Define forward propagation.
6. What is back propagation?
7. Explain gradient checking.

Further reading

You can also check out some of these resources for more information:

- Understand more about gradient descent from this amazing video: `https://www.youtube.com/watch?v=IHZwWFHWa-w`
- Learn about implementing a neural network from scratch to recognize handwritten digits: `https://github.com/sar-gupta/neural-network-from-scratch`

Getting to Know TensorFlow

In this chapter, we will learn about TensorFlow, which is one of the most popularly used deep learning libraries. Throughout this book, we will be using TensorFlow to build deep learning models from scratch. So, in this chapter, we will get the hang of TensorFlow and its functionalities. We will also learn about TensorBoard, which is a visualization tool provided by TensorFlow used for visualizing models. Moving on, we will learn how to build our first neural network, using TensorFlow to perform handwritten digit classification. Following that, we will learn about TensorFlow 2.0, which is the latest version of TensorFlow. We will learn how TensorFlow 2.0 differs from its previous versions and how it uses Keras as its high-level API.

In this chapter, we will cover the following topics:

- TensorFlow
- Computational graphs and sessions
- Variables, constants, and placeholders
- TensorBoard
- Handwritten digit classification in TensorFlow
- Math operations in TensorFlow
- TensorFlow 2.0 and Keras

What is TensorFlow?

TensorFlow is an open source software library from Google, which is extensively used for numerical computation. It is one of the most popularly used libraries for building deep learning models. It is highly scalable and runs on multiple platforms, such as Windows, Linux, macOS, and Android. It was originally developed by the researchers and engineers of the Google Brain team.

TensorFlow supports execution on everything, including CPUs, GPUs, and TPUs, which are tensor processing units, and on mobile and embedded platforms. Due to its flexible architecture and ease of deployment, it has become a popular choice of library among many researchers and scientists for building deep learning models.

In TensorFlow, every computation is represented by a data flow graph, also known as a **computational graph**, where a node represents operations, such as addition or multiplication, and an edge represents tensors. Data flow graphs can also be shared and executed on many different platforms. TensorFlow provides a visualization tool, called TensorBoard, for visualizing data flow graphs.

A **tensor** is just a multidimensional array. So, when we say TensorFlow, it is literally a flow of multidimensional arrays (tensors) in a computation graph.

You can install TensorFlow easily through `pip` by just typing the following command in your Terminal. We will install TensorFlow 1.13.1:

```
pip install tensorflow==1.13.1
```

We can check the successful installation of TensorFlow by running the following simple `Hello TensorFlow!` program:

```
import tensorflow as tf

hello = tf.constant("Hello TensorFlow!")
sess = tf.Session()
print(sess.run(hello))
```

The preceding program should print `Hello TensorFlow!`. If you get any errors, then you probably have not installed TensorFlow correctly.

Understanding computational graphs and sessions

As we have learned, every computation in TensorFlow is represented by a computational graph. They consist of several nodes and edges, where nodes are mathematical operations, such as addition and multiplication, and edges are tensors. Computational graphs are very efficient at optimizing resources and promote distributed computing.

A computational graph consists of several TensorFlow operations, arranged in a graph of nodes.

Let's consider a basic addition operation:

```
import tensorflow as tf

x = 2
y = 3
z = tf.add(x, y, name='Add')
```

The computational graph for the preceding code would look like the following:

A computational graph helps us to understand the network architecture when we work on building a really complex neural network. For instance, let's consider a simple layer, $h = \text{Relu}(WX + b)$. Its computational graph would be represented as follows:

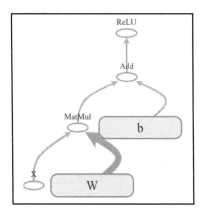

There are two types of dependency in the computational graph, called direct and indirect dependency. Say we have the b node, the input of which is dependent on the output of the a node; this type of dependency is called **direct dependency**, as shown in the following code:

```
a = tf.multiply(8,5)
b = tf.multiply(a,1)
```

When the b node doesn't depend on the a node for its input, it is called **indirect dependency**, as shown in the following code:

```
a = tf.multiply(8,5)
b = tf.multiply(4,3)
```

So, if we can understand these dependencies, we can distribute the independent computations in the available resources and reduce the computation time. Whenever we import TensorFlow, a default graph is created automatically and all of the nodes we create are associated with the default graph. We can also create our own graphs instead of using the default graph, and this is useful when building multiple models that do not depend on one another in one file. A TensorFlow graph can be created using tf.Graph(), as follows:

```
graph = tf.Graph()

with graph.as_default():
    z = tf.add(x, y, name='Add')
```

If we want to clear the default graph (that is, if we want to clear the previously defined variables and operations), then we can do that using tf.reset_default_graph().

Sessions

A computational graph with operations on its nodes and tensors to its edges will be created, and in order to execute the graph, we use a TensorFlow session.

A TensorFlow session can be created using tf.Session(), as shown in the following code, and it will allocate memory for storing the current value of the variable:

```
sess = tf.Session()
```

After creating the session, we can execute our graph, using the sess.run() method.

Every computation in TensorFlow is represented by a computational graph, so we need to run a computational graph for everything. That is, in order to compute anything on TensorFlow, we need to create a TensorFlow session.

Let's execute the following code to multiply two numbers:

```
a = tf.multiply(3,3)
print(a)
```

Instead of printing 9, the preceding code will print a TensorFlow object, `Tensor("Mul:0", shape=(), dtype=int32)`.

As we discussed earlier, whenever we import TensorFlow, a default computational graph will automatically be created and all nodes will get attached to the graph. Hence, when we print a, it just returns the TensorFlow object because the value for a is not computed yet, as the computation graph has not been executed.

In order to execute the graph, we need to initialize and run the TensorFlow session, as follows:

```
a = tf.multiply(3,3)
with tf.Session as sess:
    print(sess.run(a))
```

The preceding code will print 9.

Variables, constants, and placeholders

Variables, constants, and placeholders are fundamental elements of TensorFlow. However, there is always confusion between these three. Let's look at each element, one by one, and learn the difference between them.

Variables

Variables are containers used to store values. Variables are used as input to several other operations in a computational graph. A variable can be created using the `tf.Variable()` function, as shown in the following code:

```
x = tf.Variable(13)
```

Let's create a variable called W, using `tf.Variable()`, as follows:

```
W = tf.Variable(tf.random_normal([500, 111], stddev=0.35), name="weights")
```

As you can see in the preceding code, we create a variable, W, by randomly drawing values from a normal distribution with a standard deviation of 0.35.

What is that parameter called `name` in `tf.Variable()`?

It is used to set the name of the variable in the computational graph. So, in the preceding code, Python saves the variable as W but in the TensorFlow graph, it will be saved as weights.

We can also initialize a new variable with a value from another variable using initialized_value(). For instance, if we want to create a new variable called weights_2, using a value from the previously defined weights variable, it can be done as follows:

```
W2 = tf.Variable(weights.initialized_value(), name="weights_2")
```

However, after defining a variable, we need to initialize all of the variables in the computational graph. That can be done using tf.global_variables_initializer().

Once we create a session, first, we run the initialization operation, which will initialize all of the defined variables, and only then can we run the other operations, as shown in the following code:

```
x = tf.Variable(1212)
init = tf.global_variables_initializer()

with tf.Session() as sess:
  sess.run(init)
  print sess.run(x)
```

We can also create a TensorFlow variable using tf.get_variable(). It takes the three important parameters, which are name, shape, and initializer.

Unlike tf.Variable(), we cannot pass the value directly to tf.get_variable(); instead, we use initializer. There are several initializers available for initializing values. For example, tf.constant_initializer(value) initializes the variable with a constant value, and tf.random_normal_initializer(mean, stddev) initializes the variable by drawing values from random normal distribution with a specified mean and standard deviation.

Variables created using tf.Variable() cannot be shared, and every time we call tf.Variable(), it will create a new variable. But tf.get_variable() checks the computational graph for an existing variable with the specified parameter. If the variable already exists, then it will be reused; otherwise, a new variable will be created:

```
W3 = tf.get_variable(name = 'weights', shape = [500, 111], initializer =
random_normal_initializer()))
```

So, the preceding code checks whether there is any variable already existing with the given parameters. If yes, then it will reuse it; otherwise, it will create a new variable.

Since we are reusing variables using `tf.get_variable()`, in order to avoid name conflicts, we use `tf.variable_scope`, as shown in the following code. A variable scope is basically a name-scoping technique that just adds a prefix to the variable within the scope to avoid the naming clash:

```
with tf.variable_scope("scope"):
  a = tf.get_variable('x', [2])

with tf.variable_scope("scope", reuse = True):
  b = tf.get_variable('x', [2])
```

If you print `a.name` and `b.name`, then it will return the same name, which is `scope/x:0`. As you can see, we specified the `reuse=True` parameter in the variable scope named `scope`, which implies that the variables can be shared. If we don't set `reuse = True`, then it will give an error saying that the variable already exists.

It is recommended to use `tf.get_variable()` rather than `tf.Variable()`, because `tf.get_variable`, allows you to share variables, and it will make the code refactoring easier.

Constants

Constants, unlike variables, cannot have their values changed. That is, constants are immutable. Once they are assigned values, they cannot be changed throughout. We can create constants using the `tf.constant()`, as shown in the following code:

```
x = tf.constant(13)
```

Placeholders and feed dictionaries

We can think of placeholders as variables, where we only define the type and dimension, but do not assign the value. Values for the placeholders will be fed at runtime. We feed the data to the computational graphs using placeholders. Placeholders are defined with no values.

A placeholder can be defined using `tf.placeholder()`. It takes an optional argument called `shape`, which denotes the dimensions of the data. If `shape` is set to `None`, then we can feed data of any size at runtime. A placeholder can be defined as follows:

```
x = tf.placeholder("float", shape=None)
```

To put it in simple terms, we use `tf.Variable` to store the data and `tf.placeholder` for feeding the external data.

Let's consider a simple example to better understand placeholders:

```
x = tf.placeholder("float", None)
y = x +3

with tf.Session() as sess:
    result = sess.run(y)
    print(result)
```

If we run the preceding code, then it will return an error because we are trying to compute y, where y= x+3 and x is a placeholder whose value is not assigned. As we have learned, values for the placeholders will be assigned at runtime. We assign the values of the placeholder using the `feed_dict` parameter. The `feed_dict` parameter is basically the dictionary where the key represents the name of the placeholder, and the value represents the value of the placeholder.

As you can see in the following code, we set `feed_dict = {x:5}`, which implies that the value for the x placeholder is 5:

```
with tf.Session() as sess:
    result = sess.run(y, feed_dict={x: 5})
    print(result)
```

The preceding code returns 8.0.

What if we want to use multiple values for x? As we have not defined any shapes for the placeholders, it takes any number of values, as shown in the following code:

```
with tf.Session() as sess:
    result = sess.run(y, feed_dict={x: [3,6,9]})
    print(result)
```

It will return the following:

```
[ 6.  9. 12.]
```

Let's say we define the shape of x as [None, 2], as shown in the following code:

```
x = tf.placeholder("float", [None, 2])
```

This means that x can take a matrix of any rows but with 2 columns, as shown in the following code:

```
with tf.Session() as sess:
    x_val = [[1, 2,],
             [3,4],
             [5,6],
             [7,8],]
    result = sess.run(y, feed_dict={x: x_val})
    print(result)
```

The preceding code returns the following:

```
[[ 4.  5.]
 [ 6.  7.]
 [ 8.  9.]
 [10. 11.]]
```

Introducing TensorBoard

TensorBoard is TensorFlow's visualization tool, which can be used to visualize a computational graph. It can also be used to plot various quantitative metrics and the results of several intermediate calculations. When we are training a really deep neural network, it becomes confusing when we have to debug the network. So, if we can visualize the computational graph in TensorBoard, we can easily understand such complex models, debug them, and optimize them. TensorBoard also supports sharing.

As shown in the following screenshot, the TensorBoard panel consists of several tabs—**SCALARS, IMAGES, AUDIO, GRAPHS, DISTRIBUTIONS, HISTOGRAMS,** and **EMBEDDINGS**:

The tabs are pretty self-explanatory. The **SCALARS** tab shows useful information about the scalar variables we use in our program. For example, it shows how the value of a scalar variable called loss changes over several iterations.

The **GRAPHS** tab shows the computational graph. The **DISTRIBUTIONS** and **HISTOGRAMS** tabs show the distribution of a variable. For example, our model's weight distribution and histogram can be seen under these tabs. The **EMBEDDINGS** tab is used for visualizing high-dimensional vectors, such as word embeddings (we will learn about this in detail in Chapter 7, *Learning Text Representations*).

Let's build a basic computational graph and visualize it in TensorBoard. Let's say we have four variables, shown as follows:

```
x = tf.constant(1,name='x')
y = tf.constant(1,name='y')
a = tf.constant(3,name='a')
b = tf.constant(3,name='b')
```

Let's multiply x and y and a and b and save them as prod1 and prod2, as shown in the following code:

```
prod1 = tf.multiply(x,y,name='prod1')
prod2 = tf.multiply(a,b,name='prod2')
```

Add prod1 and prod2 and store them in sum:

```
sum = tf.add(prod1,prod2,name='sum')
```

Now, we can visualize all of these operations in TensorBoard. In order to visualize in TensorBoard, we first need to save our event files. It can be done using tf.summary.FileWriter(). It takes two important parameters, logdir and graph.

As the name suggests, logdir specifies the directory where we want to store the graph, and graph specifies which graph we want to store:

```
with tf.Session() as sess:
    writer = tf.summary.FileWriter(logdir='./graphs',graph=sess.graph)
    print(sess.run(sum))
```

In the preceding code, graphs is the directory where we are storing our event file, and sess.graph specifies the current graph in our TensorFlow session. So, we are storing the current graph in the TensorFlow session in the graphs directory.

To start TensorBoard, go to your Terminal, locate the working directory, and type the following:

```
tensorboard --logdir=graphs --port=8000
```

The logdir parameter indicates the directory where the event file is stored and port is the port number. Once you run the preceding command, open your browser and type http://localhost:8000/.

In the TensorBoard panel, under the **GRAPHS** tab, you can see the computational graph:

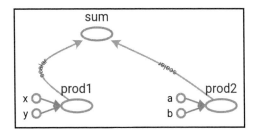

As you may notice, all of the operations we have defined are clearly shown in the graph.

Creating a name scope

Scoping is used to reduce complexity and helps us to better understand a model by grouping related nodes together. Having a name scope helps us to group similar operations in a graph. It comes in handy when we are building a complex architecture. Scoping can be created using tf.name_scope(). In the previous example, we performed two operations, Product and sum. We can simply group them into two different name scopes as Product and sum.

In the previous section, we saw how prod1 and prod2 perform multiplication and compute the result. We'll define a name scope called Product, and group the prod1 and prod2 operations, as shown in the following code:

```
with tf.name_scope("Product"):
    with tf.name_scope("prod1"):
        prod1 = tf.multiply(x,y,name='prod1')
    with tf.name_scope("prod2"):
        prod2 = tf.multiply(a,b,name='prod2')
```

Now, define the name scope for sum:

```
with tf.name_scope("sum"):
    sum = tf.add(prod1,prod2,name='sum')
```

Store the file in the graphs directory:

```
with tf.Session() as sess:
    writer = tf.summary.FileWriter('./graphs', sess.graph)
    print(sess.run(sum))
```

Visualize the graph in TensorBoard:

```
tensorboard --logdir=graphs --port=8000
```

As you may notice, now, we have only two nodes, **sum** and **Product**:

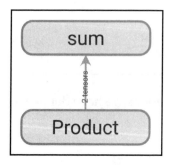

Once we double-click on the nodes, we can see how the computation is happening. As you can see, the **prod1** and **prod2** nodes are grouped under the **Product** scope, and their results are sent to the **sum** node, where they will be added. You can see how the **prod1** and **prod2** nodes compute their value:

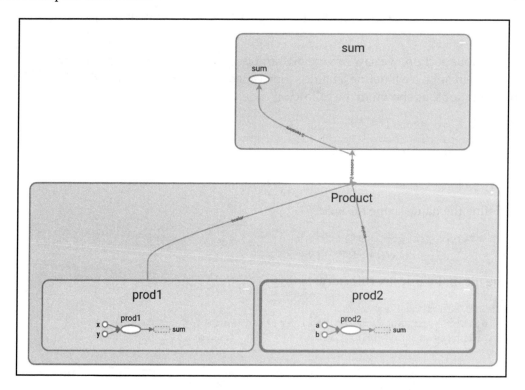

The preceding example is just a simple example. When we are working on a complex project with a lot of operations, name scoping helps us to group similar operations together and enables us to understand the computational graph better.

Handwritten digit classification using TensorFlow

Putting together all the concepts we have learned so far, we will see how we can use TensorFlow to build a neural network to recognize handwritten digits. If you have been playing around with deep learning of late, then you must have come across the MNIST dataset. It has been called the *hello world* of deep learning. It consists of 55,000 data points of handwritten digits (0 to 9).

In this section, we will see how we can use our neural network to recognize these handwritten digits, and we will get the hang of TensorFlow and TensorBoard.

Importing the required libraries

As a first step, let's import all of the required libraries:

```
import warnings
warnings.filterwarnings('ignore')

import tensorflow as tf
from tensorflow.examples.tutorials.mnist import input_data
tf.logging.set_verbosity(tf.logging.ERROR)

import matplotlib.pyplot as plt
%matplotlib inline
```

Loading the dataset

Load the dataset, using the following code:

```
mnist = input_data.read_data_sets("data/mnist", one_hot=True)
```

In the preceding code, `data/mnist` implies the location where we store the MNIST dataset, and `one_hot=True` implies that we are one-hot encoding the labels (0 to 9).

We will see what we have in our data by executing the following code:

```
print("No of images in training set {}".format(mnist.train.images.shape))
print("No of labels in training set {}".format(mnist.train.labels.shape))

print("No of images in test set {}".format(mnist.test.images.shape))
print("No of labels in test set {}".format(mnist.test.labels.shape))

No of images in training set (55000, 784)
No of labels in training set (55000, 10)
No of images in test set (10000, 784)
No of labels in test set (10000, 10)
```

We have 55000 images in the training set, each image is of size 784, and we have 10 labels, which are actually 0 to 9. Similarly, we have 10000 images in the test set.

Now, we'll plot an input image to see what it looks like:

```
img1 = mnist.train.images[0].reshape(28,28)
plt.imshow(img1, cmap='Greys')
```

Thus, our input image looks like the following:

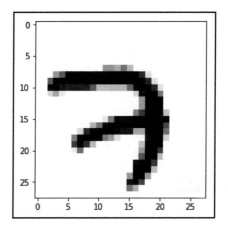

Defining the number of neurons in each layer

We'll build a four-layer neural network with three hidden layers and one output layer. As the size of the input image is 784, we set num_input to 784, and since we have 10 handwritten digits (0 to 9), we set 10 neurons in the output layer. We define the number of neurons in each layer as follows:

```
#number of neurons in input layer
num_input = 784

#num of neurons in hidden layer 1
num_hidden1 = 512

#num of neurons in hidden layer 2
num_hidden2 = 256

#num of neurons in hidden layer 3
num_hidden_3 = 128

#num of neurons in output layer
num_output = 10
```

Defining placeholders

As we have learned, we first need to define the placeholders for input and output. Values for the placeholders will be fed in at runtime through feed_dict:

```
with tf.name_scope('input'):
    X = tf.placeholder("float", [None, num_input])

with tf.name_scope('output'):
    Y = tf.placeholder("float", [None, num_output])
```

Since we have a four-layer network, we have four weights and four biases. We initialize our weights by drawing values from the truncated normal distribution with a standard deviation of 0.1. Remember, the dimensions of the weights matrix should be a *number of neurons in the previous layer* x *a number of neurons in the current layer*. For instance, the dimension of weight matrix w3 should be the *number of neurons in hidden layer 2* x *the number of neurons in hidden layer 3*.

We often define all of the weights in a dictionary, as follows:

```
with tf.name_scope('weights'):

 weights = {
 'w1': tf.Variable(tf.truncated_normal([num_input, num_hidden1],
stddev=0.1),name='weight_1'),
 'w2': tf.Variable(tf.truncated_normal([num_hidden1, num_hidden2],
stddev=0.1),name='weight_2'),
 'w3': tf.Variable(tf.truncated_normal([num_hidden2, num_hidden_3],
stddev=0.1),name='weight_3'),
 'out': tf.Variable(tf.truncated_normal([num_hidden_3, num_output],
stddev=0.1),name='weight_4'),
 }
```

The shape of the bias should be the number of neurons in the current layer. For instance, the dimension of the b2 bias is the number of neurons in hidden layer 2. We set the bias value as a constant; 0.1 in all of the layers:

```
with tf.name_scope('biases'):

    biases = {
        'b1': tf.Variable(tf.constant(0.1,
shape=[num_hidden1]),name='bias_1'),
        'b2': tf.Variable(tf.constant(0.1,
shape=[num_hidden2]),name='bias_2'),
        'b3': tf.Variable(tf.constant(0.1,
shape=[num_hidden_3]),name='bias_3'),
        'out': tf.Variable(tf.constant(0.1,
shape=[num_output]),name='bias_4')
    }
```

Forward propagation

Now we'll define the forward propagation operation. We'll use ReLU activations in all layers. In the last layers, we'll apply sigmoid activation, as shown in the following code:

```
with tf.name_scope('Model'):
    with tf.name_scope('layer1'):
        layer_1 = tf.nn.relu(tf.add(tf.matmul(X, weights['w1']),
biases['b1']) )
    with tf.name_scope('layer2'):
        layer_2 = tf.nn.relu(tf.add(tf.matmul(layer_1, weights['w2']),
biases['b2']))
    with tf.name_scope('layer3'):
        layer_3 = tf.nn.relu(tf.add(tf.matmul(layer_2, weights['w3']),
```

```
biases['b3']))
    with tf.name_scope('output_layer'):
        y_hat = tf.nn.sigmoid(tf.matmul(layer_3, weights['out']) +
biases['out'])
```

Computing loss and backpropagation

Next, we'll define our loss function. We'll use softmax cross-entropy as our loss function. TensorFlow provides the `tf.nn.softmax_cross_entropy_with_logits()` function for computing softmax cross-entropy loss. It takes two parameters as inputs, `logits` and `labels`:

- The `logits` parameter specifies the `logits` predicted by our network; for example, y_hat
- The `labels` parameter specifies the actual labels; for example, true labels, Y

We take the mean of the `loss` function using `tf.reduce_mean()`:

```
with tf.name_scope('Loss'):
    loss =
tf.reduce_mean(tf.nn.softmax_cross_entropy_with_logits(logits=y_hat,labels=
Y))
```

Now, we need to minimize the loss using backpropagation. Don't worry! We don't have to calculate the derivatives of all the weights manually. Instead, we can use TensorFlow's optimizer. In this section, we the use the Adam optimizer. It is a variant of the gradient descent optimization technique we learned about in Chapter 1, *Introduction to Deep Learning*. In Chapter 3, *Gradient Descent and Its Variants*, we will dive into the details and see how exactly the Adam optimizer and several other optimizers work. For now, let's say we use the Adam optimizer as our backpropagation algorithm:

```
learning_rate = 1e-4
optimizer = tf.train.AdamOptimizer(learning_rate).minimize(loss)
```

Computing accuracy

We calculate the accuracy of our model as follows:

- The y_hat parameter denotes the predicted probability for each class of our model. Since we have 10 classes, we will have 10 probabilities. If the probability is high at position 7, then it means that our network predicts the input image as digit 7 with high probability. The tf.argmax() function returns the index of the largest value. Thus, tf.argmax(y_hat,1) gives the index where the probability is high. Thus, if the probability is high at index 7, then it returns 7.

- The Y parameter denotes the actual labels, and they are the one-hot encoded values. That is, it consists of zeros everywhere except at the position of the actual image, where it consists of 1. For instance, if the input image is 7, then Y has 0 at all indices except at index 7, where it has 1. Thus, tf.argmax(Y,1) returns 7 because that is where we have a high value, 1.

Thus, tf.argmax(y_hat,1) gives the predicted digit, and tf.argmax(Y,1) gives us the actual digit.

The tf.equal(x, y) function takes x and y as inputs and returns the truth value of *(x == y)* element-wise. Thus, correct_pred = tf.equal(predicted_digit,actual_digit) consists of True where the actual and predicted digits are the same, and False where the actual and predicted digits are not the same. We convert the Boolean values in correct_pred into float values using TensorFlow's cast operation, tf.cast(correct_pred, tf.float32). After converting them into float values, we take the average using tf.reduce_mean().

Thus, tf.reduce_mean(tf.cast(correct_pred, tf.float32)) gives us the average correct predictions:

```
with tf.name_scope('Accuracy'):
    predicted_digit = tf.argmax(y_hat, 1)
    actual_digit = tf.argmax(Y, 1)
    correct_pred = tf.equal(predicted_digit,actual_digit)
    accuracy = tf.reduce_mean(tf.cast(correct_pred, tf.float32))
```

Creating summary

We can also visualize how the loss and accuracy of our model changes during several iterations in TensorBoard. So, we use tf.summary() to get the summary of the variable. Since the loss and accuracy are scalar variables, we use tf.summary.scalar(), as shown in the following code:

```
tf.summary.scalar("Accuracy", accuracy)
tf.summary.scalar("Loss", loss)
```

Next, we merge all of the summaries we use in our graph, using tf.summary.merge_all(). We do this because when we have many summaries, running and storing them would become inefficient, so we run them once in our session instead of running multiple times:

```
merge_summary = tf.summary.merge_all()
```

Training the model

Now, it is time to train our model. As we have learned, first, we need to initialize all of the variables:

```
init = tf.global_variables_initializer()
```

Define the batch size, number of iterations, and learning rate, as follows:

```
learning_rate = 1e-4
num_iterations = 1000
batch_size = 128
```

Start the TensorFlow session:

```
with tf.Session() as sess:
```

Initialize all the variables:

```
sess.run(init)
```

Save the event files:

```
summary_writer = tf.summary.FileWriter('./graphs', graph=sess.graph)
```

Train the model for a number of iterations:

```
for i in range(num_iterations):
```

Get a batch of data according to the batch size:

```
batch_x, batch_y = mnist.train.next_batch(batch_size)
```

Train the network:

```
sess.run(optimizer, feed_dict={ X: batch_x, Y: batch_y})
```

Print `loss` and `accuracy` for every 100th iteration:

```
if i % 100 == 0:

    batch_loss, batch_accuracy,summary = sess.run(
        [loss, accuracy, merge_summary], feed_dict={X: batch_x, Y:
batch y}
        )

    #store all the summaries
    summary_writer.add_summary(summary, i)


    print('Iteration: {}, Loss: {}, Accuracy:
{}'.format(i,batch_loss,batch_accuracy))
```

As you may notice from the following output, the loss decreases and the accuracy increases over various training iterations:

```
Iteration: 0, Loss: 2.30789709091, Accuracy: 0.1171875
Iteration: 100, Loss: 1.76062202454, Accuracy: 0.859375
Iteration: 200, Loss: 1.60075569153, Accuracy: 0.9375
Iteration: 300, Loss: 1.60388696194, Accuracy: 0.890625
Iteration: 400, Loss: 1.59523034096, Accuracy: 0.921875
Iteration: 500, Loss: 1.58489584923, Accuracy: 0.859375
Iteration: 600, Loss: 1.51407408714, Accuracy: 0.953125
Iteration: 700, Loss: 1.53311181068, Accuracy: 0.9296875
Iteration: 800, Loss: 1.57677125931, Accuracy: 0.875
Iteration: 900, Loss: 1.52060437202, Accuracy: 0.9453125
```

Visualizing graphs in TensorBoard

After training, we can visualize our computational graph in TensorBoard, as shown in the following diagram. As you can see, our **Model** takes **input**, **weights**, and **biases** as input and returns the output. We compute **Loss** and **Accuracy** based on the output of the model. We minimize the loss by calculating **gradients** and updating **weights**. We can observe all of this in the following diagram:

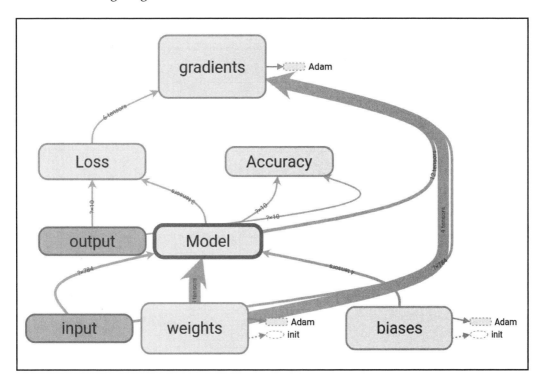

If we double-click and expand the **Model**, we can see that we have three hidden layers and one output layer:

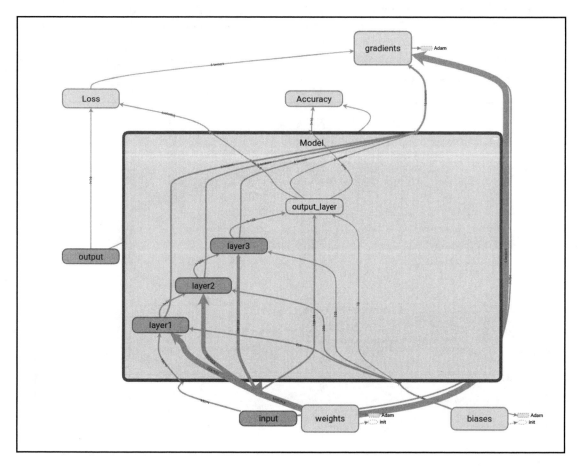

Similarly, we can double-click and see every node. For instance, if we open **weights**, we can see how the four weights are initialized using truncated normal distribution, and how it is updated using the Adam optimizer:

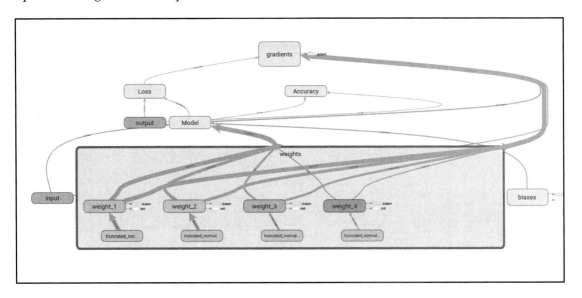

As we have learned, the computational graph helps us to understand what is happening on each node. We can see how the accuracy is being calculated by double-clicking on the **Accuracy** node:

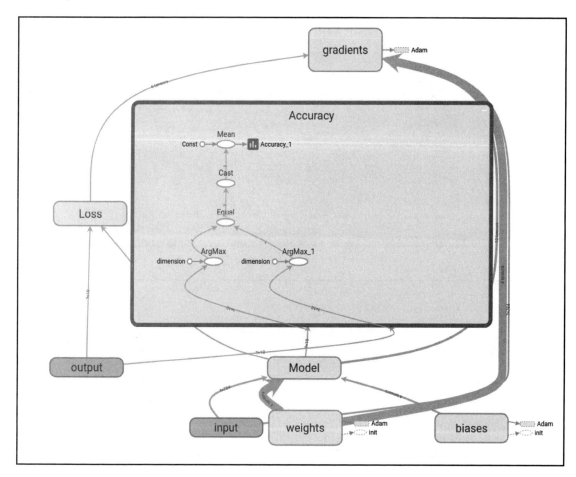

Remember that we also stored the summary of our `loss` and `accuracy` variables. We can find them under the **SCALARS** tab in TensorBoard, as shown in the following screenshot. We can see how that loss decreases over iterations, as shown in the following screenshot:

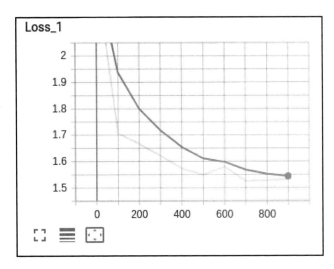

The following screenshot shows how accuracy increases over iterations:

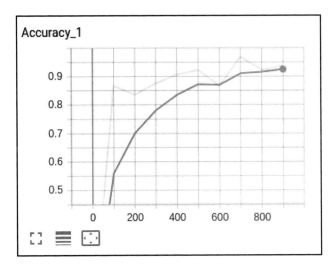

Introducing eager execution

Eager execution in TensorFlow is more Pythonic and allows for rapid prototyping. Unlike the graph mode, where we need to construct a graph every time to perform any operations, eager execution follows the imperative programming paradigm, where any operations can be performed immediately, without having to create a graph, just like we do in Python. Hence, with eager execution, we can say goodbye to sessions and placeholders. It also makes the debugging process easier with an immediate runtime error, unlike the graph mode.

For instance, in the graph mode, to compute anything, we run the session. As shown in the following code, to evaluate the value of z, we have to run the TensorFlow session:

```
x = tf.constant(11)
y = tf.constant(11)
z = x*y

with tf.Session() as sess:
    print sess.run(z)
```

With eager execution, we don't need to create a session; we can simply compute z, just like we do in Python. In order to enable eager execution, just call the `tf.enable_eager_execution()` function:

```
x = tf.constant(11)
y = tf.constant(11)
z = x*y

print z
```

It will return the following:

```
<tf.Tensor: id=789, shape=(), dtype=int32, numpy=121>
```

In order to get the output value, we can print the following:

```
z.numpy()

121
```

Math operations in TensorFlow

Now, we will explore some of the operations in TensorFlow using the eager execution mode:

```
x = tf.constant([1., 2., 3.])
y = tf.constant([3., 2., 1.])
```

Let's start with some basic arithmetic operations.

Use tf.add to add two numbers:

```
sum = tf.add(x,y)
sum.numpy()

array([4., 4., 4.], dtype=float32)
```

The tf.subtract function is used for finding the difference between two numbers:

```
difference = tf.subtract(x,y)
difference.numpy()

array([-2., 0., 2.], dtype=float32)
```

The tf.multiply function is used for multiplying two numbers:

```
product = tf.multiply(x,y)
product.numpy()

array([3., 4., 3.], dtype=float32)
```

Divide two numbers using tf.divide:

```
division = tf.divide(x,y)
division.numpy()

array([0.33333334, 1.        , 3.        ], dtype=float32)
```

The dot product can be computed as follows:

```
dot_product = tf.reduce_sum(tf.multiply(x, y))
dot_product.numpy()

10.0
```

Next, let's find the index of the minimum and maximum elements:

```
x = tf.constant([10, 0, 13, 9])
```

The index of the minimum value is computed using `tf.argmin()`:

```
tf.argmin(x).numpy()

1
```

The index of the maximum value is computed using `tf.argmax()`:

```
tf.argmax(x).numpy()

2
```

Run the following code to find the squared difference between x and y:

```
x = tf.Variable([1,3,5,7,11])
y = tf.Variable([1])

tf.math.squared_difference(x,y).numpy()

[  0,   4,  16,   36, 100]
```

Let's try typecasting; that is, converting from one data type into another.

Print the type of x:

```
print x.dtype

tf.int32
```

We can convert the type of x, which is `tf.int32`, into `tf.float32` using `tf.cast`, as shown in the following code:

```
x = tf.cast(x, dtype=tf.float32)
```

Now, check the x type. It will be `tf.float32`, as follows:

```
print x.dtype

tf.float32
```

Concatenate the two matrices:

```
x = [[3,6,9], [7,7,7]]
y = [[4,5,6], [5,5,5]]
```

Concatenate the matrices row-wise:

```
tf.concat([x, y], 0).numpy()

array([[3, 6, 9],
       [7, 7, 7],
       [4, 5, 6],
       [5, 5, 5]], dtype=int32)
```

Use the following code to concatenate the matrices column-wise:

```
tf.concat([x, y], 1).numpy()

array([[3, 6, 9, 4, 5, 6],
       [7, 7, 7, 5, 5, 5]], dtype=int32)
```

Stack the x matrix using the `stack` function:

```
tf.stack(x, axis=1).numpy()

array([[3, 7],
       [6, 7],
       [9, 7]], dtype=int32)
```

Now, let' see how to perform the `reduce_mean` operation:

```
x = tf.Variable([[1.0, 5.0], [2.0, 3.0]])

x.numpy()

array([[1., 5.],
       [2., 3.]]
```

Compute the mean value of x; that is, *(1.0 + 5.0 + 2.0 + 3.0) / 4*:

```
tf.reduce_mean(input_tensor=x).numpy()

2.75
```

Compute the mean across the row; that is, *(1.0+5.0)/2, (2.0+3.0)/2*:

```
tf.reduce_mean(input_tensor=x, axis=0).numpy()

array([1.5, 4. ], dtype=float32)
```

Compute the mean across the column; that is, *(1.0+5.0)/2.0, (2.0+3.0)/2.0*:

```
tf.reduce_mean(input_tensor=x, axis=1, keepdims=True).numpy()

array([[3. ],
       [2.5]], dtype=float32)
```

Draw random values from the probability distributions:

```
tf.random.normal(shape=(3,2), mean=10.0, stddev=2.0).numpy()

tf.random.uniform(shape = (3,2), minval=0, maxval=None,
dtype=tf.float32,).numpy()
```

Compute the softmax probabilities:

```
x = tf.constant([7., 2., 5.])

tf.nn.softmax(x).numpy()

array([0.8756006 , 0.00589975, 0.11849965], dtype=float32)
```

Now, we'll look at how to compute the gradients.

Define the `square` function:

```
def square(x):
    return tf.multiply(x, x)
```

The gradients can be computed for the preceding `square` function using `tf.GradientTape`, as follows:

```
with tf.GradientTape(persistent=True) as tape:
    print square(6.).numpy()

36.0
```

More TensorFlow operations are available in the Notebook on GitHub at `http://bit.ly/2YSYbYu`.

> TensorFlow is a lot more than this. We will learn about various important functionalities of TensorFlow as we move on through this book.

TensorFlow 2.0 and Keras

TensorFlow 2.0 has got some really cool features. It sets the eager execution mode by default. It provides a simplified workflow and uses Keras as the main API for building deep learning models. It is also backward compatible with TensorFlow 1.x versions.

To install TensorFlow 2.0, open your Terminal and type the following command:

```
pip install tensorflow==2.0.0-alpha0
```

Since TensorFlow 2.0 uses Keras as a high-level API, we will look at how Keras works in the next section.

Bonjour Keras

Keras is another popularly used deep learning library. It was developed by François Chollet at Google. It is well known for its fast prototyping, and it makes model building simple. It is a high-level library, meaning that it does not perform any low-level operations on its own, such as convolution. It uses a backend engine for doing that, such as TensorFlow. The Keras API is available in `tf.keras`, and TensorFlow 2.0 uses it as the primary API.

Building a model in Keras involves four important steps:

1. Defining the model
2. Compiling the model
3. Fitting the model
4. Evaluating the model

Defining the model

The first step is defining the model. Keras provides two different APIs to define the model:

- The sequential API
- The functional API

Defining a sequential model

In a sequential model, we stack each layer, one above another:

```
from keras.models import Sequential
from keras.layers import Dense
```

First, let's define our model as a `Sequential()` model, as follows:

```
model = Sequential()
```

Now, define the first layer, as shown in the following code:

```
model.add(Dense(13, input_dim=7, activation='relu'))
```

In the preceding code, `Dense` implies a fully connected layer, `input_dim` implies the dimension of our input, and `activation` specifies the activation function that we use. We can stack up as many layers as we want, one above another.

Define the next layer with the `relu` activation, as follows:

```
model.add(Dense(7, activation='relu'))
```

Define the output layer with the `sigmoid` activations:

```
model.add(Dense(1, activation='sigmoid'))
```

The final code block of the sequential model is shown as follows. As you can see, the Keras code is much simpler than the TensorFlow code:

```
model = Sequential()
model.add(Dense(13, input_dim=7, activation='relu'))
model.add(Dense(7, activation='relu'))
model.add(Dense(1, activation='sigmoid'))
```

Defining a functional model

A functional model provides more flexibility than a sequential model. For instance, in a functional model, we can easily connect any layer to another layer, whereas, in a sequential model, each layer is in a stack of one above another. A functional model comes in handy when creating complex models, such as directed acyclic graphs, models with multiple input values, multiple output values, and shared layers. Now, we will see how to define a functional model in Keras.

The first step is to define the input dimensions:

```
input = Input(shape=(2,))
```

Now, we'll define our first fully connected layer with 10 neurons and relu activations, using the Dense class, as shown:

```
layer1 = Dense(10, activation='relu')
```

We defined layer1, but where is the input to layer1 coming from? We need to specify the input to layer1 in a bracket notation at the end, as shown:

```
layer1 = Dense(10, activation='relu')(input)
```

We define the next layer, layer2, with 13 neurons and relu activation. The input to layer2 comes from layer1, so that is added in the bracket at the end, as shown in the following code:

```
layer2 = Dense(10, activation='relu')(layer1)
```

Now, we can define the output layer with the sigmoid activation function. Input to the output layer comes from layer2, so that is added in parentheses at the end:

```
output = Dense(1, activation='sigmoid')(layer2)
```

After defining all of the layers, we define the model using a Model class, where we need to specify inputs and outputs, as follows:

```
model = Model(inputs=input, outputs=output)
```

The complete code for the functional model is shown here:

```
input = Input(shape=(2,))
layer1 = Dense(10, activation='relu')(input)
layer2 = Dense(10, activation='relu')(layer1)
output = Dense(1, activation='sigmoid')(layer2)
model = Model(inputs=input, outputs=output)
```

Compiling the model

Now that we have defined the model, the next step is to compile it. In this phase, we set up how the model should learn. We define three parameters when compiling the model:

- The optimizer parameter: This defines the optimization algorithm we want to use; for example, the gradient descent, in this case.
- The loss parameter: This is the objective function that we are trying to minimize; for example, the mean squared error or cross-entropy loss.

- The `metrics` parameter: This is the metric through which we want to assess the model's performance; for example, `accuracy`. We can also specify more than one metric.

Run the following code to compile the model:

```
model.compile(loss='binary_crossentropy', optimizer='sgd',
metrics=['accuracy'])
```

Training the model

We defined and also compiled the model. Now, we will train the model. Training the model can be done using the `fit` function. We specify our features, x; labels, y; the number of `epochs` we want to train; and the `batch_size`, as follows:

```
model.fit(x=data, y=labels, epochs=100, batch_size=10)
```

Evaluating the model

After training the model, we will evaluate the model on the test set:

```
model.evaluate(x=data_test,y=labels_test)
```

We can also evaluate the model on the same train set, and that will help us to understand the training accuracy:

```
model.evaluate(x=data,y=labels)
```

MNIST digit classification using TensorFlow 2.0

Now, we will see how we can perform MNIST handwritten digit classification, using TensorFlow 2.0. It requires only a few lines of code compared to TensorFlow 1.x. As we have learned, TensorFlow 2.0 uses Keras as its high-level API; we just need to add `tf.keras` to the Keras code.

Let's start by loading the dataset:

```
mnist = tf.keras.datasets.mnist
```

Create a train and test set with the following code:

```
(x_train,y_train), (x_test, y_test) = mnist.load_data()
```

Normalize the train and test sets by dividing the values of x by the maximum value of x; that is, 255.0:

```
x_train, x_test = tf.cast(x_train/255.0, tf.float32), tf.cast(x_test/255.0,
tf.float32)
y_train, y_test = tf.cast(y_train,tf.int64),tf.cast(y_test,tf.int64)
```

Define the sequential model as follows:

```
model = tf.keras.models.Sequential()
```

Now, let's add layers to the model. We use a three-layer network with the relu function and softmax in the final layer:

```
model.add(tf.keras.layers.Flatten())
model.add(tf.keras.layers.Dense(256, activation="relu"))
model.add(tf.keras.layers.Dense(128, activation="relu"))
model.add(tf.keras.layers.Dense(10, activation="softmax"))
```

Compile the model by running the following line of code:

```
model.compile(optimizer='sgd', loss='sparse_categorical_crossentropy',
metrics=['accuracy'])
```

Train the model:

```
model.fit(x_train, y_train, batch_size=32, epochs=10)
```

Evaluate the model:

```
model.evaluate(x_test, y_test)
```

That's it! Writing code with the Keras API is that simple.

Should we use Keras or TensorFlow?

We have learned that TensorFlow 2.0 uses Keras as a high-level API. Using a high-level API enables rapid prototyping. But we can't use high-level APIs when we want to build a model on a low level, or if we want to build something that a high-level API doesn't provide.

In addition to this, writing code from scratch strengthens our knowledge of algorithms and helps us to understand and learn concepts better than directly diving into high-level APIs. That's why, in this book, we will code most of the algorithms from scratch using TensorFlow, without using high-level APIs such as Keras. We will be using TensorFlow version 1.13.1.

Summary

We started off this chapter by learning about TensorFlow and how it uses computational graphs. We learned that every computation in TensorFlow is represented by a computational graph, which consists of several nodes and edges, where nodes are mathematical operations, such as addition and multiplication, and edges are tensors.

We learned that variables are containers used to store values and they are used as input to several other operations in a computational graph. Later, we learned that placeholders are like variables, where we only define the type and dimension but will not assign the values, and values for the placeholders will be fed at runtime.

Going forward, we learned about TensorBoard, which is TensorFlow's visualization tool and can be used to visualize a computational graph. It can also be used to plot various quantitative metrics and the results of several intermediate calculations.

We also learned about eager execution, which is more Pythonic, and allows for rapid prototyping. We understood that, unlike the graph mode, where we need to construct a graph every time to perform any operations, eager execution follows the imperative programming paradigm, where any operations can be performed immediately, without having to create a graph, just like we do in Python.

In the next chapter, we will learn about gradient descent and the variants of gradient descent algorithms.

Questions

Assess your knowledge about TensorFlow by answering the following questions:

1. Define a computational graph.
2. What are sessions?
3. How do we create a session in TensorFlow?
4. What is the difference between variables and placeholders?
5. Why do we need TensorBoard?
6. What is the name scope and how is it created?
7. What is eager execution?

Further reading

You can learn more about TensorFlow by checking out the official documentation at `https://www.tensorflow.org/tutorials`.

Section 2: Fundamental Deep Learning Algorithms

In this section, we will explore all the fundamental deep learning algorithms. We first understand each algorithm intuitively, then we will deep dive into the underlying math. We will also learn to implement each algorithm in TensorFlow.

The following chapters are included in this section:

3
Gradient Descent and Its Variants

Gradient descent is one of the most popular and widely used optimization algorithms, and is a first-order optimization algorithm. First-order optimization means that we calculate only the first-order derivative. As we saw in `Chapter 1`, *Introduction to Deep Learning*, we used gradient descent and calculated the first-order derivative of the loss function with respect to the weights of the network to minimize the loss.

Gradient descent is not only applicable to neural networks—it is also used in situations where we need to find the minimum of a function. In this chapter, we will go deeper into gradient descent, starting with the basics, and learn several variants of gradient descent algorithms. There are various flavors of gradient descent that are used for training neural networks. First, we will understand **Stochastic Gradient Descent** (**SGD**) and mini-batch gradient descent. Then, we'll explore how momentum is used to speed up gradient descent to attain convergence. Later in this chapter, we will learn about how to perform gradient descent in an adaptive manner by using various algorithms, such as Adagrad, Adadelta, RMSProp, Adam, Adamax, AMSGrad, and Nadam. We will take a simple linear regression equation and see how we can find the minimum of a linear regression's cost function using various types of gradient descent algorithms.

In this chapter, we will learn about the following topics:

- Demystifying gradient descent
- Gradient descent versus stochastic gradient descent
- Momentum and Nesterov accelerated gradient
- Adaptive methods of gradient descent

Demystifying gradient descent

Before we get into the details, let's understand the basics. What is a function in mathematics? A function represents the relation between input and output. We generally use f to denote a function. For instance, $f(x) = x^2$ implies a function that takes x as an input and returns x^2 as an output. It can also be represented as $y = x^2$.

Here, we have a function, $y = x^2$, and we can plot and see what our function looks like:

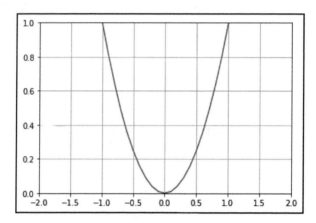

The smallest value of a function is called the **minimum of a function**. As you can see in the preceding plot, the minimum of the x^2 function lies at 0. The previous function is called a **convex function**, and is where we have only one minimum value. A function is called a **non-convex function** when there is more than one minimum value. As we can see in the following diagram, a non-convex function can have many local minima and one global minimum value, whereas a convex function has only one global minimum value:

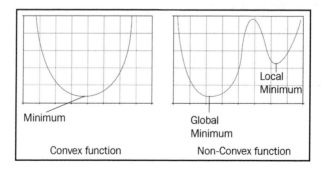

By looking at the graph of the x^2 function, we can easily say that it has its minimum value at $x = 0$. But how can we find the minimum value of a function mathematically? First, let's assume $x = 0.7$. Thus, we are at a position where $x = 0.7$, as shown in the following graph:

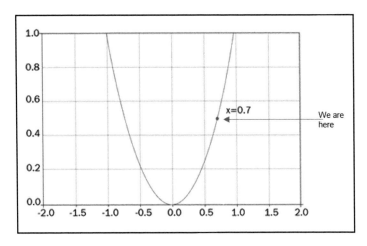

Now, we need to go to zero, which is our minimum value, but how can we reach it? We can reach it by calculating the derivative of the function, $y = x^2$. So, the derivative of the function, y, with respect to x, is as follows:

$$y = x^2$$

$$\frac{dy}{dx} = 2x$$

Since we are at $x = 0.7$ and substituting this in the previous equation, we get the following equation:

$$\frac{dy}{dx} = 2(0.7) = 1.4$$

After calculating the derivative, we update our position of x according to the following update rule:

$$x = x - \frac{dy}{dx}$$

$$x = 0.7 - 1.4$$

$$x = -0.7$$

As we can see in the following graph, we were at *x = 0.7* initially, but, after computing the gradient, we are now at the updated position of *x = -0.7*. However, this is something we don't want because we missed our minimum value, which is *x = 0*, and reached somewhere else:

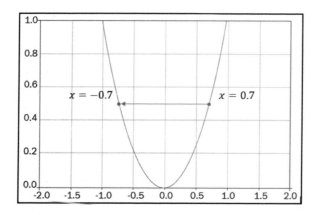

To avoid this, we introduce a new parameter called learning rate, α, in the update rule. It helps us to slow down our gradient steps so that we won't miss out the minimal point. We multiply the gradients by the learning rate and update the x value, as follows:

$$x = x - \alpha \frac{dy}{dx}$$

Let's say that $\alpha = 0.15$; now, we can write the following:

$$x = 0.7 - (0.15 * 1.4)$$

$$x = 0.49$$

As we can see in the following graph, after multiplying the gradients by the learning rate with the updated *x* value, we descended from the initial position, *x = 0.7*, to *x = 0.49*:

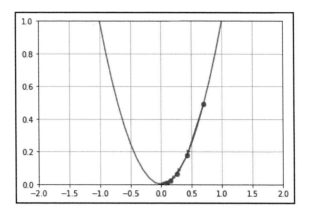

However, this still isn't our optimal minimum value. We need to go further down until we reach the minimum value; that is, $x = 0$. So, for some n number of iterations, we have to repeat the same process until we reach the minimal point. That is, for some n number of iterations, we update the value of x using the following update rule until we reach the minimal point:

$$x = x - \alpha. \frac{dy}{dx}$$

Okay – why is there a minus in the preceding equation? That is, why we are subtracting $\alpha. \frac{dy}{dx}$ from x? Why can't we add them and have our equation as $x = x + \alpha. \frac{dy}{dx}$?

This is because we are finding the minimum of a function, so we need to go downward. If we add x to $\alpha. \frac{dy}{dx}$, then we go upward on every iteration, and we cannot find the minimum value, as shown in the following graphs:

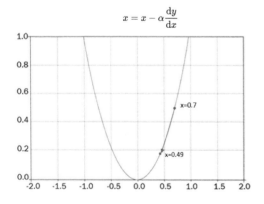
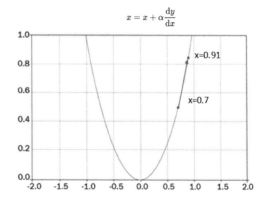

Thus, on every iteration, we compute gradients of y with respect to x, that is, $\frac{dy}{dx}$, multiply the gradients by the learning rate, that is, $\alpha \cdot \frac{dy}{dx}$, and subtract it from the x value to get the updated x value, as follows:

$$x = x - \alpha . \frac{dy}{dx}$$

By repeating this step on every iteration, we go downward from the cost function and reach the minimum point. As we can see in the following graph, we moved downward from the initial position of 0.7 to 0.49, and then, from there, we reached 0.2.

Then, after several iterations, we reach the minimum point, which is 0.0:

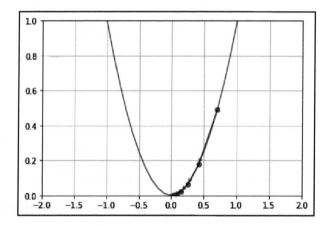

We say we attained **convergence** when we reach the minimum of the function. But the question is: how do we know that we attained convergence? In our example, $y = x^2$, we know that the minimum value is 0. So, when we reach 0, we can say that we found the minimum value that we attained convergence. But how can we mathematically say that 0 is the minimum value of the function, $y = x^2$?

Let's take a closer look at the following graph, which shows how the value of x changes on every iteration. As you may notice, the value of x is 0.009 in the fifth iteration, 0.008 in the sixth iteration, and 0.007 in the seventh iteration. As you can see, there's not much difference between the fifth, sixth, and seventh iterations. When there is little change in the value of x over iterations, then we can conclude that we have attained convergence:

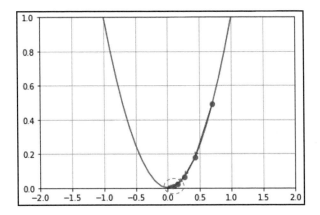

Okay, but what is the use of all this? Why are we trying to find the minimum of a function? When we're training a model, our goal is to minimize the loss function of the model. Thus, with gradient descent, we can find the minimum of the cost function. Finding the minimum of cost function gives us an optimal parameter of the model with which we can obtain the minimal loss. In general, we denote the parameters of the model by θ. The following equation is called the parameter update rule or weight update rule:

$$\theta = \theta - \alpha \cdot \nabla_\theta J(\theta) \tag{1}$$

Here, we have the following:

- θ is the parameter of the model
- α is the learning rate
- $\nabla_\theta J(\theta)$ is the gradient

We update the parameter of the model for several iterations according to the parameter update rule until we attain convergence.

Performing gradient descent in regression

So far, we have understood how the gradient descent algorithm finds the optimal parameters of the model. In this section, we will understand how we can use gradient descent in linear regression and find the optimal parameter.

The equation of a simple linear regression can be expressed as follows:

$$\hat{y} = mx + b$$

Thus, we have two parameters, m and b. Now, we will see how can we use gradient descent and find the optimal values for these two parameters.

Importing the libraries

First, we need to import the required libraries:

```
import warnings
warnings.filterwarnings('ignore')

import random
import math
import numpy as np
from matplotlib import pyplot as plt
%matplotlib inline
```

Preparing the dataset

Next, we will generate some random data points with 500 rows and 2 columns (x and y) and use them for training:

```
data = np.random.randn(500, 2)
```

As you can see, our data has two columns:

```
print data[0]
```

```
array([-0.08575873,  0.45157591])
```

The first column indicates the x value:

```
print data[0,0]
```

```
-0.08575873243708057
```

The second column indicates the y value:

```
print data[0,1]
```

```
0.4515759149158441
```

We know that the equation of a simple linear regression is expressed as follows:

$$\hat{y} = mx + b \tag{2}$$

Thus, we have two parameters, m and b. We store both of these parameters in an array called `theta`. First, we initialize `theta` with zeros, as follows:

```
theta = np.zeros(2)
```

The `theta[0]` function represents the value of m, while the `theta[1]` function represents the value of b:

```
print theta

array([0., 0.])
```

Defining the loss function

The **mean squared error** (**MSE**) of regression is given as follows:

$$J = \frac{1}{N} \sum_{i=1}^{N} (y - \hat{y})^2 \tag{3}$$

Here, N is the number of training samples, y is the actual value, and $\hat{y}$ is the predicted value.

The implementation of the preceding loss function is shown here. We feed the `data` and the model parameter, `theta`, to the loss function, which returns the MSE. Remember that `data[,0]` has an x value and that `data[,1]` has a y value. Similarly, `theta [0]` has a value of m and `theta[1]` has a value of b.

Let's define the loss function:

```
def loss_function(data,theta):
```

Now, we need to get the value of m and b:

```
    m = theta[0]
    b = theta[1]

    loss = 0
```

We do this for each iteration:

```
for i in range(0, len(data)):
```

Now, we get the value of x and y:

```
x = data[i, 0]
y = data[i, 1]
```

Then, we predict the value of $\hat{y}$:

```
y_hat = (m*x + b)
```

Here, we compute the loss as given in equation (3):

```
loss = loss + ((y - (y_hat)) ** 2)
```

Then, we compute the mean squared error:

```
mse = loss / float(len(data))

return mse
```

When we feed our randomly initialized `data` and model parameter, `theta`, `loss_function` returns the mean squared loss, as follows:

```
loss_function(data, theta)

1.0253548008165727
```

Now, we need to minimize this loss. In order to minimize the loss, we need to calculate the gradient of the loss function, J, with respect to the model parameters, m and b, and update the parameter according to the parameter update rule. First, we will calculate the gradients of the loss function.

Computing the gradients of the loss function

The gradients of the loss function, J, with respect to the parameter m, are given as follows:

$$\frac{dJ}{dm} = \frac{2}{N} \sum_{i=1}^{N} -x_i \left(y_i - (mx_i + b) \right) \tag{4}$$

The gradients of the loss function, J, with respect to the parameter b, are given as follows:

$$\frac{dJ}{db} = \frac{2}{N} \sum_{i=1}^{N} - (y_i - (mx_i + b)) \qquad (5)$$

We define a function called `compute_gradients`, which takes the parameters, `data` and `theta` as input and returns the computed gradients:

```
def compute_gradients(data, theta):
```

Now, we need to initialize the gradients:

```
gradients = np.zeros(2)
```

Then, we need to save the total number of data points in `N`:

```
N = float(len(data))
```

Now, we can get the value of m and b:

```
m = theta[0]
b = theta[1]
```

We do the same for each iteration:

```
for i in range(0, len(data)):
```

Then, we get the value of x and y:

```
x = data[i, 0]
y = data[i, 1]
```

Now, we compute the gradient of the loss with respect to m, as given in equation *(4)*:

```
gradients[0] += - (2 / N) * x * (y - (( m* x) + b))
```

Then, we compute the gradient of the loss with respect to b, as given in equation *(5)*:

```
gradients[1] += - (2 / N) * (y - ((theta[0] * x) + b))
```

We need to add `epsilon` to avoid division by zero error:

```
epsilon = 1e-6
gradients = np.divide(gradients, N + epsilon)

return gradients
```

When we feed our randomly initialized `data` and `theta` model parameter, the `compute_gradients` function returns the gradients with respect to m, that is, $\frac{dJ}{dm}$, and gradients with respect to b, that is, $\frac{dJ}{db}$, as follows:

```
compute_gradients(data,theta)

array([-9.08423989e-05,  1.05174511e-04])
```

Updating the model parameters

Now that we've computed the gradients, we need to update our model parameters according to our update rule, as follows:

$$m = m - \alpha \frac{dJ}{dm} \qquad (6)$$

$$b = b - \alpha \frac{dJ}{db} \qquad (7)$$

Since we stored m in `theta[0]` and b in `theta[1]`, we can write our update equation as follows:

$$\theta = \theta - \alpha \frac{dJ}{d\theta} \qquad (8)$$

As we learned in the previous section, updating gradients on just one iteration will not lead us to convergence, that is, the minimum of the cost function, so we need to compute gradients and update the model parameter for several iterations.

First, we need to set the number of iterations:

```
num_iterations = 50000
```

Now, we need to define the learning rate:

```
lr = 1e-2
```

Next, we will define a list called `loss` for storing the loss on every iteration:

```
loss = []
```

On each iteration, we will calculate and update the gradients according to our parameter update rule from equation *(8)*:

```
theta = np.zeros(2)

for t in range(num_iterations):
    #compute gradients
    gradients = compute_gradients(data, theta)
    #update parameter
    theta = theta - (lr*gradients)
    #store the loss
    loss.append(loss_function(data,theta))
```

Now, we need to plot the `loss` (`Cost`) function:

```
plt.plot(loss)
plt.grid()
plt.xlabel('Training Iterations')
plt.ylabel('Cost')
plt.title('Gradient Descent')
```

The following plot shows how the loss (**Cost**) decreases over the training iterations:

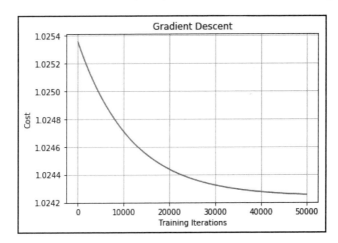

Thus, we learned that gradient descent can be used to find the optimal parameters of the model, which we can then use to minimize the loss. In the next section, we will learn about several variants of the gradient descent algorithm.

Gradient descent versus stochastic gradient descent

We update the parameter of the model multiple times with our parameter update equation *(1)* until we find the optimal parameter value. In gradient descent, to perform a single parameter update, we iterate through all the data points in our training set. So, every time we update the parameters of the model, we iterate through all the data points in the training set. Updating the parameters of the model only after iterating through all the data points in the training set makes gradient descent very slow and it will increase the training time, especially when we have a large dataset.

Let's say we have a training set with 1 million data points. We know that we update the parameters of the model multiple times to find the optimal parameter value. So, even to perform a single parameter update, we go through all 1 million data points in our training set and then update the model parameters. This will definitely make the training slow. This is because we can't just find the optimal parameter with a single update; we need to update the parameters of the model several times to find the optimal value. So, if we iterate through all 1 million data points in our training set for every parameter update, it will definitely slow down our training.

Thus, to combat this, we introduce **stochastic gradient descent (SGD)**. Unlike gradient descent, we don't have to wait to update the parameter of the model after iterating all the data points in our training set; we just update the parameters of the model after iterating through every single data point in our training set.

Since we update the parameters of the model in SGD after iterating every single data point, it will learn the optimal parameter of the model faster compared to gradient descent, and this will minimize the training time.

How is SGD useful? When we have a huge dataset, by using the vanilla gradient descent method, we update the parameters only after iterating through all the data points in that huge dataset. So, after many iterations over the whole dataset, we reach convergence and, apparently, it takes a long time. But, in SGD, we update the parameters after iterating through every single training sample. That is, we are learning to find the optimal parameters right from the first training sample, which helps to attain convergence faster compared to the vanilla gradient descent method.

We know that the epoch specifies the number of times the neural network sees the whole training data. Thus, in gradient descent, on each epoch, we perform the parameter update. This means that, after every epoch, the neural networks see the whole training data. We perform the parameter update for each epoch as follows:

$$\theta = \theta - \alpha \cdot \nabla_\theta J(\theta)$$

However, in stochastic gradient descent, we don't have to wait until the completion of each epoch to update the parameters. That is, we don't have to wait until the neural network sees the whole training data to update the parameters. Instead, we update the parameters of the network right from seeing a single training sample for each epoch:

$$\theta = \theta - \alpha \cdot \nabla_\theta J(\theta)$$

The following contour plot shows how gradient descent and stochastic gradient descent perform the parameter updates and find the minimum cost. The star symbol in the center of the plot denotes the position where we have a minimum cost. As you can see, SGD reaches convergence faster than vanilla gradient descent. You can also observe the oscillations in the gradient steps on SGD; this is because we are updating the parameter for every training sample, so, the gradient step in SGD changes frequently compared to the vanilla gradient descent:

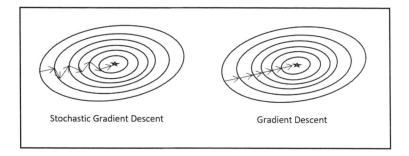

There is also another variant of gradient descent called **mini-batch gradient descent**. It takes the pros of both vanilla gradient descent and stochastic gradient descent. In SGD, we saw that we update the parameter of the model for every training sample. However, in mini-batch gradient descent, instead of updating the parameters after iterating each training sample, we update the parameters after iterating some batches of data points. Let's say the batch size is 50, which means that we update the parameter of the model after iterating through 50 data points instead of updating the parameter after iterating through each individual data point.

The following diagram shows the contour plot of SGD and mini-batch gradient descent:

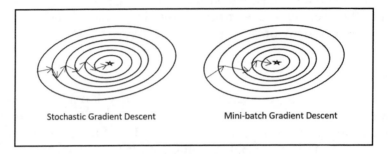

Here are the differences between these types of gradient descent in a nutshell:

- **Gradient descent**: Updates the parameters of the model after iterating through all the data points in the training set
- **Stochastic gradient descent**: Updates the parameter of the model after iterating through every single data point in the training set
- **Mini-batch gradient descent**: Updates the parameters of the model after iterating n number of data points in the training set

 Mini-batch gradient descent is preferred over vanilla gradient descent and SGD for large datasets since mini-batch gradient descent outperforms the other two.

The code for mini-batch gradient descent is as follows.

First, we need to define the `minibatch` function:

```
def minibatch(data, theta, lr = 1e-2, minibatch_ratio = 0.01,
num_iterations = 1000):
```

Next, we will define `minibatch_size` by multiplying the length of data by `minibatch_ratio`:

```
minibatch_size = int(math.ceil(len(data) * minibatch_ratio))
```

Now, on each iteration, we perform the following:

```
for t in range(num_iterations):
```

Next, select `sample_size`:

```
sample_size = random.sample(range(len(data)), minibatch_size)
np.random.shuffle(data)
```

Now, sample the data based on `sample_size`:

```
sample_data = data[0:sample_size[0], :]
```

Compute the gradients for `sample_data` with respect to `theta`:

```
grad = compute_gradients(sample_data, theta)
```

After computing the gradients for the sampled data with the given mini-batch size, we update the model parameter, `theta`, as follows:

```
theta = theta - (lr * grad)
```

```
return theta
```

Momentum-based gradient descent

In this section, we will learn about two new variants of gradient descent, called **momentum** and Nesterov accelerated gradient.

Gradient descent with momentum

We have a problem with SGD and mini-batch gradient descent due to the oscillations in the parameter update. Take a look at the following plot, which shows how mini-batch gradient descent is attaining convergence. As you can see, there are oscillations in the gradient steps. The oscillations are shown by the dotted line. As you may notice, it is making a gradient step toward one direction, and then taking a different direction, and so on, until it reaches convergence:

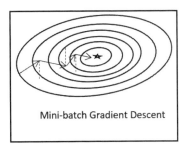

Mini-batch Gradient Descent

This oscillation occurs because, since we update the parameters after iterating every n number of data points, the direction of the update will have some variance, and this leads to oscillations in every gradient step. Due to this oscillation, it is hard to reach convergence, and it slows down the process of attaining it.

To alleviate this, we'll introduce a new technique called **momentum**. If we can understand what the right direction is for the gradient steps to attain convergence faster, then we can make our gradient steps navigate in that direction and reduce the oscillation in the irrelevant directions; that is, we can reduce taking directions that do not lead us to convergence.

So, how can we do this? We basically take a fraction of the parameter update from the previous gradient step and add it to the current gradient step. In physics, momentum keeps an object moving after a force is applied. Here, the momentum keeps our gradient moving toward the direction that leads to convergence.

If you take a look at the following equation, you can see we are basically taking the parameter update from the previous step, v_{t-1}, and adding it to the current gradient step, $\nabla_\theta J(\theta)$. How much information we want to take from the previous gradient step depends on the factor, that is, γ, and the learning rate, which is denoted by η:

$$v_t = \gamma v_{t-1} + \eta \nabla_\theta J(\theta)$$

In the preceding equation, v_t is called velocity, and it accelerates gradients in the direction that leads to convergence. It also reduces oscillations in an irrelevant direction by adding a fraction of a parameter update from the previous step to the current step.

Thus, the parameter update equation with momentum is expressed as follows:

$$\boxed{\theta = \theta - v_t}$$

By doing this, performing mini-batch gradient descent with momentum helps us to reduce oscillations in gradient steps and attain convergence faster.

Now, let's look at the implementation of momentum.

First, we define the momentum function, as follows:

```
def momentum(data, theta, lr = 1e-2, gamma = 0.9, num_iterations = 1000):
```

Then, we initialize `vt` with zeros:

```
vt = np.zeros(theta.shape[0])
```

The following code is executed to cover the range for each iteration:

```
for t in range(num_iterations):
```

Now, we compute `gradients` with respect to `theta`:

```
gradients = compute_gradients(data, theta)
```

Next, we update `vt` to be $v_t = \gamma v_{t-1} + \eta \nabla_\theta J(\theta)$:

```
vt = gamma * vt + lr * gradients
```

Now, we update the model parameter, `theta`, as $\theta = \theta - v_t$:

```
theta = theta - vt
```

```
    return theta
```

Nesterov accelerated gradient

One problem with momentum is that it might miss out the minimum value. That is, as we move closer toward convergence (the minimum point), the value for momentum will be high. When the value of momentum is high while we are near to attaining convergence, then the momentum actually pushes the gradient step high and it might miss out on the actual minimum value; that is, it might overshoot the minimum value when the momentum is high when we are near to convergence, as shown in the following diagram:

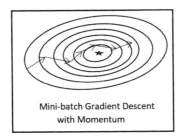

Mini-batch Gradient Descent
with Momentum

To overcome this, Nesterov introduced a new method called **Nesterov accelerated gradient** (**NAG**).

The fundamental motivation behind Nesterov momentum is that, instead of calculating the gradient at the current position, we calculate gradients at the position where the momentum would take us to, and we call that position the lookahead position.

What does this mean, though? In the *Gradient descent with momentum* section, we learned about the following equation:

$$v_t = \gamma v_{t-1} + \eta \nabla_\theta J(\theta) \tag{9}$$

The preceding equation tells us that we are basically pushing the current gradient step, $\nabla_\theta J(\theta)$, to a new position using a fraction of the parameter update from the previous step, γv_{t-1}, which will help us to attain convergence. However, when the momentum is high, this new position will actually overshoot the minimum value.

Thus, before making a gradient step with momentum and reaching a new position, if we understand which position the momentum will take us to, then we can avoid overshooting the minimum value. If we find out that momentum will take us to the position that actually misses the minimum value, then we can slow down the momentum and try to reach the minimum value.

But how can we find the position that the momentum will take us to? In equation (2), instead of calculating gradients with respect to the current gradient step, $\nabla_\theta J(\theta)$, we calculate gradients with respect to $\nabla_\theta J(\theta - \gamma v_{t-1})$. The term, $\theta - \gamma v_{t-1}$, basically tells us the approximate position of where our next gradient step is going to be. We call this the lookahead position. This gives us an idea of where our next gradient step is going to be.

So, we can rewrite our v_t equation according to NAG as follows:

$$v_t = \gamma v_{t-1} + \eta \nabla_\theta J(\theta - \gamma v_{t-1})$$

We update our parameter as follows:

$$\boxed{\theta = \theta - v_t}$$

Updating the parameters with the preceding equation prevents us from missing the minimum value by slowing down the momentum when the gradient steps are near to convergence. The Nesterov accelerated method is implemented as follows.

First, we define the NAG function:

```
def NAG(data, theta, lr = 1e-2, gamma = 0.9, num_iterations = 1000):
```

Then, we initialize the value of `vt` with zeros:

```
vt = np.zeros(theta.shape[0])
```

For every iteration, we perform the following steps:

```
for t in range(num_iterations):
```

Now, we need to compute the gradients with respect to $\theta - \gamma v_{t-1}$:

```
gradients = compute_gradients(data, theta - gamma * vt)
```

Then, we update `vt` as $v_t = \gamma v_{t-1} + \eta \nabla_\theta J(\theta - \gamma v_{t-1})$:

```
vt = gamma * vt + lr * gradients
```

Now, we update the model parameter, `theta`, to $\theta = \theta - v_t$:

```
theta = theta - vt

return theta
```

Adaptive methods of gradient descent

In this section, we will learn about several adaptive versions of gradient descent.

Setting a learning rate adaptively using Adagrad

When we build a deep neural network, we have many parameters. Parameters are basically the weights of the network, so when we build a network with many layers, we will have many weights, say, $\theta^1, \theta^2, \theta^3 .. \theta^i .. \theta^n$. Our goal is to find the optimal values for all these weights. In all of the previous methods we learned about, the learning rate was a common value for all the parameters of the network. However **Adagrad** (short for **adaptive gradient**) adaptively sets the learning rate according to a parameter.

Parameters that have frequent updates or high gradients will have a slower learning rate, while a parameter that has an infrequent update or small gradients will also have a slower learning rate. But why do we have to do this? It is because parameters that have infrequent updates implies that they are not trained enough, so we set a high learning rate for them, and parameters that have frequent updates implies that they are trained enough, so we set their learning rate to a low value so that we don't overshoot the minimum.

Now, let's see how Adagrad adaptively changes the learning rate. Previously, we represented the gradient with $\nabla_\theta J(\theta)$. For simplicity, from now on in this chapter, we'll represent gradients with g. So, the gradient of a parameter, θ^i, at an iteration, t, can be represented as follows:

$$g_t^i = \nabla_\theta J(\theta_t^i)$$

Therefore, we can rewrite our update equation with g as the gradient notation as follows:

$$\theta_t^i = \theta_{t-1}^i - \eta \cdot g_t^i$$

Now, for every iteration, t, to update a parameter, θ^i, we divide the learning rate by the sum of squares of all previous gradients of the parameter, θ^i, as follows:

$$\theta_t^i = \theta_{t-1}^i - \frac{\eta}{\sqrt{\sum_{\tau=1}^t (g_\tau^i)^2 + \epsilon}} \cdot g_t^i$$

Here, $\sqrt{\sum_{\tau=1}^t (g_\tau^i)^2 + \epsilon}$ implies the sum of squares of all previous gradients of the parameter θ^i. We added ϵ just to avoid the division by zero error. We typically set the value of ϵ to a small number. The question that arises here is, why are we dividing the learning rate by a sum of squares of all the previous gradients?

We learned that parameters that have frequent updates or high gradients will have a slower learning rate, while parameters that have an infrequent update or small gradients will also have a high learning rate.

The sum, $\sqrt{\sum_{\tau=1}^t (g_\tau^i)^2 + \epsilon}$, actually scales our learning rate. That is, when the sum of the squared past gradients has a high value, we are basically dividing the learning rate by a high value, so our learning rate will become less. Similarly, if the sum of the squared past gradients has a low value, we are dividing the learning rate by a lower value, so our learning rate value will become high. This implies that the learning rate is inversely proportional to the sum of the squares of all the previous gradients of the parameter.

Here, our update equation is expressed as follows:

$$\theta_t^i = \theta_{t-1}^i - \frac{\eta}{\sqrt{\sum_{\tau=1}^t (g_\tau^i)^2 + \epsilon}} \cdot g_t^i$$

In a nutshell, in Adagrad, we set the learning rate to a low value when the previous gradient value is high, and to a high value when the past gradient value is lower. This means that our learning rate value changes according to the past gradient updates of the parameters.

Now that we have learned how the Adagrad algorithm works, let's strengthen our knowledge by implementing it. The code for the Adagrad algorithm is given as follows.

First, define the `AdaGrad` function:

```
def AdaGrad(data, theta, lr = 1e-2, epsilon = 1e-8, num_iterations =
10000):
```

Define the variable called `gradients_sum` to hold the sum of gradients and initialize them with zeros:

```
gradients_sum = np.zeros(theta.shape[0])
```

For every iteration, we perform the following steps:

```
for t in range(num_iterations):
```

Then, we compute the `gradients` of loss with respect to `theta`:

```
gradients = compute_gradients(data, theta)
```

Now, we calculate the sum of the gradients squared, that is, $\sum_{\tau=1}^{t}(g_\tau^i)^2$:

```
gradients_sum += gradients ** 2
```

Afterward, we compute the gradient updates, that is, $\frac{g_t^i}{\sqrt{\sum_{\tau=1}^t (g_\tau^i)^2 + \epsilon}}$:

```
gradient_update = gradients / (np.sqrt(gradients_sum + epsilon))
```

Now, update the `theta` model parameter so that it's $\theta_t^i = \theta_{t-1}^i - \dfrac{\eta}{\sqrt{\sum_{\tau=1}^{t}(g_\tau^i)^2 + \epsilon}} \cdot g_t^i$:

```
theta = theta - (lr * gradient_update)

return theta
```

Again, there is a shortcoming associated with the Adagrad method. For every iteration, we are accumulating and summing all the past squared gradients. So, on every iteration, our sum of the squared past gradients value will increase. When the sum of the squared past gradient value is high, we will have a large number in the denominator. When we divide the learning rate by a very large number, then the learning rate will become very small. So, over several iterations, the learning rate starts decaying and becomes an infinitesimally small number – that is, our learning rate will be monotonically decreasing. When the learning rate reaches a very low value, then it takes a long time to attain convergence.

In the next section, we will see how Adadelta tackles this shortcoming.

Doing away with the learning rate using Adadelta

Adadelta is an enhancement of the Adagrad algorithm. In Adagrad, we noticed the problem of the learning rate diminishing to a very low number. Although Adagrad learns the learning rate adaptively, we still need to set the initial learning rate manually. However, in Adadelta, we don't need the learning rate at all. So how does the Adadelta algorithm learn?

In Adadelta, instead of taking the sum of all the squared past gradients, we can set a window of size w and take the sum of squared past gradients only from that window. In Adagrad, we took the sum of all the squared past gradients and it led to the learning rate diminishing to a low number. To avoid that, we take the sum of the squared past gradients only from a window.

If w is the window size, then our parameter update equation becomes the following:

$$\theta_t^i = \theta_{t-1}^i - \frac{\eta}{\sqrt{\sum_{\tau=t-w+1}^{t}(g_\tau^i)^2 + \epsilon}} \cdot g_t^i$$

However, the problem is that, although we are taking gradients only from within a window, w, squaring and storing all the gradients from the window in each iteration is inefficient. So, instead of doing that, we can take the running average of gradients.

We compute the running average of gradients at an iteration, t, $E[g^2]_t$, by adding the previous running average of gradients, $E[g^2]_{t-1}$, and current gradients, g_t^2:

$$E[g^2]_t = E[g^2]_{t-1} + g_t^2$$

Instead of just taking the running average, we take the exponentially decaying running average of gradients, as follows:

$$E[g^2]_t = \gamma E[g^2]_{t-1} + (1 - \gamma)g_t^2 \tag{10}$$

Here, γ is called the exponential decaying rate and is similar to the one we saw in momentum – that is, it is used for deciding how much information from the previous running average of gradients should be added.

Now, our update equation becomes the following:

$$\theta_t^i = \theta_{t-1}^i - \frac{\eta}{\sqrt{E[g^2]_t + \epsilon}} \cdot g_t^i$$

For notation simplicity, let's denote $- \frac{\eta}{\sqrt{E[g^2]_t + \epsilon}} \cdot g_t^i$ as $\nabla\theta_t$ so that we can rewrite the previous update equation as follows:

$$\theta_t^i = \theta_{t-1}^i + \nabla\theta_t \tag{11}$$

From the previous equation, we can infer the following:

$$\nabla\theta_t = -\frac{\eta}{\sqrt{E[g^2]_t + \epsilon}} \cdot g_t^i \tag{12}$$

If you look at the denominator in the previous equation, we are basically computing the root mean squared of gradients up to an iteration, t, so we can simply write that in shorthand as follows:

$$RMS[g_t] = \sqrt{E[g^2]_t + \epsilon} \tag{13}$$

By substituting equation *(13)* in equation *(12)*, we can write the following:

$$\nabla\theta_t = -\frac{\eta}{RMS[g_t]} \cdot g_t^i \tag{14}$$

However, we still have the learning rate, η, term in our equation. How can we do away with that? We can do so by making the units of the parameter update in accordance with the parameter. As you may have noticed, the units of θ_t and $\nabla\theta_t$ don't really match. To combat this, we compute the exponentially decaying average of the parameter updates, $\nabla\theta_t$, as we computed an exponentially decaying average of gradients, g_t, in equation *(10)*. So, we can write the following:

$$E[\Delta\theta^2]_t = \gamma E[\Delta\theta^2]_{t-1} + (1-\gamma)\Delta\theta_t^2$$

It's like the RMS of gradients, $RMS[g_t]$, which is similar to equation *(13)*. We can write the RMS of the parameter update as follows:

$$RMS[\Delta\theta]_t = \sqrt{E[\Delta\theta^2]_t + \epsilon}$$

However, the RMS value for the parameter update, $\Delta\theta_t$ is not known, that is, $RMS[\Delta\theta]_t$ is not known, so we can just approximate it by considering until the previous update, $RMS[\Delta\theta]_{t-1}$.

Now, we just replace our learning rate with the RMS value of the parameter updates. That is, we replace η with $RMS[\Delta\theta]_{t-1}$ in equation *(14)* and write the following:

$$\nabla\theta_t = -\frac{RMS[\Delta\theta]_{t-1}}{RMS[g_t]} \cdot g_t^i \tag{15}$$

Substituting equation *(15)* in equation *(11)*, our final update equation becomes the following:

$$\theta_t^i = \theta_{t-1}^i - \frac{RMS[\Delta\theta]_{t-1}}{RMS[g_t]} \cdot g_t^i$$

$$\boxed{\theta_t^i = \theta_{t-1}^i + \nabla\theta_t}$$

Now, let's understand the Adadelta algorithm by implementing it.

First, we define the `AdaDelta` function:

```
def AdaDelta(data, theta, gamma = 0.9, epsilon = 1e-5, num_iterations =
1000):
```

Then, we initialize the `E_grad2` variable with zero for storing the running average of gradients, and `E_delta_theta2` with zero for storing the running average of the parameter update, as follows:

```
# running average of gradients
E_grad2 = np.zeros(theta.shape[0])

#running average of parameter update
E_delta_theta2 = np.zeros(theta.shape[0])
```

For every iteration, we perform the following steps:

```
for t in range(num_iterations):
```

Now, we need to compute the `gradients` with respect to `theta`:

```
gradients = compute_gradients(data, theta)
```

Then, we can compute the running average of gradients:

```
E_grad2 = (gamma * E_grad2) + ((1. - gamma) * (gradients ** 2))
```

Here, we will compute `delta_theta`, that is,
$$\nabla\theta_t = -\frac{RMS[\Delta\theta]_{t-1}}{RMS[g_t]} \cdot g_t^i$$:

```
delta_theta = - (np.sqrt(E_delta_theta2 + epsilon)) /
(np.sqrt(E_grad2 + epsilon)) * gradients
```

Now, we can compute the running average of the parameter update,
$E[\Delta\theta^2]_t = \gamma E[\Delta\theta^2]_{t-1} + (1-\gamma)\Delta\theta_t^2$:

```
E_delta_theta2 = (gamma * E_delta_theta2) + ((1. - gamma) *
(delta_theta ** 2))
```

Next, we will update the parameter of the model, `theta`, so that it's $\theta_t^i = \theta_{t-1}^i + \nabla\theta_t$:

```
theta = theta + delta_theta

return theta
```

Overcoming the limitations of Adagrad using RMSProp

Similar to Adadelta, RMSProp was introduced to combat the decaying learning rate problem of Adagrad. So, in RMSProp, we compute the exponentially decaying running average of gradients as follows:

$$E[g^2]_t = \gamma E[g^2]_{t-1} + (1 - \gamma)g_t^2$$

Instead of taking the sum of the square of all the past gradients, we use this running average of gradients. This means that our update equation becomes the following:

$$\theta_t^i = \theta_{t-1}^i - \frac{\eta}{\sqrt{E[g^2]_t + \epsilon}} \cdot g_t^i$$

It is recommended to assign a value of learning η to 0.9. Now, we will learn how to implement RMSProp in Python.

First, we need to define the `RMSProp` function:

```
def RMSProp(data, theta, lr = 1e-2, gamma = 0.9, epsilon = 1e-6,
num_iterations = 1000):
```

Now, we need to initialize the `E_grad2` variable with zeros to store the running average of gradients:

```
E_grad2 = np.zeros(theta.shape[0])
```

For every iteration, we perform the following steps:

```
for t in range(num_iterations):
```

Then, we compute the `gradients` with respect to `theta`:

```
gradients = compute_gradients(data, theta)
```

Next, we compute the running average of the gradients, that is, $E[g^2]_t = \gamma E[g^2]_{t-1} + (1 - \gamma)g_t^2$:

```
E_grad2 = (gamma * E_grad2) + ((1. - gamma) * (gradients ** 2))
```

Now, we update the parameter of the model, theta, so that it's $\theta_t^i = \theta_{t-1}^i - \dfrac{\eta}{\sqrt{E[g^2]_t + \epsilon}} \cdot g_t^i$:

```
    theta = theta - (lr / (np.sqrt(E_grad2 + epsilon)) * gradients)
return theta
```

Adaptive moment estimation

Adaptive moment estimation, known as **Adam** for short, is one of the most popularly used algorithms for optimizing a neural network. While reading about RMSProp, we learned that we compute the running average of squared gradients to avoid the diminishing learning rate problem:

$$E[g^2]_t = \gamma E[g^2]_{t-1} + (1 - \gamma)g_t^2$$

The final updated equation of RMSprop is given as follows:

$$\theta_t^i = \theta_{t-1}^i - \frac{\eta}{\sqrt{E[g^2]_t + \epsilon}} \cdot g_t^i$$

Similar to this, in Adam, we also compute the running average of the squared gradients. However, along with computing the running average of the squared gradients, we also compute the running average of the gradients.

The running average of gradients is given as follows:

$$E[g]_t = \beta_1 E[g]_{t-1} + (1 - \beta_1)g_t \qquad (16)$$

The running average of squared gradients is given as follows:

$$E[g^2]_t = \beta_2 E[g^2]_{t-1} + (1 - \beta_2)g_t^2 \qquad (17)$$

Since a lot of literature and libraries represent the decaying rate in Adam as β instead of γ, we'll also use β to represent the decaying rate in Adam. Thus, β_1 and β_2 in equations *(16)* and *(17)* denote the exponential decay rates for the running average of the gradients and the squared gradients, respectively.

So, our updated equation becomes the following:

$$\theta_t^i = \theta_{t-1}^i - \frac{\eta}{\sqrt{E[g^2]_t} + \epsilon} \cdot E[g]_t$$

The running average of the gradients and running average of the squared gradients are basically the first and second moments of those gradients. That is, they are the mean and uncentered variance of our gradients, respectively. So, for notation simplicity, let's denote $E[g]_t$ as m_t and $E[g^2]_t$ as v_t.

Therefore, we can rewrite equations *(16)* and *(17)* as follows:

$$m_t = \beta_1 m_{t-1} + (1 - \beta_1) g_t$$

$$v_t = \beta_2 v_{t-1} + (1 - \beta_2) g_t^2$$

We begin by setting the initial moments estimates to zero. That is, we initialize m_t and v_t with zeros. When the initial estimates are set to 0, they remain very small, even after many iterations. This means that they would be biased toward 0, especially when β_1 and β_2 are close to 1. So, to combat this, we compute the bias-corrected estimates of m_t and v_t by just dividing them by $1 - \beta^t$, as follows:

$$\hat{m}_t = \frac{m_t}{1 - \beta_1^t}$$

$$\hat{v}_t = \frac{v_t}{1 - \beta_2^t}$$

Here, $\hat{m}_t$ and $\hat{v}_t$ are the bias-corrected estimates of m_t and v_t, respectively.

So, our final update equation is given as follows:

$$\theta_t = \theta_{t-1} - \frac{\eta}{\sqrt{\hat{v}_t} + \epsilon} \hat{m}_t$$

Now, let's understand how to implement Adam in Python.

First, let's define the `Adam` function, as follows:

```
def Adam(data, theta, lr = 1e-2, beta1 = 0.9, beta2 = 0.9, epsilon = 1e-6,
num_iterations = 1000):
```

Then, we initialize the first moment, `mt`, and the second moment, `vt`, with `zeros`:

```
mt = np.zeros(theta.shape[0])
vt = np.zeros(theta.shape[0])
```

For every iteration, we perform the following steps:

```
for t in range(num_iterations):
```

Next, we compute the `gradients` with respect to `theta`:

```
gradients = compute_gradients(data, theta)
```

Then, we update the first moment, `mt`, so that it's $m_t = \beta_1 m_{t-1} + (1 - \beta_1) g_t$:

```
mt = beta1 * mt + (1. - beta1) * gradients
```

Next, we update the second moment, `vt`, so that it's $v_t = \beta_2 v_{t-1} + (1 - \beta_2) g_t^2$:

```
vt = beta2 * vt + (1. - beta2) * gradients ** 2
```

Now, we compute the bias-corrected estimate of `mt`, that is, $\hat{m}_t = \dfrac{m_t}{1 - \beta_1^t}$:

```
mt_hat = mt / (1. - beta1 ** (t+1))
```

Next, we compute the bias-corrected estimate of `vt`, that is, $\hat{v}_t = \dfrac{v_t}{1 - \beta_2^t}$:

```
vt_hat = vt / (1. - beta2 ** (t+1))
```

Finally, we update the model parameter, `theta`, so that it's $\theta_t = \theta_{t-1} - \dfrac{\eta}{\sqrt{\hat{v}_t} + \epsilon} \hat{m}_t$:

```
theta = theta - (lr / (np.sqrt(vt_hat) + epsilon)) * mt_hat

return theta
```

Adamax – Adam based on infinity-norm

Now, we will look at a small variant of the Adam algorithm called **Adamax**. Let's recall the equation of the second-order moment in Adam:

$$v_t = \beta_2 v_{t-1} + (1 - \beta_2) g_t^2$$

As you may have noticed from the preceding equation, we scale the gradients inversely proportional to the L^2 norm of the current and past gradients (L^2 norm basically means the square of values):

$$v_t = \beta_2 v_{t-1} + (1 - \beta_2)|g_t|^2$$

Instead of having just L^2, can we generalize it to the L^p norm? In general, when we have a large p for norm, our update would become unstable. However, when we set the p value to ∞, that is, when L^∞, the v_t equation becomes simple and stable. Instead of just parameterizing the gradients, g_t, alone, we also parameterize the decay rate, β_2. Thus, we can write the following:

$$v_t = \beta_2^\infty v_{t-1} + (1 - \beta_2^\infty)|g_t|^\infty$$

When we set the limits, p tends to reach infinity, and then we get the following final equation:

$$v_t = max(\beta_2^{t-1}|g_1|, \beta_2^{t-2}|g_2|\ldots\beta_2|g_{t-1}|, |g_t|)$$

You can check the paper listed in the *Further reading* section at the end of this chapter to see how exactly this is derived.

We can rewrite the preceding equation as a simple recursive equation, as follows:

$$v_t = max(\beta_2 \cdot v_{t-1}, |g_t|)$$

Computing m_t is similar to what we saw in the *Adaptive moment estimation* section, so we can write the following directly:

$$m_t = \beta_1 m_{t-1} + (1 - \beta_1)g_t$$

By doing this, we can compute the bias-corrected estimate of m_t:

$$\hat{m}_t = \frac{m_t}{1 - \beta_1^t}$$

Therefore, the final update equation becomes the following:

$$\boxed{\theta_t = \theta_{t-1} - \frac{\eta}{v_t}\hat{m}_t}$$

To better understand the Adamax algorithm, let's code it, step by step.

First, we define the `Adamax` function, as follows:

```
def Adamax(data, theta, lr = 1e-2, beta1 = 0.9, beta2 = 0.999, epsilon =
1e-6, num_iterations = 1000):
```

Then, we initialize the first moment, `mt`, and the second moment, `vt`, with zeros:

```
mt = np.zeros(theta.shape[0])
vt = np.zeros(theta.shape[0])
```

For every iteration, we perform the following steps:

```
for t in range(num_iterations):
```

Now, we can compute the gradients with respect to `theta`, as follows:

```
gradients = compute_gradients(data, theta)
```

Then, we compute the first moment, `mt`, as $m_t = \beta_1 m_{t-1} + (1 - \beta_1)g_t$:

```
mt = beta1 * mt + (1. - beta1) * gradients
```

Next, we compute the second moment, `vt`, as $v_t = max(\beta_2 \cdot v_{t-1}, |g_t|)$:

```
vt = np.maximum(beta2 * vt, np.abs(gradients))
```

Now, we can compute the bias-corrected estimate of `mt`; that is, $\hat{m}_t = \frac{m_t}{1 - \beta_1^t}$:

```
mt_hat = mt / (1. - beta1 ** (t+1))
```

Update the model parameter, `theta`, so that it's $\theta_t = \theta_{t-1} - \frac{\eta}{v_t}\hat{m}_t$:

```
theta = theta - ((lr / (vt + epsilon)) * mt_hat)
return theta
```

Adaptive moment estimation with AMSGrad

One problem with the Adam algorithm is that it sometimes fails to attain optimal convergence, or it reaches a suboptimal solution. It has been noted that, in some settings, Adam fails to attain convergence or reach the suboptimal solution instead of a global optimal solution. This is due to exponentially moving the averages of gradients. Remember when we used the exponential moving averages of gradients in Adam to avoid the problem of learning rate decay?

However, the problem is that since we are taking an exponential moving average of gradients, we miss out information about the gradients that occur infrequently.

To resolve this issue, the authors of AMSGrad made a small change to the Adam algorithm. Recall the second-order moment estimates we saw in Adam, as follows:

$$v_t = \beta_2 v_{t-1} + (1 - \beta_2) g_t^2$$

In AMSGrad, we use a slightly modified version of v_t. Instead of using v_t directly, we take the maximum value of v_t until the previous step, as follows:

$$\hat{v}_t = \max(\hat{v}_{t-1}, v_t)$$

This will retain the informative gradients instead of being phased out due to the exponential moving average.

So, our final update equation becomes the following:

$$\theta_t = \theta_{t-1} - \frac{\eta}{\sqrt{\hat{v}_t} + \epsilon} \hat{m}_t$$

Now, let's understand how to code AMSGrad in Python.

First, we define the AMSGrad function, as follows:

```
def AMSGrad(data, theta, lr = 1e-2, beta1 = 0.9, beta2 = 0.9, epsilon =
1e-6, num_iterations = 1000):
```

Then, we initialize the first moment, mt, the second moment, vt, and the modified version of vt, that is, vt_hat, with zeros, as follows:

```
mt = np.zeros(theta.shape[0])
vt = np.zeros(theta.shape[0])
vt_hat = np.zeros(theta.shape[0])
```

For every iteration, we perform the following steps:

```
for t in range(num_iterations):
```

Now, we can compute the gradients with respect to `theta`:

```
gradients = compute_gradients(data, theta)
```

Then, we compute the first moment, `mt`, as $m_t = \beta_1 m_{t-1} + (1 - \beta_1)g_t$:

```
mt = beta1 * mt + (1. - beta1) * gradients
```

Next, we update the second moment, `vt`, as $v_t = \beta_2 v_{t-1} + (1 - \beta_2)g_t^2$:

```
vt = beta2 * vt + (1. - beta2) * gradients ** 2
```

In AMSGrad, we use a slightly modified version of v_t. Instead of using v_t directly, we take the maximum value of v_t until the previous step. Thus, $\hat{v}_t = \max(\hat{v}_{t-1}, v_t)$ is implemented as follows:

```
vt_hat = np.maximum(vt_hat, vt)
```

Here, we will compute the bias-corrected estimate of `mt`, that is, $\hat{m}_t = \dfrac{m_t}{1 - \beta_1^t}$:

```
mt_hat = mt / (1. - beta1 ** (t+1))
```

Now, we can update the model parameter, `theta`, so that it's $\theta_t = \theta_{t-1} - \dfrac{\eta}{\sqrt{\hat{v}_t} + \epsilon}\hat{m}_t$:

```
theta = theta - (lr / (np.sqrt(vt_hat) + epsilon)) * mt_hat

return theta
```

Nadam – adding NAG to ADAM

Nadam is another small extension of the Adam method. As the name suggests, here, we incorporate NAG into Adam. First, let's recall what we learned about in Adam.

We calculated the first and second moments as follows:

$$m_t = \beta_1 m_{t-1} + (1 - \beta_1)g_t$$

$$v_t = \beta_2 v_{t-1} + (1 - \beta_2)g_t^2$$

Then, we calculated the bias-corrected estimates of the first and second moments, as follows:

$$\hat{m}_t = \frac{m_t}{1 - \beta_1^t}$$

$$\hat{v}_t = \frac{v_t}{1 - \beta_2^t}$$

Our final update equation of Adam is expressed as follows:

$$\theta_t = \theta_{t-1} - \frac{\eta}{\sqrt{v_t} + \epsilon} m_t$$

Now, we will see how Nadam modifies Adam to use Nesterov momentum. In Adam, we compute the first moment as follows:

$$m_t = \beta_1 m_{t-1} + (1 - \beta_1) g_t$$

We change this first moment so that it's Nesterov accelerated momentum. That is, instead of using the previous momentum, we use the current momentum and use that as a lookahead:

$$\tilde{m}_t = \beta_1^{t+1} m_t + (1 - \beta_1^t) g_t$$

We can't compute the bias-corrected estimates in the same way as we computed them in Adam because, here, g_t comes from the current step, and m_t comes from the subsequent step. Therefore, we change the bias-corrected estimate step, as follows:

$$\hat{m}_t = \frac{m_t}{1 - \prod_{i=1}^{t+1} \beta_1^i}$$

$$\hat{g}_t = \frac{g_t}{1 - \prod_{i=1}^{t} \beta_1^i}$$

Thus, we can rewrite our first-moment equation as follows:

$$\tilde{m}_t = \beta_1^{t+1} \hat{m}_t + (1 - \beta_1^t) \hat{g}_t$$

Therefore, our final update equation becomes the following:

$$\theta_t = \theta_{t-1} - \frac{\eta}{\sqrt{v_t} + \epsilon} \tilde{m}_t$$

Now let's see how we can implement the Nadam algorithm in Python.

First, we define the `nadam` function:

```
def nadam(data, theta, lr = 1e-2, beta1 = 0.9, beta2 = 0.999, epsilon =
1e-6, num_iterations = 500):
```

Then, we initialize the first moment, `mt`, and the second moment, `vt`, with zeros:

```
mt = np.zeros(theta.shape[0])
vt = np.zeros(theta.shape[0])
```

Next, we set `beta_prod` to 1:

```
beta_prod = 1
```

For every iteration, we perform the following steps:

```
for t in range(num_iterations):
```

Then, we compute the gradients with respect to `theta`:

```
gradients = compute_gradients(data, theta)
```

Afterward, we compute the first moment, `mt`, so that it's $m_t = \beta_1 m_{t-1} + (1 - \beta_1)g_t$:

```
mt = beta1 * mt + (1. - beta1) * gradients
```

Now, we can update the second moment, `vt`, so that its $v_t = \beta_2 v_{t-1} + (1 - \beta_2)g_t^2$:

```
vt = beta2 * vt + (1. - beta2) * gradients ** 2
```

Now, we compute `beta_prod`; that is, $\prod_{i=1}^{t+1} \beta_1^i$:

```
beta_prod = beta_prod * (beta1)
```

Next, we compute the bias-corrected estimate of `mt` so that it's $\hat{m}_t = \dfrac{m_t}{1 - \prod_{i=1}^{t+1} \beta_1^i}$:

```
mt_hat = mt / (1. - beta_prod)
```

Then, we compute the bias-corrected estimate of `gt` so that it's $\hat{g}_t = \dfrac{g_t}{1 - \prod_{i=1}^{t} \beta_1^i}$:

```
g_hat = grad / (1. - beta_prod)
```

From here, we compute the bias-corrected estimate of `vt` so that it's $\hat{v}_t = \dfrac{v_t}{1 - \beta_2^t}$:

```
vt_hat = vt / (1. - beta2 ** (t))
```

Now, we compute `mt_tilde` so that it's $\tilde{m}_t = \beta_1^{t+1} \hat{m}_t + (1 - \beta_1^t) \hat{g}_t$:

```
mt_tilde = (1-beta1**t+1) * mt_hat + ((beta1**t)* g_hat)
```

Finally, we update the model parameter, `theta`, by using $\theta_t = \theta_{t-1} - \dfrac{\eta}{\sqrt{v_t} + \epsilon} \tilde{m}_t$:

```
        theta = theta - (lr / (np.sqrt(vt_hat) + epsilon)) * mt_hat
    return theta
```

By doing this, we have learned about various popular variants of gradient descent algorithms that are used for training neural networks. The complete code to perform regression with all the variants of regression is available as a Jupyter Notebook at `http://bit.ly/2XoW0vH`.

Summary

We started off this chapter by learning about what convex and non-convex functions are. Then, we explored how we can find the minimum of a function using gradient descent. We learned how gradient descent minimizes a loss function by computing optimal parameters through gradient descent. Later, we looked at SGD, where we update the parameters of the model after iterating through each and every data point, and then we learned about mini-batch SGD, where we update the parameters after iterating through a batch of data points.

Going forward, we learned how momentum is used to reduce oscillations in gradient steps and attain convergence faster. Following this, we understood Nesterov momentum, where, instead of calculating the gradient at the current position, we calculate the gradient at the position the momentum will take us to.

We also learned about the Adagrad method, where we set the learning rate low for parameters that have frequent updates, and high for parameters that have infrequent updates. Next, we learned about the Adadelta method, where we completely do away with the learning rate and use an exponentially decaying average of gradients. We then learned about the Adam method, where we use both first and second momentum estimates to update gradients.

Following this, we explored variants of Adam, such as Adamax, where we generalized the L^2 norm of Adam to L^∞, and AMSGrad, where we combated the problem of Adam reaching a suboptimal solution. At the end of this chapter, we learned about Nadam, where we incorporated Nesterov momentum into the Adam algorithm.

In the next chapter, we will learn about one of the most widely used deep learning algorithms, called **recurrent neural networks** (**RNNs**), and how to use them to generate song lyrics.

Questions

Let's recap on gradient descent by answering the following questions:

1. How does SGD differ from vanilla gradient descent?
2. Explain mini-batch gradient descent.
3. Why do we need momentum?
4. What is the motivation behind NAG?
5. How does Adagrad set the learning rate adaptively?
6. What is the update rule of Adadelta?
7. How does RMSProp overcome the limitations of Adagrad?
8. Define the update equation of Adam.

Further reading

For further information, refer to the following links:

- *Adaptive Subgradient Methods for Online Learning and Stochastic Optimization*, by John Duchi et al., http://www.jmlr.org/papers/volume12/duchi11a/duchi11a.pdf
- *Adadelta: An Adaptive Learning Rate Method*, by Matthew D. Zeiler, https://arxiv.org/pdf/1212.5701.pdf
- *Adam: A Method For Stochastic Optimization*, by Diederik P. Kingma and Jimmy Lei Ba, https://arxiv.org/pdf/1412.6980.pdf
- *On the Convergence of Adam and Beyond*, by Sashank J. Reddi, Satyen Kale, and Sanjiv Kumar, https://openreview.net/pdf?id=ryQu7f-RZ
- *Incorporating Nesterov Momentum into Adam*, by Timothy Dozat, http://cs229.stanford.edu/proj2015/054_report.pdf

4
Generating Song Lyrics Using RNN

In a normal feedforward neural network, each input is independent of other input. But with a sequential dataset, we need to know about the past input to make a prediction. A sequence is an ordered set of items. For instance, a sentence is a sequence of words. Let's suppose that we want to predict the next word in a sentence; to do so, we need to remember the previous words. A normal feedforward neural network cannot predict the correct next word, as it will not remember the previous words of the sentence. Under such circumstances (in which we need to remember the previous input), to make predictions, we use **recurrent neural networks** (**RNNs**).

In this chapter, we will describe how an RNN is used to model sequential datasets and how it remembers the previous input. We will begin by investigating how an RNN differs from a feedforward neural network. Then, we will inspect how forward propagation works in an RNN.

Moving on, we will examine the **backpropagation through time** (**BPTT**) algorithm, which is used for training RNNs. Later, we will look at the vanishing and exploding gradient problem, which occurs while training recurrent networks. You will also learn how to generate song lyrics using an RNN in TensorFlow.

At the end of the chapter, will we examine the different types of RNN architectures, and how they are used for various applications.

In this chapter, we will learn about the following topics:

- Recurrent neural networks
- Forward propagation in RNNs
- Backpropagation through time
- The vanishing and exploding gradient problem
- Generating song lyrics using RNNs
- Different types of RNN architectures

Introducing RNNs

The Sun rises in the ____ .

If we were asked to predict the blank term in the preceding sentence, we would probably say east. Why would we predict that the word east would be the right word here? Because we read the whole sentence, understood the context, and predicted that the word east would be an appropriate word to complete the sentence.

If we use a feedforward neural network to predict the blank, it would not predict the right word. This is due to the fact that in feedforward networks, each input is independent of other input and they make predictions based only on the current input, and they don't remember previous input.

Thus, the input to the network will just be the word preceding the blank, which is the word *the*. With this word alone as an input, our network cannot predict the correct word, because it doesn't know the context of the sentence, which means that it doesn't know the previous set of words to understand the context of the sentence and to predict an appropriate next word.

Here is where we use RNNs. They predict output not only based on the current input, but also on the previous hidden state. Why do they have to predict the output based on the current input and the previous hidden state? Why can't they just use the current input and the previous input?

This is because the previous input will only store information about the previous word, while the previous hidden state will capture the contextual information about all the words in the sentence that the network has seen so far. Basically, the previous hidden state acts like a memory and it captures the context of the sentence. With this context and the current input, we can predict the relevant word.

For instance, let's take the same sentence, *The sun rises in the* ____. As shown in the following figure, we first pass the word *the* as an input, and then we pass the next word, *sun*, as input; but along with this, we also pass the previous hidden state, h_0. So, every time we pass the input word, we also pass a previous hidden state as an input.

In the final step, we pass the word *the*, and also the previous hidden state h_3, which captures the contextual information about the sequence of words that the network has seen so far. Thus, h_3 acts as the memory and stores information about all the previous words that the network has seen. With h_3 and the current input word (*the*), we can predict the relevant next word:

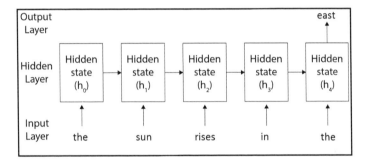

In a nutshell, an RNN uses the previous hidden state as memory which captures and stores the contextual information (input) that the network has seen so far.

RNNs are widely applied for use cases that involve sequential data, such as time series, text, audio, speech, video, weather, and much more. They have been greatly used in various **natural language processing** (**NLP**) tasks, such as language translation, sentiment analysis, text generation, and so on.

The difference between feedforward networks and RNNs

A comparison between an RNN and a feedforward network is shown in the following diagram:

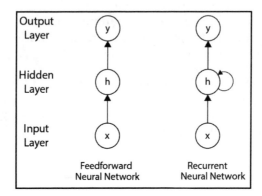

As you can observe in the preceding diagram, the RNN contains a looped connection in the hidden layer, which implies that we use the previous hidden state along with the input to predict the output.

Still confused? Let's look at the following unrolled version of an RNN. But wait; what is the unrolled version of an RNN?

It means that we roll out the network for a complete sequence. Let's suppose that we have an input sentence with T words; then, we will have 0 to $T-1$ layers, one for each word, as shown in the following figure:

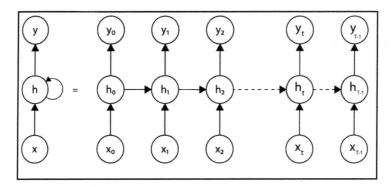

As you can see in the preceding figure, at the time step $t = 1$, the output y_1 is predicted based on the current input x_1 and the previous hidden state h_0. Similarly, at time step $t = 2$, y_2 is predicted using the current input x_2 and the previous hidden state h_1. This is how an RNN works; it takes the current input and the previous hidden state to predict the output.

Forward propagation in RNNs

Let's look at how an RNN uses forward propagation to predict the output; but before we jump right in, let's get familiar with the notations:

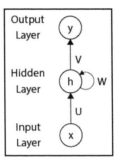

The preceding figure illustrates the following:

- U represents the input to hidden layer weight matrix
- W represents the hidden to hidden layer weight matrix
- V represents the hidden to output layer weight matrix

The hidden state h at a time step t can be computed as follows:

$$h_t = \tanh(Ux_t + Wh_{t-1})$$

That is, *hidden state at a time step, t = tanh([input to hidden layer weight x input] + [hidden to hidden layer weight x previous hidden state]).*

The output at a time step t can be computed as follows:

$$\hat{y}_t = \text{softmax}(Vh_t)$$

That is, *output at a time step, t = softmax (hidden to output layer weight x hidden state at a time t).*

We can also represent RNNs as shown in the following figure. As you can see, the hidden layer is represented by an RNN block, which implies that our network is an RNN, and previous hidden states are used in predicting the output:

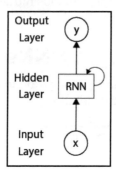

The following diagram shows how forward propagation works in an unrolled version of an RNN:

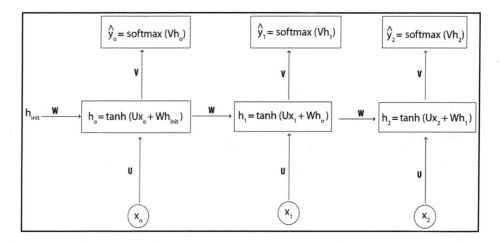

We initialize the initial hidden state h_{init} with random values. As you can see in the preceding figure, the output, $\hat{y}_0$, is predicted based on the current input, x_0 and the previous hidden state, which is an initial hidden state, h_{init}, using the following formula:

$$h_0 = \tanh(Ux_0 + Wh_{init})$$

$$\hat{y}_0 = \text{softmax}(Vh_0)$$

Similarly, look at how the output, $\hat{y}_1$, is computed. It takes the current input, x_1, and the previous hidden state, h_0:

$$h_1 = \tanh(Ux_1 + Wh_0)$$

$$\hat{y}_1 = \text{softmax}(Vh_1)$$

Thus, in forward propagation to predict the output, RNN uses the current input and the previous hidden state.

To achieve clarity, let's look at how to implement forward propagation in RNN to predict the output:

1. Initialize all the weights, U, W, and V, by randomly drawing from the uniform distribution:

   ```
   U = np.random.uniform(-np.sqrt(1.0 / input_dim), np.sqrt(1.0 /
   input_dim), (hidden_dim, input_dim))

   W = np.random.uniform(-np.sqrt(1.0 / hidden_dim), np.sqrt(1.0 /
   hidden_dim), (hidden_dim, hidden_dim))

   V = np.random.uniform(-np.sqrt(1.0 / hidden_dim), np.sqrt(1.0 /
   hidden_dim), (input_dim, hidden_dim))
   ```

2. Define the number of time steps, which will be the length of our input sequence, x:

   ```
   num_time_steps = len(x)
   ```

3. Define the hidden state:

   ```
   hidden_state = np.zeros((num_time_steps + 1, hidden_dim))
   ```

4. Initialize the initial hidden state, h_{init}, with zeros:

   ```
   hidden_state[-1] = np.zeros(hidden_dim)
   ```

5. Initialize the output:

   ```
   YHat = np.zeros((num_time_steps, output_dim))
   ```

6. For every time step, we perform the following:

```
for t in np.arange(num_time_steps):

    #h_t = tanh(UX + Wh_{t-1})
    hidden_state[t] = np.tanh(U[:, x[t]] + W.dot(hidden_state[t -
1])))

    # yhat_t = softmax(vh)
    YHat[t] = softmax(V.dot(hidden_state[t]))
```

Backpropagating through time

We just learned how forward propagation works in RNNs and how it predicts the output. Now, we compute the loss, L, at each time step, t, to determine how well the RNN has predicted the output. We use the cross-entropy loss as our loss function. The loss L at a time step t can be given as follows:

$$L_t = -y_t \log(\hat{y}_t)$$

Here, y_t is the actual output, and $\hat{y}_t$ is the predicted output at a time step t.

The final loss is a sum of the loss at all the time steps. Suppose that we have $T-1$ layers; then, the final loss can be given as follows:

$$L = \sum_{j=0}^{T-1} L_j$$

As shown in the following figure, the final loss is obtained by the sum of loss at all the time steps:

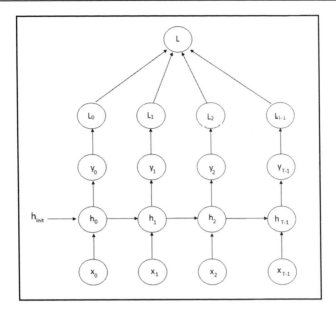

We computed the loss, now our goal is to minimize the loss. How can we minimize the loss? We can minimize the loss by finding the optimal weights of the RNN. As we learned, we have three weights in RNNs: input to hidden, U, hidden to hidden, W, and hidden to output, V.

We need to find optimal values for all of these three weights to minimize the loss. We can use our favorite gradient descent algorithm to find the optimal weights. We begin by calculating the gradients of the loss function with respect to all the weights; then, we update the weights according to the weight update rule as follows:

$$V = V - \alpha \frac{\partial L}{\partial V}$$

$$W = W - \alpha \frac{\partial L}{\partial W}$$

$$U = U - \alpha \frac{\partial L}{\partial U}$$

You can skip the upcoming sections if you don't want to understand the math behind the gradient calculation. However, it will help you to better understand how BPTT works in RNN better.

First, we calculate the gradients of loss with respect to the final layer $\hat{y}_t$, that is $\frac{\partial L}{\partial \hat{y}_t}$, so that we can use it in the upcoming steps.

As we have learned, the loss L at a time step t can be given as follows:

$$L_t = -y_t log(\hat{y}_t)$$

Since we know:

$$\frac{d}{dx}(\log x) = \frac{1}{x}$$

We can write:

$$\frac{\partial L_t}{\partial \hat{y}_t} = -y_t \frac{1}{\hat{y}_t}$$

Thus, the gradient of the loss L with respect to $\hat{y}_t$ becomes:

$$\frac{\partial L_t}{\partial \hat{y}_t} = -\frac{y_t}{\hat{y}_t} \tag{1}$$

Now, we will learn how to calculate the gradient of the loss with respect to all the weights one by one.

Gradients with respect to the hidden to output weight, V

First, let's recap the steps involved in the forward propagation:

$$h_t = \tanh(Ux_t + Wh_{t-1})$$

$$\hat{y}_t = \text{softmax}(Vh_t) \tag{2}$$

$$L_t = -y_t log(\hat{y}_t)$$

$$h_t = \tanh(Ux_t + Wh_{t-1})$$

$$\hat{y}_t = \text{softmax}(z_t)$$

$$L_t = -y_t log(\hat{y}_t)$$

After predicting the output $\hat{y}$, we are in the final layer of the network. Since we are backpropagating, that is, going from the output layer to the input layer, our first weight would be V, which is the hidden to output layer weight.

We have seen that the final loss is the sum of the loss over all the time steps, and similarly, the final gradient is the sum of gradients over all the time steps:

$$\frac{\partial L}{\partial V} = \frac{\partial L_0}{\partial V} + \frac{\partial L_1}{\partial V} + \frac{\partial L_2}{\partial V} + \dots \frac{\partial L_{T-1}}{\partial V}$$

Hence, we can write:

$$\frac{\partial L}{\partial V} = \sum_{j=0}^{T-1} \frac{\partial L_j}{\partial V}$$

Recall our loss function, $L_t = -y_t log(\hat{y}_t)$; we cannot calculate the gradient with respect to V directly from L_t, as there are no V terms in it. So, we apply the chain rule. Recall the forward propagation equation; there is a V term in $\hat{y}_t$:

$$\hat{y}_t = \sigma(z_t) \text{ where } z_t = Vh_t$$

First, we calculate a partial derivative of the loss with respect to $\hat{y}_t$, and then, from $\hat{y}_t$, we will calculate the partial derivative with respect to z_t. From z_t, we can calculate the derivative with respect to V.

Thus, our equation becomes the following:

$$\frac{\partial L}{\partial V} = \sum_{j=0}^{T-1} \frac{\partial L_j}{\partial \hat{y}_j} \frac{\partial \hat{y}_j}{\partial z_j} \frac{\partial z_j}{\partial V} \tag{3}$$

As we know that, $\hat{y}_t = \text{softmax}(z_t)$, the gradient of loss with respect to z_j can be calculated as follows:

$$\frac{\partial L_j}{\partial z_j} = \frac{\partial L_j}{\partial \hat{y}_j} \frac{\partial \hat{y}_j}{\partial z_j} \tag{4}$$

Substituting equation *(4)* in equation *(3)*, we can write the following:

$$\frac{\partial L}{\partial V} = \sum_{j=0}^{T-1} \frac{\partial L_j}{\partial z_j} \frac{\partial z_j}{\partial V} \tag{5}$$

For better understanding, let's take each of the terms from the preceding equation and compute them one by one:

$$\frac{\partial L}{\partial z_t} = \sum_{j=0}^{T-1} \frac{\partial L_j}{\partial \hat{y}_j} \frac{\partial \hat{y}_j}{\partial z_t} \tag{6}$$

From equation *(1)*, we can substitute the value of $\frac{\partial L_j}{\partial \hat{y}_j}$ in the preceding equation *(6)* as follows:

$$\frac{\partial L}{\partial z_t} = \sum_{j=0}^{T-1} -\frac{y_j}{\hat{y}_j} \frac{\partial \hat{y}_j}{\partial z_t} \tag{7}$$

Now, we will compute the term $\frac{\partial \hat{y}_j}{\partial z_t}$. Since we know, $\hat{y}_t = \text{softmax}(z_t)$, computing $\frac{\partial \hat{y}_j}{\partial z_t}$ gives us the derivative of the softmax function:

$$\frac{\partial y_j}{\partial z_t} = \text{softmax}'$$

The derivative of the softmax function can be represented as follows:

$$\frac{\partial \hat{y}_j}{\partial z_t} = \begin{cases} -\hat{y}_j \hat{y}_t, & j \neq t \\ \hat{y}_t \left(1 - \hat{y}_t \right), & j = t \end{cases} \tag{8}$$

Substituting equation *(8)* into equation *(7)*, we can write the following:

$$\frac{\partial L}{\partial z_t} = -\frac{y_t}{\hat{y}_t}\hat{y}_t(1-\hat{y}_t) + \sum_{j\neq t}^{T}\frac{y_j}{\hat{y}_j}\hat{y}_j\hat{y}_t$$

$$= -\frac{y_t}{\hat{y}_t}\hat{y}_t(1-\hat{y}_t) + \sum_{j\neq t}^{T}\frac{y_j}{\hat{y}_j}\hat{y}_j\hat{y}_t$$

$$= -y_t(1-\hat{y}_t) + \sum_{j\neq t}^{T}y_j\hat{y}_t$$

$$= -y_t + \hat{y}_t y_t + \hat{y}_t\sum_{j\neq t}^{T}y_j$$

$$= -y_t + \hat{y}_t\sum_{j=1}^{T}y_j$$

$$say\ \sum_{j=1}^{T}y_j = 1, then\ we\ can\ write$$

$$= -y_t + \hat{y}_t$$

$$= \hat{y}_t - y_t$$

Thus, the final equation becomes:

$$\frac{\partial L}{\partial z_t} = \hat{y}_t - y_t \tag{9}$$

Now, we can substitute equation *(9)* into equation *(5)*:

$$\frac{\partial L}{\partial V} = \sum_{j=0}^{T-1}(\hat{y}_j - y_j)\frac{\partial z_j}{\partial V} \tag{10}$$

Since we know that $z_t = Vh_t$, we can write:

$$\frac{\partial z_t}{\partial V} = h_t$$

Substituting the preceding equation into equation *(10)*, we get our final equation, that is, gradient of the loss function with respect to V, as follows:

$$\frac{\partial L}{\partial V} = \sum_{j=0}^{T-1} (\hat{y}_j - y_j) \otimes h_j$$

Gradients with respect to hidden to hidden layer weights, W

Now, we will compute the gradients of loss with respect to hidden to hidden layer weights, W. Similar to V, the final gradient is the sum of the gradients at all time steps:

$$\frac{\partial L}{\partial W} = \frac{\partial L_0}{\partial W} + \frac{\partial L_1}{\partial W} + \frac{\partial L_2}{\partial W} + \dots \frac{\partial L_{T-1}}{\partial W}$$

So, we can write:

$$\frac{\partial L}{\partial W} = \sum_{j=0}^{T-1} \frac{\partial L_j}{\partial W}$$

First, let's compute gradient of loss, L_0 with respect to W, that is, $\frac{\partial L_0}{\partial W}$.

We cannot compute derivative of L_0 with respect to W directly from as there are no W terms in it. So, we use the chain rule to compute the gradients of loss with respect to W. Let's recollect the forward propagation equation:

$$L_0 = -y_0 \, log(\hat{y}_0)$$

$$y_0 = softmax(h_0 V)$$

$$h_0 = tanh(Ux_0 + Wh_{init})$$

First, we calculate the partial derivative of loss L_0 with respect to y_0; then, from y_0, we calculate the partial derivative with respect to h_o; then, from h_o, we can calculate the derivative with respect to W, as follows:

$$\frac{\partial L_0}{\partial W} = \frac{\partial L_0}{\partial y_0} \frac{\partial y_0}{\partial h_0} \frac{\partial h_0}{\partial W}$$

Now, let's compute the gradient of loss, L_1 with respect to W, that is, $\frac{\partial L_1}{\partial W}$. Thus, again, we will apply the chain rule and get the following:

$$\frac{\partial L_1}{\partial W} = \frac{\partial L_1}{\partial y_1} \frac{\partial y_1}{\partial h_1} \frac{\partial h_1}{\partial W}$$

If you look at the preceding equation, how can we calculate the term $\frac{\partial h_1}{\partial W}$? Let's recall the equation of h_1:

$$h_1 = tanh(Ux_1 + Wh_0)$$

As you can see in the preceding equation, computing h_1 depends on h_0 and W, but h_0 is not a constant; it is a function again. So, we need to calculate the derivative with respect to that, as well.

The equation then becomes the following:

$$\frac{\partial L_1}{\partial W} = \frac{\partial L_1}{\partial y_1} \frac{\partial y_1}{\partial h_1} \frac{\partial h_1}{\partial W} + \frac{\partial L_1}{\partial y_1} \frac{\partial y_1}{\partial h_1} \frac{\partial h_1}{\partial h_0} \frac{\partial h_0}{\partial W}$$

The following figure shows computing $\frac{\partial L_1}{\partial W}$; we can notice how h_1 is dependent on h_0:

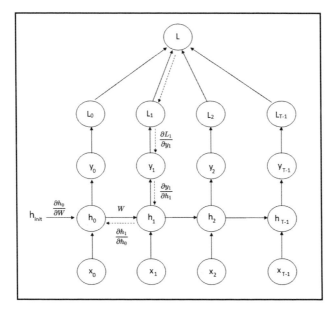

Now, let's compute the gradient of loss, L_2 with respect to W, that is, $\frac{\partial L_2}{\partial W}$. Thus, again, we will apply the chain rule and get the following:

$$\frac{\partial L_2}{\partial W} = \frac{\partial L_2}{\partial y_2} \frac{\partial y_2}{\partial h_2} \frac{\partial h_2}{\partial W}$$

In the preceding equation, we can't compute $\frac{\partial h_2}{\partial W}$ directly. Recall the equation of h_2:

$$h_2 = tanh(Ux_2 + Wh_1)$$

As you observe, computing h_2, depends on a function h_1, whereas h_1 is again a function which depends on function h_0. As shown in the following figure, to compute the derivative with respect to h_2, we need to traverse until h_0, as each function is dependent on one another:

$$\frac{\partial L_2}{\partial W} = \frac{\partial L_2}{\partial y_2} \frac{\partial y_2}{\partial h_2} \frac{\partial h_2}{\partial W} + \frac{\partial L_2}{\partial y_2} \frac{\partial y_2}{\partial h_2} \frac{\partial h_2}{\partial h_1} \frac{\partial h_1}{\partial W} + \frac{\partial L_2}{\partial y_2} \frac{\partial y_2}{\partial h_2} \frac{\partial h_2}{\partial h_0} \frac{\partial h_0}{\partial W}$$

This can be pictorially represented as follows:

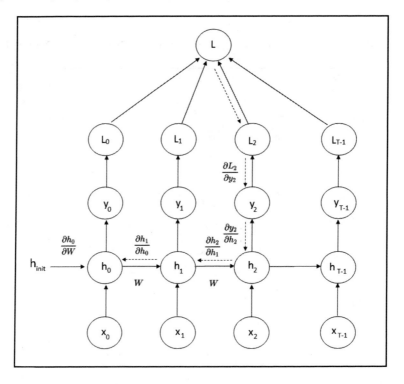

This applies for the loss at any time step; say, $L_0, L_1, L_2, \ldots L_j, \ldots L_{T-1}$. So, we can say that to compute any loss L_j, we need to traverse all the way to h_0, as shown in the following figure:

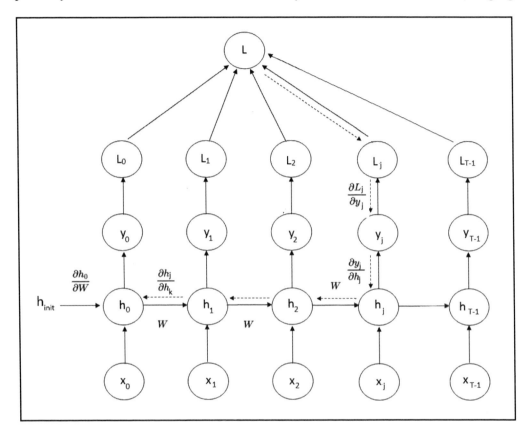

This is because in RNNs, the hidden state at a time t is dependent on a hidden state at a time $t-1$, which implies that the current hidden state is always dependent on the previous hidden state.

So, any loss L_j can be computed as shown in the following figure:

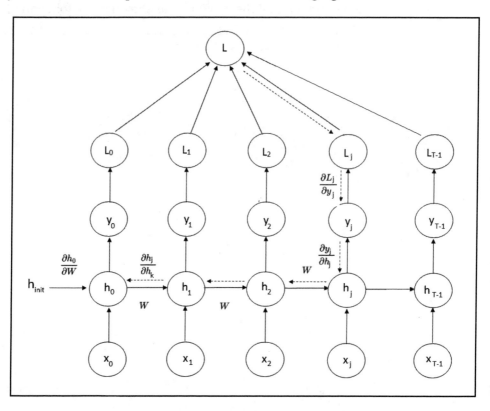

Thus, we can write, gradient of loss L_j with respect to W becomes:

$$\frac{\partial L_j}{\partial W} = \sum_{k=0}^{j} \frac{\partial L_j}{\partial y_j} \frac{\partial y_j}{\partial h_j} \frac{\partial h_j}{\partial h_k} \frac{\partial h_k}{\partial W} \qquad (11)$$

The sum $\sum_{k=0}^{j}$ in the previous equation implies the sum over all the hidden states h_k. In the preceding equation, $\frac{\partial h_j}{\partial h_k}$ can be computed using the chain rule. So, we can say:

$$\frac{\partial h_j}{\partial h_k} = \prod_{m=k+1}^{j} \frac{\partial h_m}{\partial h_{m-1}} \qquad (12)$$

Assume that $j=3$ and $k=0$; then, the preceding equation becomes:

$$\frac{\partial h_3}{\partial h_0} = \frac{\partial h_3}{\partial h_2} \frac{\partial h_2}{\partial h_1} \frac{\partial h_1}{\partial h_0}$$

Substituting equation *(12)* into equation *(11)* will give us the following:

$$\frac{\partial L_j}{\partial W} = \sum_{k=0}^{j} \frac{\partial L_j}{\partial y_j} \frac{\partial y_j}{\partial h_j} \left(\prod_{m=k+1}^{j} \frac{\partial h_m}{\partial h_{m-1}} \right) \frac{\partial h_k}{\partial W} \qquad (13)$$

We know that final loss is the sum of loss across all the time steps:

$$\frac{\partial L}{\partial W} = \sum_{j=0}^{T-1} \frac{\partial L_j}{\partial W}$$

Substituting equation *(13)* into the preceding equation, we get the following:

$$\frac{\partial L}{\partial W} = \sum_{j=0}^{T-1} \sum_{k=0}^{j} \frac{\partial L_j}{\partial y_j} \frac{\partial y_j}{\partial h_j} \left(\prod_{m=k+1}^{j} \frac{\partial h_m}{\partial h_{m-1}} \right) \frac{\partial h_k}{\partial W} \qquad (14)$$

We have two summations in the preceding equation, where:

- $\sum_{j=0}^{T-1}$ implies the sum of loss across all the time steps

- $\sum_{k=0}^{j}$ is the summation over hidden states

So, our final equation for computing gradient of loss with respect to W, is given as:

$$\frac{\partial L}{\partial W} = \sum_{j=0}^{T-1} \sum_{k=0}^{j} \frac{\partial L_j}{\partial y_j} \frac{\partial y_j}{\partial h_j} \left(\prod_{m=k+1}^{j} \frac{\partial h_m}{\partial h_{m-1}} \right) \frac{\partial h_k}{\partial W} \qquad (15)$$

Now, we will look at how to compute each of the terms in the preceding equation, one by one. From equation *(4)* and equation *(9)*, we can say that:

$$\frac{\partial L_j}{\partial y_j} \frac{\partial y_j}{\partial h_j} = \hat{y}_j - y_j$$

Let's look at the next term:

$$\prod_{m=k+1}^{j} \frac{\partial h_m}{\partial h_{m-1}}$$

We know that the hidden state h_m is computed as:

$$h_m = tanh(Ux_m + Wh_{m-1})$$

The derivative of $tanh(x)$ is $1 - tanh^2(x)$, so we can write:

$$\frac{\partial h_m}{\partial h_{m-1}} = W^T diag(1 - tanh^2(Wh_{m-1} + Ux_m))$$

Let's look at the final term $\frac{h_k}{\partial w}$. We know that the hidden state h_k is computed as, $h_k = tanh(Ux_k + Wh_{k-1})$. Thus, the derivation of h_k with respect to W becomes:

$$\frac{\partial h_k}{\partial W} = h_k - 1$$

Substituting all of the calculated terms into equation *(15)*, we get our final equation for gradient of loss L with respect to W as follows:

$$\boxed{\frac{\partial L}{\partial W} = \sum_{j=0}^{T-1}\sum_{k=0}^{j}(\hat{y}_j - y_j)\prod_{m=k+1}^{j} W^T diag(1 - tanh^2(Wh_{m-1} + Ux_m)) \otimes h_{k-1}}$$

Gradients with respect to input to the hidden layer weight, U

Computing the gradients of the loss function with respect to U is the same as W, since here also we take the sequential derivative of h_t. Similar to W, to compute the derivative of any loss L_j with respect to U, we need to traverse all the way back to h_0.

The final equation for computing the gradient of the loss with respect to U is given as follows. As you may notice, it is basically the same as the equation *(15)*, except that we have the term $\frac{\partial h_k}{\partial u}$ instead of $\frac{\partial h_k}{\partial w}$ shown as follows:

$$\frac{\partial L}{\partial U} = \sum_{j=0}^{T-1} \sum_{k=0}^{j} \frac{\partial L_j}{\partial \hat{y}_j} \frac{\partial \hat{y}_j}{\partial h_j} \left(\prod_{m=k+1}^{j} \frac{\partial h_m}{\partial h_{m-1}} \right) \frac{\partial h_k}{\partial U}$$

We have already seen how to compute to the first two terms in the previous section.

Let's look at the final term $\frac{h_k}{\partial u}$. We know that the hidden state h_k is computed as, $h_k = \tanh(Ux_k + Wh_{k-1})$. Thus, the derivation of h_k with respect to U becomes:

$$\frac{\partial h_k}{\partial U} = x_k$$

So, our final equation for a gradient of the loss L, with respect to U, can be written as follows:

$$\frac{\partial L}{\partial U} = \sum_{j=0}^{T-1} \sum_{k=0}^{j} (\hat{y}_j - y_j) \prod_{m=k+1}^{j} W^T diag(1 - \tanh^2(Wh_{m-1} + Ux_m)) \otimes x_k$$

Vanishing and exploding gradients problem

We just learned how BPTT works, and we saw how the gradient of loss can be computed with respect to all the weights in RNNs. But here, we will encounter a problem called the **vanishing and exploding gradients**.

While computing the derivatives of loss with respect to W and U, we saw that we have to traverse all the way back to the first hidden state, as each hidden state at a time t is dependent on its previous hidden state at a time $t - 1$.

For instance, the gradient of loss L_2 with respect to W is given as:

$$\frac{\partial L_2}{\partial W} = \frac{\partial L_2}{\partial y_2} \frac{\partial y_2}{\partial h_2} \frac{\partial h_2}{\partial W}$$

If you look at the term $\frac{\partial h_2}{\partial W}$ from the preceding equation, we can't calculate the derivative of h_2 with respect to W directly. As we know, $h_2 = tanh(Ux_2 + Wh_1)$ is a function that is dependent on h_1 and W. So, we need to calculate the derivative with respect to h_1, as well. Even $h_1 = tanh(Ux_2 + Wh_0)$ is a function that is dependent on h_0 and W. Thus, we need to calculate the derivative with respect to h_0, as well.

As shown in the following figure, to compute the derivative of L_2, we need to go all the way back to the initial hidden state h_0, as each hidden state is dependent on its previous hidden state:

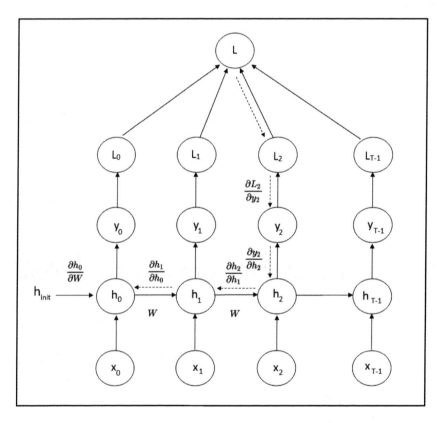

So, to compute any loss L_j, we need to traverse all the way back to the initial hidden state h_0, as each hidden state is dependent on its previous hidden state. Suppose that we have a deep recurrent network with 50 layers. To compute the loss L_{50}, we need to traverse all the way back to h_0, as shown in the following figure:

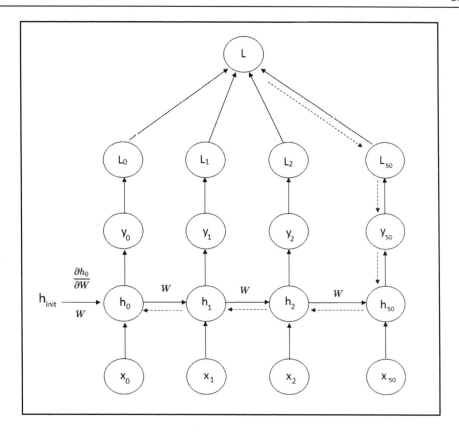

So, what is the problem here, exactly? While backpropagating towards the initial hidden state, we lose information, and the RNN will not backpropagate perfectly.

Remember $h_t = \tanh(Ux_t + Wh_{t-1})$? Every time we move backward, we compute the derivative of h_t. A derivative of tanh is bounded to 1. We know that any two values between 0 and 1, when multiplied with each other will give us a smaller number. We usually initialize the weights of the network to a small number. Thus, when we multiply the derivatives and weights while backpropagating, we are essentially multiplying smaller numbers.

So, when we multiply smaller numbers at every step while moving backward, our gradient becomes infinitesimally small and leads to a number that the computer can't handle; this is called the **vanishing gradient problem.**

Recall the equation of gradient of the loss with respect to the W that we saw in the *Gradients with respect to hidden to hidden layer weights, W* section:

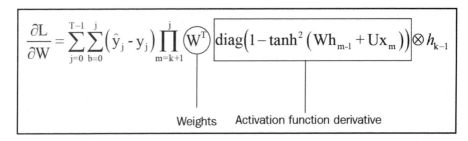

$$\frac{\partial L}{\partial W} = \sum_{j=0}^{T-1}\sum_{b=0}^{j}\left(\hat{y}_j - y_j\right)\prod_{m=k+1}^{j}\left(W^T\right)\left[\text{diag}\left(1-\tanh^2\left(Wh_{m-1}+Ux_m\right)\right)\right]\otimes h_{k-1}$$

Weights   Activation function derivative

As you can observe, we are multiplying the weights and derivative of the tanh function at every time step. Repeated multiplication of these two leads to a small number and causes the vanishing gradients problem.

The vanishing gradients problem occurs not only in RNN but also in other deep networks where we use sigmoid or tanh as the activation function. So, to overcome this, we can use ReLU as an activation function instead of tanh.

However, we have a variant of the RNN called the **long short-term memory (LSTM)** network, which can solve the vanishing gradient problem effectively. We will look at how it works in `Chapter 5`, *Improvements to the RNN*.

Similarly, when we initialize the weights of the network to a very large number, the gradients will become very large at every step. While backpropagating, we multiply a large number together at every time step, and it leads to infinity. This is called the **exploding gradient problem.**

Gradient clipping

We can use gradient clipping to bypass the exploding gradient problem. In this method, we normalize the gradients according to a vector norm (say, *L2*) and clip the gradient value to a certain range. For instance, if we set the threshold as 0.7, then we keep the gradients in the -0.7 to +0.7 range. If the gradient value exceeds -0.7, then we change it to -0.7, and similarly, if it exceeds 0.7, then we change it to +0.7.

Let's assume $\hat{g}$ is the gradient of loss L with respect to W:

$$\hat{g} = \frac{\partial L}{\partial W}$$

First, we normalize the gradients using the L2 norm, that is, $||\hat{g}||$. If the normalized gradient exceeds the defined threshold, we update the gradient, as follows:

$$\hat{g} = \frac{threshold}{||\hat{g}||} \cdot \hat{g}$$

Generating song lyrics using RNNs

We have learned enough about RNNs; now, we will look at how to generate song lyrics using RNNs. To do this, we simply build a character-level RNN, meaning that on every time step, we predict a new character.

Let's consider a small sentence, *What a beautiful d*.

At the first time step, the RNN predicts a new character as *a*. The sentence will be updated to, *What a beautiful da*.

At the next time step, it predicts a new character as *y*, and the sentence becomes, *What a beautiful day*.

In this manner, we predict a new character at each time step and generate a song. Instead of predicting a new character every time, we can also predict a new word every time, which is called **word level RNN**. For simplicity, let's start with a character level RNN.

But how does RNN predicts a new character on each time step? Let's suppose at a time step t=0, we feed an input character say *x*. Now the RNN predicts the next character based on the given input character *x*. To predict the next character, it predicts the probability of all the characters in our vocabulary to be the next character. Once we have this probability distribution we randomly select the next character based on this probability. Confusing? Let us better understand this with an example.

For instance, as shown in the following figure, let's suppose that our vocabulary contains four characters *L, O, V,* and *E*; when we feed the character *L* as an input, RNN computes the probability of all the words in the vocabulary to be the next character:

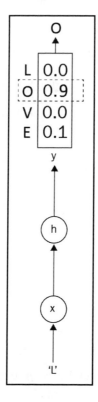

So, we have the probabilities as **[0.0, 0.9, 0.0, 0.1]**, corresponding to the characters in the vocabulary *[L,O,V,E]*. With this probability distribution, we select *O* as the next character 90% of the time, and *E* as the next character 10% of the time. Predicting the next character by sampling from this probability distribution adds some randomness to the output.

On the next time step, we feed the predicted character from the previous time step and the previous hidden state as an input to predict the next character, as shown in the following figure:

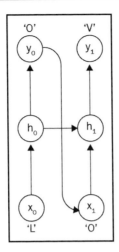

So, on each time step, we feed the predicted character from the previous time step and the previous hidden state as input and predict the next character shown as follows:

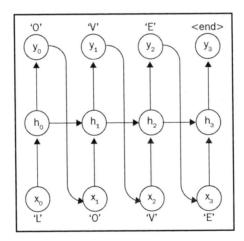

As you can see in the preceding figure, at time step *t=2*, *V* is passed as an input, and it predicts the next character as *E*. But this does not mean that every time character *V* is sent as an input it should always return *E* as output. Since we are passing input along with the previous hidden state, the RNN has the memory of all the characters it has seen so far.

So, the previous hidden state captures the essence of the previous input characters, which are *L* and *O*. Now, with this previous hidden state and the input *V*, the RNN predicts the next character as *E*.

Implementing in TensorFlow

Now, we will look at how to build the RNN model in TensorFlow to generate song lyrics. The dataset and also the complete code used in this section with step by step explanation is available on GitHub at http://bit.ly/2QJttyp. After downloading, unzip the archive, and place the songdata.csv in the data folder.

Import the required libraries:

```
import warnings
warnings.filterwarnings('ignore')

import random
import numpy as np
import tensorflow as tf

tf.logging.set_verbosity(tf.logging.ERROR)

import warnings
warnings.filterwarnings('ignore')
```

Data preparation

Read the downloaded input dataset:

```
df = pd.read_csv('data/songdata.csv')
```

Let's see what we have in our dataset:

```
df.head()
```

The preceding code generates the following output:

	artist	song	link	text
0	ABBA	Ahe's My Kind Of Girl	/a/abba/ahes+my+kind+of+girl_20598417.html	Look at her face, it's a wonderful face \nAnd...
1	ABBA	Andante, Andante	/a/abba/andante+andante_20002708.html	Take it easy with me, please \nTouch me gentl...
2	ABBA	As Good As New	/a/abba/as+good+as+new_20003033.html	I'll never know why I had to go \nWhy I had t...
3	ABBA	Bang	/a/abba/bang_20598415.html	Making somebody happy is a question of give an...
4	ABBA	Bang-A-Boomerang	/a/abba/bang+a+boomerang_20002668.html	Making somebody happy is a question of give an...

Our dataset consists of about 57,650 song lyrics:

```
df.shape[0]
```

```
57650
```

We have song lyrics from about 643 artists:

```
len(df['artist'].unique())
```

```
643
```

The number of songs from each artist is shown as follows:

```
df['artist'].value_counts()[:10]
```

```
Donna Summer          191
Gordon Lightfoot      189
George Strait         188
Bob Dylan             188
Loretta Lynn          187
Cher                  187
Alabama               187
Reba Mcentire         187
Chaka Khan            186
Dean Martin           186
Name: artist, dtype: int64
```

On average, we have about 89 songs from each artist:

```
df['artist'].value_counts().values.mean()
```

```
89
```

We have song lyrics in the column text, so we combine all the rows of that column and save it as a text in a variable called data, as follows:

```
data = ', '.join(df['text'])
```

Let's see a few lines of a song:

```
data[:369]
```

```
"Look at her face, it's a wonderful face  \nAnd it means something special
to me  \nLook at the way that she smiles when she sees me  \nHow lucky can
one fellow be?  \n  \nShe's just my kind of girl, she makes me feel fine
\nWho could ever believe that she could be mine?  \nShe's just my kind of
girl, without her I'm blue  \nAnd if she ever leaves me what could I do,
what co"
```

Since we are building a char-level RNN, we will store all the unique characters in our dataset into a variable called `chars`; this is basically our vocabulary:

```
chars = sorted(list(set(data)))
```

Store the vocabulary size in a variable called `vocab_size`:

```
vocab_size = len(chars)
```

Since the neural networks only accept the input in numbers, we need to convert all the characters in the vocabulary to a number.

We map all the characters in the vocabulary to their corresponding index that forms a unique number. We define a `char_to_ix` dictionary, which has a mapping of all the characters to their index. To get the index by a character, we also define the `ix_to_char` dictionary, which has a mapping of all the indices to their respective characters:

```
char_to_ix = {ch: i for i, ch in enumerate(chars)}
ix_to_char = {i: ch for i, ch in enumerate(chars)}
```

As you can see in the following code snippet, the character `'s'` is mapped to an index `68` in the `char_to_ix` dictionary:

```
print char_to_ix['s']
```

```
68
```

Similarly, if we give `68` as an input to the `ix_to_char`, then we get the corresponding character, which is `'s'`:

```
print ix_to_char[68]
```

```
's'
```

Once we obtain the character to integer mapping, we use one-hot encoding to represent the input and output in vector form. A **one-hot encoded vector** is basically a vector full of 0s, except, 1 at a position corresponding to a character index.

For example, let's suppose that the vocabSize is 7, and the character *z* is in the fourth position in the vocabulary. Then, the one-hot encoded representation for the character *z* can be represented as follows:

```
vocabSize = 7
char_index = 4

print np.eye(vocabSize)[char_index]

array([0., 0., 0., 0., 1., 0., 0.])
```

As you can see, we have a 1 at the corresponding index of the character, and the rest of the values are 0s. This is how we convert each character into a one-hot encoded vector.

In the following code, we define a function called one_hot_encoder, which will return the one-hot encoded vectors, given an index of the character:

```
def one_hot_encoder(index):
    return np.eye(vocab_size)[index]
```

Defining the network parameters

Next, we define all the network parameters:

1. Define the number of units in the hidden layer:

```
hidden_size = 100
```

2. Define the length of the input and output sequence:

```
seq_length = 25
```

3. Define the learning rate for gradient descent:

```
learning_rate = 1e-1
```

4. Set the seed value:

```
seed_value = 42
tf.set_random_seed(seed_value)
random.seed(seed_value)
```

Defining placeholders

Now, we will define the TensorFlow placeholders:

1. The `placeholders` for the input and output is defined as:

```
inputs = tf.placeholder(shape=[None, vocab_size],dtype=tf.float32,
name="inputs")
targets = tf.placeholder(shape=[None, vocab_size],
dtype=tf.float32, name="targets")
```

2. Define the `placeholder` for the initial hidden state:

```
init_state = tf.placeholder(shape=[1, hidden_size],
dtype=tf.float32, name="state")
```

3. Define an `initializer` for initializing the weights of the RNN:

```
initializer = tf.random_normal_initializer(stddev=0.1)
```

Defining forward propagation

Let's define the forward propagation involved in the RNN, which is mathematically given as follows:

$$h_t = \tanh(Ux_t + Wh_{t-1} + bh)$$

$$\hat{y} = \text{softmax}(Vh_t + bv)$$

The bh and bv are the biases of the hidden and output layers, respectively. For simplicity, we haven't added them to our equations in the previous sections. Forward propagation can be implemented as follows:

```
with tf.variable_scope("RNN") as scope:
    h_t = init_state
    y_hat = []

    for t, x_t in enumerate(tf.split(inputs, seq_length, axis=0)):
        if t > 0:
            scope.reuse_variables()

        #input to hidden layer weights
        U = tf.get_variable("U", [vocab_size, hidden_size],
initializer=initializer)

        #hidden to hidden layer weights
```

```
        W = tf.get_variable("W", [hidden_size, hidden_size],
initializer=initializer)

        #output to hidden layer weights
        V = tf.get_variable("V", [hidden_size, vocab_size],
initializer=initializer)

        #bias for hidden layer
        bh = tf.get_variable("bh", [hidden_size], initializer=initializer)

        #bias for output layer
        by = tf.get_variable("by", [vocab_size], initializer=initializer)

        h_t = tf.tanh(tf.matmul(x_t, U) + tf.matmul(h_t, W) + bh)

        y_hat_t = tf.matmul(h_t, V) + by

        y_hat.append(y_hat_t)
```

Apply `softmax` on the output and get the probabilities:

```
output_softmax = tf.nn.softmax(y_hat[-1])
outputs = tf.concat(y_hat, axis=0)
```

Compute the cross-entropy loss:

```
loss =
tf.reduce_mean(tf.nn.softmax_cross_entropy_with_logits(labels=targets,
logits=outputs))
```

Store the final hidden state of the RNN in `hprev`. We use this final hidden state for making predictions:

```
hprev = h_t
```

Defining BPTT

Now, we will perform the BPTT, with Adam as our optimizer. We will also perform gradient clipping to avoid the exploding gradients problem:

1. Initialize the Adam optimizer:

    ```
    minimizer = tf.train.AdamOptimizer()
    ```

2. Compute the gradients of the loss with the Adam optimizer:

    ```
    gradients = minimizer.compute_gradients(loss)
    ```

3. Set the threshold for the gradient clipping:

```
threshold = tf.constant(5.0, name="grad_clipping")
```

4. Clip the gradients that exceed the threshold and bring it to the range:

```
clipped_gradients = []
for grad, var in gradients:
    clipped_grad = tf.clip_by_value(grad, -threshold, threshold)
    clipped_gradients.append((clipped_grad, var))
```

5. Update the gradients with the clipped gradients:

```
updated_gradients = minimizer.apply_gradients(clipped_gradients)
```

Start generating songs

Start the TensorFlow session and initialize all the variables:

```
sess = tf.Session()

init = tf.global_variables_initializer()

sess.run(init)
```

Now, we will look at how to generate the song lyrics using an RNN. What should the input and output to the RNN? How does it learn? What is the training data? Let's understand this an explanation, along with the code, step by step.

We know that in RNNs, the output predicted at a time step t will be sent as the input to the next time step; that is, on every time step, we need to feed the predicted character from the previous time step as input. So, we prepare our dataset in the same way.

For instance, look at the following table. Let's suppose that each row is a different time step; on a time step $t = 0$, the RNN predicted a new character, g, as the output. This will be sent as the input to the next time step, $t = 1$.

However, if you notice the input in the time step $t = 1$, we removed the first character from the input o and added the newly predicted character g at the end of our sequence. Why are we removing the first character from the input? Because we need to maintain the sequence length.

Let's suppose that our sequence length is eight; adding a newly predicted character to our sequence increases the sequence length to nine. To avoid this, we remove the first character from the input, while adding a newly predicted character from the previous time step.

Similarly, in the output data, we also remove the first character on each time step, because once it predicts the new character, the sequence length increases. To avoid this, we remove the first character from the output on each time step, as shown in the following table:

	Input	Output
t=0	On a br(i) (12345678)	n a bri(g) (12345678)
t=1	n a bri(g) (12345678)	a brig(h) (12345678)
t=2	a brig(h) (12345678)	a brigh(t) (12345678)

Now, we will look at how we can prepare our input and output sequence similarly to the preceding table.

Define a variable called `pointer`, which points to the character in our dataset. We will set our `pointer` to 0, which means it points to the first character:

```
pointer = 0
```

Define the input data:

```
input_sentence = data[pointer: pointer + seq_length]
```

What does this mean? With the pointer and the sequence length, we slice the data. Consider that the `seq_length` is 25 and the pointer is 0. It will return the first 25 characters as input. So, `data[pointer:pointer + seq_length]` returns the following output:

```
"Look at her face, it's a "
```

Define the output, as follows:

```
output_sentence = data[pointer + 1: pointer + seq_length + 1]
```

We slice the output data with one character ahead moved from input data. So, `data[pointer + 1:pointer + seq_length + 1]` returns the following:

```
"ook at her face, it's a w"
```

As you can see, we added the next character in the preceding sentence and removed the first character. So, on every iteration, we increment the pointer and traverse the entire dataset. This is how we obtain the input and output sentence for training the RNN.

As we have learned, an RNN accepts only numbers as input. Once we have sliced the input and output sequence, we get the indices of the respective characters, using the `char_to_ix` dictionary that we defined:

```
input_indices = [char_to_ix[ch] for ch in input_sentence]
target_indices = [char_to_ix[ch] for ch in output_sentence]
```

Convert the indices into one-hot encoded vectors by using the `one_hot_encoder` function we defined previously:

```
input_vector = one_hot_encoder(input_indices)
target_vector = one_hot_encoder(target_indices)
```

This `input_vector` and `target_vector` become the input and output for training the RNN. Now, Let's start training.

The `hprev_val` variable stores the last hidden state of our trained RNN model which we use for making predictions, and we store the loss in a variable called `loss_val`:

```
hprev_val, loss_val, _ = sess.run([hprev, loss, updated_gradients],
feed_dict={inputs: input_vector,targets: target_vector,init_state:
hprev_val})
```

We train the model for *n* iterations. After training, we start making predictions. Now, we will look at how to make predictions and generate song lyrics using our trained RNN. Set the `sample_length`, that is, the length of the sentence (song) we want to generate:

```
sample_length = 500
```

Randomly select the starting index of the input sequence:

```
random_index = random.randint(0, len(data) - seq_length)
```

Select the input sentence with the randomly selected index:

```
sample_input_sent = data[random_index:random_index + seq_length]
```

As we know, we need to feed the input as numbers; convert the selected input sentence to indices:

```
sample_input_indices = [char_to_ix[ch] for ch in sample_input_sent]
```

Remember, we stored the last hidden state of the RNN in `hprev_val`. We used that for making predictions. We create a new variable called `sample_prev_state_val` by copying values from `hprev_val`.

The `sample_prev_state_val` is used as an initial hidden state for making predictions:

```
sample_prev_state_val = np.copy(hprev_val)
```

Initialize the list for storing the predicted output indices:

```
predicted_indices = []
```

Now, for *t* in range of `sample_length`, we perform the following and generate the song for the defined `sample_length`.

Convert the `sampled_input_indices` to the one-hot encoded vectors:

```
sample_input_vector = one_hot_encoder(sample_input_indices)
```

Feed the `sample_input_vector`, and also the hidden state `sample_prev_state_val`, as the initial hidden state to the RNN, and get the predictions. We store the output probability distribution in `probs_dist`:

```
probs_dist, sample_prev_state_val = sess.run([output_softmax, hprev],
    feed_dict={inputs: sample_input_vector,init_state: sample_prev_state_val})
```

Randomly select the index of the next character with the probability distribution generated by the RNN:

```
ix = np.random.choice(range(vocab_size), p=probs_dist.ravel())
```

Add this newly predicted index, `ix`, to the `sample_input_indices`, and also remove the first index from `sample_input_indices` to maintain the sequence length. This will form the input for the next time step:

```
sample_input_indices = sample_input_indices[1:] + [ix]
```

Store all the predicted `chars` indices in the `predicted_indices` list:

```
predicted_indices.append(ix)
```

Convert all the `predicted_indices` to their characters:

```
predicted_chars = [ix_to_char[ix] for ix in predicted_indices]
```

Combine all the `predicted_chars` and save it as `text`:

```
text = ''.join(predicted_chars)
```

Print the predicted text on every 50,000th iteration:

```
print ('\n')
print (' After %d iterations' %(iteration))
print('\n %s \n' % (text,))
print('-'*115)
```

Increment the `pointer` and `iteration`:

```
pointer += seq_length
iteration += 1
```

On the initial iteration, you can see that the RNN has generated the random characters. But at the 50,000th iteration, it has started to generate meaningful text:

```
After 0 iterations

 Y?a6C.-eMSfk0pHD v!74YNeI 3YeP,h-
h6AADuANJJv:HA(QXNeKzwCjBnAShbavSrGw7:ZcSv[!?dUno Qt?OmE-PdY
wrqhSu?Yvxdek?5Rn'Pj!n5:32a?cjue  ZIj
Xr6qn.scqpa7)[MSUjG-Sw8n3ZexdUrLXDQ:MOXBMX
EiuKjGudcznGMkF:Y6)ynj0Hiajj?d?n2Iapmfc?WYd BWVyB-GAxe.Hq0PaEce5H!u5t:
AkO?F(oz0Ma!BUMtGtSsAP]Oh, 1nHf5tZCwU(F?X5CDzhOgSNH(4Cl-Ldk? HO7
WD9boZyPIDghWUfY B:r5z9Muzdw2'WWtf4srCgyX?hS!,BL
GZHqgTY:K3!wn:aZGoxr?zmayANhMKJsZhGjpbgiwSw5Z:oatGAL4Xenk]jE3zJ?ymB6v?j7(mL
[3DFsO['Hw-d7htzMn?nm20o'?6gfPZhBa
NlOjnBd2n0 T"d'e1k?OY6Wwnx6d!F


----------------------------------------------------------------------
-------------------


After 50000 iterations

 Hem-:]
[Ex" what
Akn'lise
[Grout his bring bear.
Gnow ourd?
Thelf
As cloume
```

```
That hands, Havi Musking me Mrse your leallas, Froking the cluse (have:
mes.
I slok and if a serfres me the sky withrioni flle rome.....Ba tut get make
ome
But it lives I dive.
[Lett it's to the srom of and a live me it's streefies
And is.
As it and is me dand a serray]
[zrtye:"
Chay at your hanyer
[Every rigbthing with farclets
[Brround.
Mad is trie
[Chare's a day-Mom shacke?


, I


-----------------------------------------------------------------------
----------------------
```

Different types of RNN architectures

Now that we have learned how an RNN works, we will look at a different type of RNN architecture that's based on numbers of input and output.

One-to-one architecture

In a **one-to-one architecture**, a single input is mapped to a single output, and the output from the time step *t* is fed as an input to the next time step. We have already seen this architecture in the last section for generating songs using RNNs.

For instance, for a text generation task, we take the output generated from a current time step and feed it as the input to the next time step to generate the next word. This architecture is also widely used in stock market predictions.

The following figure shows the one-to-one RNN architecture. As you can see, output predicted at the time step t is sent as the input to the next time step:

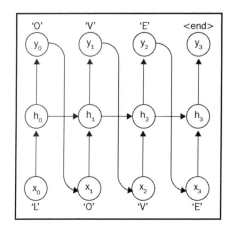

One-to-many architecture

In a **one-to-many architecture**, a single input is mapped to multiple hidden states and multiple output values, which means RNN takes a single input and maps it to an output sequence. Although we have a single input value, we share the hidden states across time steps to predict the output. Unlike the previous one-to-one architecture, here, we only share the previous hidden states across time steps and not the previous outputs.

One such application of this architecture is image caption generation. We pass a single image as an input, and the output is the sequence of words constituting a caption of the image.

As shown in the following figure, a single image is passed as an input to the RNN, and at the first time step, t_0, the word *Horse* is predicted; on the next time step, t_1, the previous hidden state h_0 is used to predict the next word which is *standing*. Similarly, it continues for a sequence of steps and predicts the next word until the caption is generated:

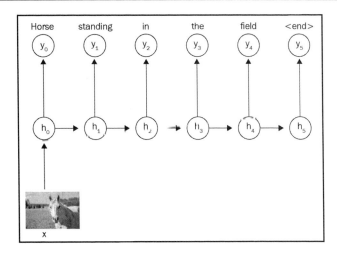

Many-to-one architecture

A **many-to-one architecture,** as the name suggests, takes a sequence of input and maps it to a single output value. One such popular example of a many-to-one architecture is **sentiment classification**. A sentence is a sequence of words, so on each time step, we pass each word as input and predict the output at the final time step.

Let's suppose that we have a sentence: *Paris is a beautiful city.* As shown in the following figure, at each time step, a single word is passed as an input, along with the previous hidden state; and, at the final time step, it predicts the sentiment of the sentence:

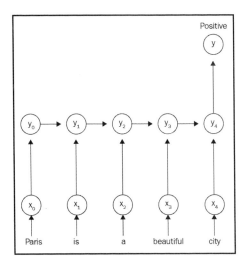

Many-to-many architecture

In **many-to-many architectures**, we map a sequence of input of arbitrary length to a sequence of output of arbitrary length. This architecture has been used in various applications. Some of the popular applications of many-to-many architectures include language translation, conversational bots, and audio generation.

Let's suppose that we are converting a sentence from English to French. Consider our input sentence: *What are you doing?* It would be mapped to, *Que faites vous* as shown in the following figure:

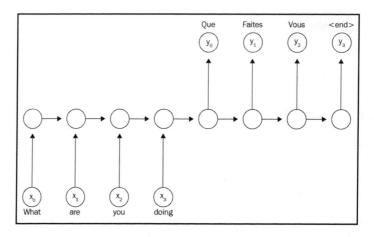

Summary

We started off the chapter by covering what an RNN is and how an RNN differs from a feedforward network. We learned that an RNN is a special type of neural network that is widely applied over sequential data; it predicts output based on not only the current input but also on the previous hidden state, which acts as the memory and stores the sequence of information that the network has seen so far.

We learned how forward propagation works in RNNs, and then we explored a detailed derivation of the BPTT algorithm, which is used for training RNN. Then, we explored RNNs by implementing them in TensorFlow to generate song lyrics. At the end of the chapter, we learned about the different architectures of RNNs, such as one-to-one, one-to-many, many-to-one, and many-to-many, which are used for various applications.

In the next chapter, we will learn about the LSTM cell, which solves the vanishing gradient problem in RNNs. We will also learn about different variants of RNNs.

Questions

Try answering the following questions:

1. What is the difference between an RNN and a feedforward neural network?
2. How is the hidden state computed in an RNN?
3. What is the use of a recurrent network?
4. How does the vanishing gradient problem occur?
5. What is the exploding gradient problem?
6. How gradient clipping mitigates the exploding gradient problem?
7. What are the different types of RNN architectures?

Further reading

Refer to the following links to learn more about RNNs:

- *Fundamentals of Recurrent Neural Network (RNN) and Long Short-Term Memory (LSTM) Network*, by Alex Sherstinsky, `https://arxiv.org/pdf/1808.03314.pdf`
- Handwriting generation by RNN with TensorFlow, based on *Generating Sequences With Recurrent Neural Networks* by Alex Graves, `https://github.com/snowkylin/rnn-handwriting-generation`

5
Improvements to the RNN

The drawback of a **recurrent neural network (RNN)** is that it will not retain information for a long time in memory. We know that an RNN stores sequences of information in its hidden state but when the input sequence is too long, it cannot retain all the information in its memory due to the vanishing gradient problem, which we discussed in the previous chapter.

To combat this, we introduce a variant of RNN called a **long short-term memory (LSTM)** cell, which resolves the vanishing gradient problem by using a special structure called a **gate**. Gates keep the information in memory as long as it is required. They learn what information to keep and what information to discard from the memory.

We will start the chapter by exploring LSTM and exactly how LSTM overcomes the shortcomings of RNN. Later, we will learn how to perform forward and backward propagation with an LSTM cell. Going ahead, we will explore how to implement LSTM cells in TensorFlow and how to use them to predict a Bitcoin's price.

Moving forward, we will get the hang of **gated recurrent unit (GRU)** cells, which act as a simplified version of the LSTM cell. We will learn how forward and backward propagation works in a GRU cell. Next, we will get a basic understanding of bidirectional RNNs and how they make use of past and future information to make predictions; we will also understand how deep RNNs work.

At the end of the chapter, we will learn about the sequence-to-sequence model, which maps the input of varying length to the output of varying length. We will dive into the architecture of the sequence-to-sequence model and the attention mechanism.

To ensure a clear understanding of these topics, the chapter has been organized as follows:

- LSTM to the rescue
- Forward and backward propagation in LSTM
- Predicting Bitcoin price using LSTM
- GRUs
- Forward and backward propagation in GRUs
- Bidirectional RNNs
- Deep RNNs
- Sequence-to-sequence model

LSTM to the rescue

While backpropagating an RNN, we discovered a problem called **vanishing gradients**. Due to the vanishing gradient problem, we cannot train the network properly, and this causes the RNN to not retain long sequences in the memory. To understand what we mean by this, let's consider a small sentence:

The sky is __.

An RNN can easily predict the blank as *blue* based on the information it has seen, but it cannot cover the long-term dependencies. What does that mean? Let's consider the following sentence to understand the problem better:

Archie lived in China for 13 years. He loves listening to good music. He is a fan of comics. He is fluent in ____.

Now, if we were asked to predict the missing word in the preceding sentence, we would predict it as *Chinese*, but how did we predict that? We simply remembered the previous sentences and understood that Archie lived for 13 years in China. This led us to the conclusion that Archie might be fluent in Chinese. An RNN, on the other hand, cannot retain all of this information in its memory to say that Archie is fluent in Chinese. Due to the vanishing gradient problem, it cannot recollect/remember information for a long time in its memory. That is, when the input sequence is long, the RNN memory (hidden state) cannot hold all the information. To alleviate this, we use an LSTM cell.

LSTM is a variant of an RNN that resolves the vanishing gradient problem and retains information in the memory as long as it is required. Basically, RNN cells are replaced with LSTM cells in the hidden units, as shown in the following diagram:

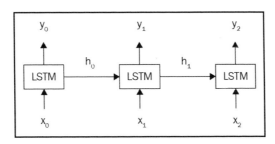

Understanding the LSTM cell

What makes LSTM cells so special? How do LSTM cells achieve long-term dependency? How does it know what information to keep and what information to discard from the memory?

This is all achieved by special structures called **gates**. As shown in the following diagram, a typical LSTM cell consists of three special gates called the input gate, output gate, and forget gate:

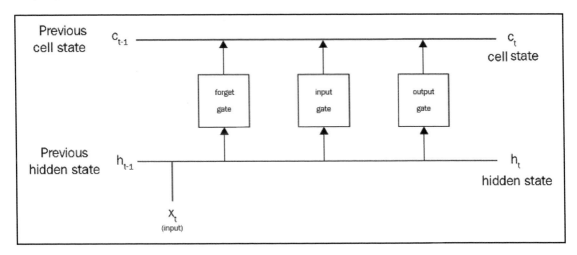

These three gates are responsible for deciding what information to add, output, and forget from the memory. With these gates, an LSTM cell effectively keeps information in the memory only as long as required.

In an RNN cell, we used the hidden state, h_t, for two purposes: one for storing the information and the other for making predictions. Unlike RNN, in the LSTM cell we break the hidden states into two states, called the **cell state** and the **hidden state**:

- The cell state is also called internal memory and is where all the information will be stored
- The hidden state is used for computing the output, that is, for making predictions

Both the cell state and hidden state are shared across every time step. Now, we will deep dive into the LSTM cell and see exactly how these gates are used and how the hidden state predicts the output.

Forget gate

The forget gate, f_t, is responsible for deciding what information should be removed from the cell state (memory). Consider the following sentence:

Harry is a good singer. He lives in New York. Zayn is also a good singer.

As soon as we start talking about Zayn, the network will understand that the subject has been changed from Harry to Zayn, and the information about Harry is no longer required. Now, the forget gate will remove/forget information about Harry from the cell state.

The forget gate is controlled by a sigmoid function. At time step t, we pass input x_t, and the previous hidden state, h_{t-1}, to the forget gate. It return 0 if the particular information from the cell state should be removed and returns 1 if the information should not be removed. The forget gate, f, at a time step, t, is expressed as follows:

$$f_t = \sigma(U_f x_t + W_f h_{t-1} + b_f)$$

Here, the following applies:

- U_f is the input-to-hidden layer weights of the forget gate
- W_f is the hidden-to-hidden layer weights of the forget gate
- b_f is the bias of the forget gate

The following diagram shows the forget gate. As you can see, input x_t is multiplied with U_f and the previous hidden state, h_{t-1}, is multiplied with W_f, then both will be added together and sent to the sigmoid function, which returns f_t, as follows:

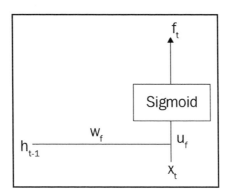

Input gate

The input gate is responsible for deciding what information should be stored in the cell state. Let's consider the same example:

Harry is a good singer. He lives in New York. Zayn is also a good singer.

After the forget gate removes information from the cell state, the input gate decides what information it has to keep in the memory. Here, since the information about Harry is removed from the cell state by the forget gate, the input gate decides to update the cell state with information about *Zayn*.

Similar to the forget gate, the input gate is controlled by a sigmoid function that returns output in the range of 0 to 1. If it returns 1, then the particular information will be stored/updated to the cell state, and if it returns 0, we will not store the information to the cell state. The input gate i at time step t is expressed as follows:

$$i_t = \sigma(U_i x_t + W_i h_{t-1} + b_i)$$

Here, the following applies:

- U_i is the input-to-hidden weights of the input gate
- W_i is the hidden-to-hidden weights of the input gate
- b_i is the bias of the input gate

The following diagram shows the input gate:

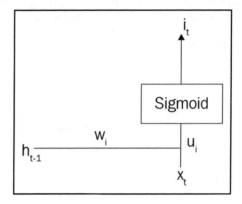

Output gate

We will have a lot of information in the cell state (memory). The output gate is responsible for deciding what information should be taken from the cell state to give as an output. Consider the following sentence:

Zayn's debut album was a huge success. Congrats ____.

The output gate will look up all the information in the cell state and select the correct information to fill the blank. Here, `congrats` is an adjective that is used to describe a noun. So, the output gate will predict *Zayn* (noun) to fill the blank. Similar to other gates, it is also controlled by a sigmoid function. The output gate o at time step t is expressed as follows:

$$o_t = \sigma(U_o x_t + W_o h_{t-1} + b_o)$$

Here, the following applies:

- U_o is the input-to-hidden weights of the output gate
- W_o is the hidden-to-hidden weights of the output gate
- b_o is the bias of the output gate

The output gate is shown in the following diagram:

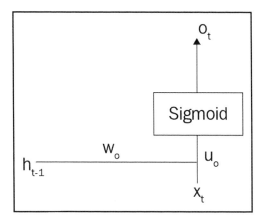

Updating the cell state

We just learned how all three gates work in an LSTM network, but the question is, how can we actually update the cell state by adding relevant new information and deleting information that is not required from the cell state with the help of the gates?

First, we will see how to add new relevant information to the cell state. To hold all the new information that can be added to the cell state (memory), we create a new vector called g_t. It is called a **candidate state** or **internal state vector**. Unlike gates that are regulated by the sigmoid function, the candidate state is regulated by the tanh function, but why? The sigmoid function returns values in the range of o to 1, that is, it is always positive. We need to allow the values of g_t to be either positive or negative. So, we use the tanh function, which returns values in the range of −1 to +1.

The candidate state, g, at time t is expressed as follows:

$$g_t = tanh(U_g x_t + W_g h_{t-1} + b_g)$$

Here, the following applies:

- U_g is the input-to-hidden weights of the candidate state
- W_g is the hidden-to-hidden weights of the candidate state
- b_g is the bias of the candidate state

Thus, the candidate state holds all the new information that can be added to the cell state (memory). The following diagram shows the candidate state:

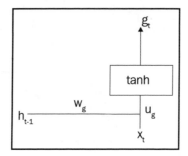

How do we decide whether the information in the candidate state is relevant? How do we decide whether to add new information or not from the candidate state to the cell state? We learned that the input gate is responsible for deciding whether to add new information or not, so if we multiply g_t and i_t, we only get relevant information that should be added to the memory.

That is, we know the input gate returns 0 if the information is not required and 1 if the information is required. Say $i_t = 0$, then multiplying g_t and i_t gives us 0, which means the information in g_t is not required and we don't want to update the cell state with g_t. When $i_t = 1$, then multiplying g_t and i_t gives us g_t, which implies we can update the information in g_t to the cell state.

Adding the new information to the cell state with input gate i_t and candidate state g_t is shown in the following diagram:

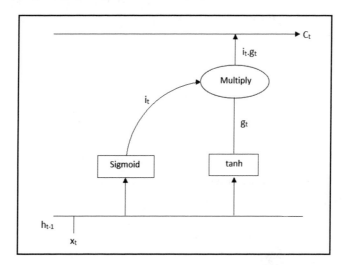

Now, we will see how to remove information from the previous cell state that is no longer required.

We learned that the forget gate is used for removing information that is not required in the cell state. So, if we multiply the previous cell state, c_{t-1}, and forget gate, f_t, then we retain only relevant information in the cell state.

Say $f_t = 0$, then multiplying c_{t-1} and f_t gives us 0, which means the information in cell state, c_{t-1}, is not required and should be removed (forgotten). When $f_t = 1$, then multiplying c_{t-1} and f_t gives c_{t-1}, which implies that information in the previous cell state is required and should not be removed.

Removing information from the previous cell state, c_{t-1}, with the forget gate, f_t, is shown in the following diagram:

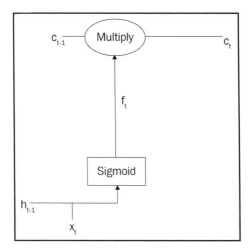

Thus, in a nutshell, we update our cell state by multiplying g_t and i_t to add new information, and multiplying c_{t-1} ,and f_t to remove information. We can express the cell state equation as follows:

$$c_t = f_t c_{t-1} + i_t g_t$$

Updating hidden state

We just learned how the information in the cell state will be updated. Now, we will see how the information in the hidden state will be updated. We learned that the hidden state, h_t, is used for computing the output, but how can we compute the output?

We know that the output gate is responsible for deciding what information should be taken from the cell state to give as output. Thus, multiplying o_t and tanh (to squash between -1 and +1) of cell state, $tanh(c_t)$, gives us the output.

Thus, the hidden state, h_t, is expressed as follows:

$$h_t = o_t tanh(c_t)$$

The following diagram shows how the hidden state, h_t, is computed by multiplying o_t and $tanh(c_t)$:

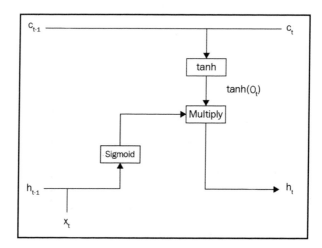

Finally, once we have the hidden state value, we can apply the softmax function and compute $\hat{y}_t$ as follows:

$$\hat{y}_t = softmax(Vh_t)$$

Here, V is the hidden-to-output layer weights.

Forward propagation in LSTM

Putting it all together, the final LSTM cell with all the operations is shown in the following diagram. Cell state and hidden states are shared across time steps, meaning that the LSTM computes the cell state, c_t, and hidden state, h_t, at time step t, and sends it to the next time step:

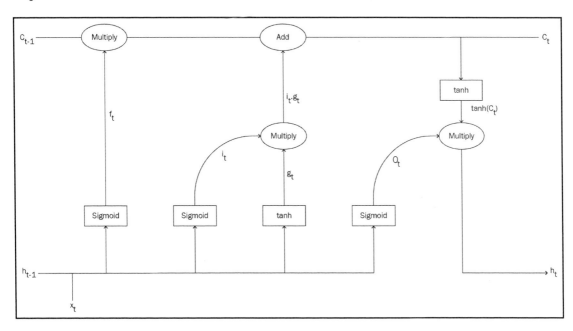

The complete forward propagation steps in the LSTM cell can be given as follows:

1. **Input gate**: $i_t = \sigma(U_i x_t + W_f h_{t-1} + b_i)$
2. **Forget gate**: $f_t = \sigma(U_f x_t + W_f h_{t-1} + b_f)$
3. **Output gate**: $o_t = \sigma(U_o x_t + W_o h_{t-1} + b_o)$
4. **Candidate state**: $g_t = tanh(U_g x_t + W_g h_{t-1} + b_g)$
5. **Cell state**: $c_t = f_t c_{t-1} + i_t g_t$
6. **Hidden state**: $h_t = o_t tanh(c_t)$
7. **Output**: $\hat{y}_t = softmax(V h_t)$

Backpropagation in LSTM

We compute the loss at each time step to determine how well our LSTM model is predicting the output. Say we use cross-entropy as a loss function, then the loss, L, at time step t is given by the following equation:

$$L_t = -y_t \log(\hat{y}_t)$$

Here, y_t is the actual output and $\hat{y}_t$ is the predicted output at time step t.

Our final loss is the sum of loss at all time steps, and can be given as follows:

$$L = \sum_{j=0}^{T} L_j$$

We minimize the loss using gradient descent. We find the derivative of loss with respect to all of the weights used in the network and find the optimal weights to minimize the loss:

- We have four inputs-to-hidden layer weights, U_i, U_f, U_o, U_g, which are the input-to-hidden layer weights of the input gate, forget gate, output gate, and candidate state, respectively
- We have four hidden-to-hidden layer weights, W_i, W_f, W_o, W_g , which implies hidden-to-hidden layer weights of input gate, forget gate, output gate, and candidate state, respectively
- We have one hidden-to-output layer weight, V

We find the optimal values for all these weights through gradient descent and update the weights according to the weight update rule. The weight update rule is given by the following equation:

$$\text{weight} = \text{weight} - \alpha \frac{\partial Loss}{\text{weight}}$$

In the next section, we'll look at how to compute gradients of loss with respect to all of the weights used in the LSTM cell step by step.

 You can skip the upcoming section if you are not interested in deriving gradients for all of the weights. However, it will strengthen your understanding of the LSTM cell.

Gradients with respect to gates

Calculating gradients of loss with respect to all the weights used in the LSTM cell requires the gradients of all the gates and the candidate state. So, in this section, we will learn how to compute the gradient of the loss function with respect to all of the gates and the candidate state.

Before we begin, let's recollect the following two things:

- The derivative of a sigmoid function is expressed as follows:

$$\sigma'(x) = \sigma(x)(1 - \sigma(x))$$

- The derivative of a tanh function is expressed as follows:

$$tanh'(x) = 1 - tanh^2(x)$$

In the upcoming calculations, we will be using gradients of loss with respect to hidden state h_t and cell state c_t at multiple places. Thus, first, we will see how to compute gradients of loss with respect to hidden state h_t and cell state c_t.

First, let's see how to calculate the **gradients of loss with respect to a hidden state,** h_t.

We know that output $\hat{y}_t$ is computed as follows:

$$\hat{y}_t = softmax(Vh_t)$$

Let's say $z_t = Vh_t$. We have the h_t term in z_t, so by the chain rule, we can write the following:

$$\frac{\partial L}{\partial h_t} = \frac{\partial L}{\partial z_t}\frac{\partial z_t}{\partial h_t}$$

$$\frac{\partial L}{\partial h_t} = \frac{\partial L}{\partial z_t}V$$

We have already seen how to compute $\frac{\partial L}{\partial z_t}$ in Chapter 4, *Generating Song Lyrics Using RNN,* so directly from the equation *(9)* of Chapter 4, *Generating Song Lyrics Using RNN,* we can write the following:

$$\frac{\partial L}{\partial h_t} = (\hat{y}_t - y_t)V$$

Now, let's see how to calculate the **gradient of the loss with respect to cell state,** c_t.

To calculate the gradients of loss with respect to the cell state, look at the equations of forward propagation and find out which equation has the c_t term. On the equation of the hidden state, we have the c_t term as follows:

$$h_t = o_t tanh(c_t)$$

So, by the chain rule, we can write the following:

$$\frac{\partial L}{\partial c_t} = \frac{\partial L}{\partial h_t}\frac{\partial h_t}{\partial c_t}$$
$$= \frac{\partial L}{\partial h_t}(o_t \cdot tanh'(c_t))$$

We know that the derivative of tanh is $tanh'(x) = 1 - tanh^2(x)$, thus we can write the following:

$$\frac{\partial L}{\partial c_t} = \frac{\partial L}{\partial h_t}(o_t \cdot 1 - tanh^2(c_t))$$

Now that we have calculated the gradients of loss with respect to the hidden state and the cell state, let's see how to calculate the gradient of loss with respect to all the gates one by one.

First, we will see how to calculate the **gradient of loss with respect to the output gate,** o_t.

To calculate the gradients of loss with respect to the output gate, look at the equations of forward propagation and find out in which equation we have the o_t term. In the equation of the hidden state we have the o_t term as follows:

$$h_t = o_t tanh(c_t)$$

So, by the chain rule, we can write the following:

$$\frac{\partial L}{\partial o_t} = \frac{\partial L}{\partial h_t} \frac{\partial h_t}{\partial o_t}$$

$$\frac{\partial L}{\partial o_t} = \frac{\partial L}{\partial h_t} tanh(c_t) \tag{1}$$

Now we will see how to calculate the **gradient of loss with respect to the input gate,** i_t.

We have the i_t term in the cell state equation for c_t:

$$c_t = f_t c_{t-1} + i_t g_t$$

By the chain rule, we can write the following:

$$\frac{\partial L}{\partial i_t} = \frac{\partial L}{\partial c_t} \frac{\partial c_t}{\partial i_t}$$

$$\frac{\partial L}{\partial i_t} = \frac{\partial L}{\partial c_t} g_t \tag{2}$$

And now we learn how to calculate the **gradient of loss with respect to the forget gate,** f_t.

We also have the f_t term in the cell state equation for c_t:

$$c_t = f_t c_{t-1} + i_t g_t$$

By the chain rule, we can write the following:

$$\frac{\partial L}{\partial f_t} = \frac{\partial L}{\partial c_t} \frac{\partial c_t}{\partial f_t}$$

$$\frac{\partial L}{\partial f_t} = \frac{\partial L}{\partial c_t} c_{t-1} \tag{3}$$

Finally, we learn how to calculate the **gradient of loss with respect to the candidate state,** g_t.

We also have the g_t term in the cell state equation for c_t:

$$c_t = f_t c_{t-1} + i_t g_t$$

Thus, by the chain rule, we can write the following:

$$\frac{\partial L}{\partial g_t} = \frac{\partial L}{\partial c_t}\frac{\partial c_t}{\partial g_t}$$

$$\frac{\partial L}{\partial g_t} = \frac{\partial L}{\partial c_t} i_t \tag{4}$$

Thus, we have calculated the gradients of loss with respect to all the gates and the candidate state. In the next section, we'll see how to calculate the gradients of loss with respect to all of the weights used in the LSTM cell.

Gradients with respect to weights

Now let's see how to calculate gradients with respect to all the weights used in the LSTM cell.

Gradients with respect to V

After predicting the output, $\hat{y}$, we are in the final layer of the network. Since we are backpropagating; that is, going from the output layer to the input layer, our first weight will be V, which is hidden-to-output layer weight.

We have learned throughout that the final loss is the sum of the loss over all the time steps. In a similar manner, our final gradient is the sum of gradients at all time steps as follows:

$$\frac{\partial L}{\partial V} = \frac{\partial L_0}{\partial V} + \frac{\partial L_1}{\partial V} + \frac{\partial L_2}{\partial V} + \dots \frac{\partial L_{T-1}}{\partial V}$$

If we have $T-1$ layers, then we can write the gradient of loss with respect to V as follows:

$$\frac{\partial L}{\partial V} = \sum_{j=0}^{T-1} \frac{\partial L_j}{\partial V}$$

Since the final equation of LSTM, that is, $\hat{y}_t = softmax(Vh_t)$, is the same as RNN, calculating gradients of loss with respect to V is exactly the same as what we computed in the RNN. Thus, we can directly write the following:

$$\boxed{\frac{\partial L}{\partial V} = \sum_{j=1}^{T} (\hat{y}_j - y_j) \otimes h_j}$$

Gradients with respect to W

Now we will see how to calculate the gradients of loss with respect to hidden-to-hidden layer weights, W, for all the gates and the candidate state.

Let's calculate **gradients of loss with respect to** W_i.

Recall the equation of the input gate, which is given as follows:

$$i_t = \sigma(U_i x_t + W_i h_{t-1} + b_i)$$

Thus, by the chain rule, we can write the following:

$$\frac{\partial L}{\partial W_i} = \sum_{j=0}^{T-1} \frac{\partial L_j}{\partial i_j} \frac{\partial i_j}{\partial W_i}$$

Let's calculate each of the terms in the preceding equation.

We have already seen how to compute the first term, the gradient of loss with respect to the input gate, $\frac{\partial L_j}{\partial i_j}$, in the *Gradients with respect to gates* section. Refer to equation *(2)*.

So, let's look at the second term:

$$\frac{\partial i_j}{\partial W_i} = \sigma'(i_j) \otimes h_{j-1}$$

Since we know the derivative of the sigmoid function, $\sigma'(x) = \sigma(x)(1 - \sigma(x))$, we can write the following:

$$\sigma'(i_j) = \sigma(i_j)(1 - \sigma(i_j))$$

But i_j is already a result of sigmoid, that is, $i_j = \sigma(U_i x_j + W_i h_{j-1} + b_i)$, so we can just write $\sigma'(i_j) = i_j(1 - i_j)$, Thus, our equation becomes the following:

$$\frac{\partial i_j}{\partial W_i} = i_j(1 - i_j) \otimes h_{j-1}$$

Thus, our final equation for calculating the gradient of loss with respect to W_i becomes the following:

$$\boxed{\frac{\partial L}{\partial W_i} = \sum_{j=0}^{T-1} \frac{\partial L_j}{\partial i_j} \cdot i_j(1 - i_j) \otimes h_{j-1}}$$

Now, let's find out the **gradients of loss with respect to** W_f.

Recall the equation of the forget gate, which is given as follows:

$$f_t = \sigma(U_f x_t + W_f h_{t-1} + b_f)$$

Thus, by the chain rule, we can write the following:

$$\frac{\partial L}{\partial W_f} = \sum_{j=0}^{T-1} \frac{\partial L_j}{\partial f_j} \frac{\partial f_j}{\partial W_f}$$

We have already seen how to compute $\frac{\partial L_j}{\partial f_j}$ in the gradients with respect to the gates section. Refer to equation *(3)*. So, let's look at computing the second term:

$$\frac{\partial f_j}{\partial W_f} = \sigma(f_j)' \otimes h_{j-1}$$
$$= f_j(1 - f_j) \otimes h_{j-1}$$

Thus, our final equation for calculating the gradient of loss with respect to W_f becomes the following:

$$\boxed{\frac{\partial L}{\partial W_f} = \sum_{j=0}^{T-1} \frac{\partial L_j}{\partial f_j} \cdot f_j(1 - f_j) \otimes h_{j-1}}$$

Let's calculate the **gradients of loss with respect to** W_o.

Recall the equation of the output gate, which is given as follows:

$$o_t = \sigma(U_o x_t + W_o h_{t-1} + b_o)$$

So, by the chain rule, we can write the following:

$$\frac{\partial L}{\partial W_o} = \sum_{j=0}^{T-1} \frac{\partial L_j}{\partial o_j} \frac{\partial o_j}{\partial W_o}$$

Check equation *(1)* for the first term. The second term can be computed as follows:

$$\frac{\partial o_j}{\partial W_o} = \sigma'(o_j) \otimes h_{j-1}$$
$$= o_j(1 - o_j) \otimes h_{j-1}$$

Thus, our final equation for calculating the gradient of loss with respect to W_o becomes the following:

$$\boxed{\frac{\partial L}{\partial W_o} = \sum_{j=0}^{T-1} \frac{\partial L_j}{\partial o_j} \cdot o_j(1 - o_j) \otimes h_{j-1}}$$

Let's move on to calculating the **gradient with respect to** W_g.

Recall the candidate state equation:

$$g_t = tanh(U_g x_t + W_g h_{t-1} + b_g)$$

Thus, by the chain rule, we can write the following:

$$\frac{\partial L}{\partial W_g} = \sum_{j=1}^{T} \frac{\partial L_j}{\partial g_j} \frac{\partial g_j}{\partial W_g}$$

Refer to equation *(4)* for the first term. The second term can be computed as follows:

$$\frac{\partial g_j}{\partial W_f} = tanh'(g_j) \otimes h_{j-1}$$

We know that derivative of tanh is $tanh(x)' = 1 - tanh^2(x)$, so we can write the following:

$$\frac{\partial g_j}{\partial W_f} = (1 - g_j^2) \otimes h_{j-1}$$

Thus, our final equation for calculating the gradient of loss with respect to W_g becomes the following:

$$\boxed{\frac{\partial L}{\partial W_g} = \sum_{j=0}^{T-1} \frac{\partial L_j}{\partial g_j} \cdot (1 - g_j^2) \otimes h_{j-1}}$$

Gradients with respect to U

Let's calculate the gradients of loss with respect to hidden-to-input layer weights U for all the gates and the candidate state. Computing gradients of loss with respect to U is exactly the same as the gradients we computed with respect to W, except that the last term will be x_j instead of h_{j-1}. Let's examine what we mean by that.

Let's find out the **gradients of loss with respect to** U_i.

The input gate equation is as follows:

$$i_t = \sigma(U_i x_t + W_f h_{t-1} + b_i)$$

Thus, using the chain rule, we can write the following:

$$\frac{\partial L}{\partial U_i} = \sum_{j=1}^{T} \frac{\partial L_j}{\partial i_j} \frac{\partial i_j}{\partial U}$$

Let's calculate each of the terms in the preceding equation. We already know the first term from equation *(2)*. So, the second term can be computed as follows:

$$\frac{\partial i_j}{\partial U_i} = \sigma'(i_j) \otimes x_j$$
$$= i_j(1 - i_j) \otimes x_j$$

Thus, our final equation for calculating the gradient of loss with respect to U_i becomes the following:

$$\frac{\partial L}{\partial U_i} = \sum_{j=0}^{T-1} \frac{\partial L_j}{\partial i_j} \cdot i_j (1 - i_j) \otimes x_j$$

As you can see, the preceding equation is exactly the same as $\frac{\partial L}{\partial W_i}$, except that the last term is x_j instead of h_{j-1}. This applies for all other weights, so we can directly write the equations as follows:

- Gradients of loss with respect to U_f:

$$\frac{\partial L}{\partial U_f} = \sum_{j=0}^{T-1} \frac{\partial L_j}{\partial f_j} \cdot f_j (1 - f_j) \otimes x_j$$

- Gradients of loss with respect to U_o:

$$\frac{\partial L}{\partial U_o} = \sum_{j=0}^{T-1} \frac{\partial L_j}{\partial o_j} \cdot o_j (1 - o_j) \otimes x_j$$

- Gradients of loss with respect to U_g:

$$\frac{\partial L}{\partial U_g} = \sum_{j=0}^{T-1} \frac{\partial L_j}{\partial g_j} \cdot g_j (1 - g_j^2) \otimes x_j$$

After the computing gradients, with respect to all of these weights, we update them using the weight update rule and minimize the loss.

Predicting Bitcoin prices using LSTM model

We have learned that LSTM models are widely used for sequential datasets, that is, datasets in which order matters. In this section, we will learn how we can use LSTM networks for time series analysis. We will learn how to predict Bitcoin prices using an LSTM network.

First, we import the required libraries as follows:

```
import numpy as np
import pandas as pd
from sklearn.preprocessing import StandardScaler

import matplotlib.pyplot as plt
%matplotlib inline
plt.style.use('ggplot')

import tensorflow as tf
tf.logging.set_verbosity(tf.logging.ERROR)

import warnings
warnings.filterwarnings('ignore')
```

Data preparation

Now, we will see how we can prepare our dataset in a way that our LSTM network needs. First, we read the input dataset as follows:

```
df = pd.read_csv('data/btc.csv')
```

Then we display a few rows of the dataset:

```
df.head()
```

The preceding code generates the following output:

	Date	Symbol	Open	High	Low	Close	Volume From	Volume To
0	5/26/2018	BTCUSD	7459.11	7640.46	7380.00	7520.00	2722.80	2.042265e+07
1	5/25/2018	BTCUSD	7584.15	7661.85	7326.94	7459.11	8491.93	6.342069e+07
2	5/24/2018	BTCUSD	7505.00	7734.99	7269.00	7584.15	11033.72	8.293137e+07
3	5/23/2018	BTCUSD	7987.70	8030.00	7433.19	7505.00	14905.99	1.148104e+08
4	5/22/2018	BTCUSD	8393.44	8400.00	7950.00	7987.70	6589.43	5.389753e+07

As shown in the preceding data frame, the `Close` column represents the closing price of Bitcoin. We need only the `Close` column to make predictions, so we take that particular column alone:

```
data = df['Close'].values
```

Next, we standardize the data and bring it to the same scale:

```
scaler = StandardScaler()
data = scaler.fit_transform(data.reshape(-1, 1))
```

We then plot and observe the trend of how the Bitcoin price changes. Since we scaled the price, it is not a bigger number:

```
plt.plot(data)
plt.xlabel('Days')
plt.ylabel('Price')
plt.grid()
```

The following plot is generated:

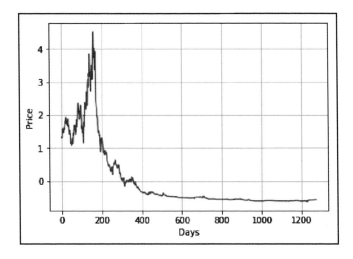

Now, we define a function called the `get_data` function, which generates the input and output. It takes the data and `window_size` as an input and generates the input and target column.

What is the window size here? We move the x values `window_size` times ahead and get the y values. For instance, as shown in the following table with `window_size` equal to 1, the y values are just one time step ahead of the x values:

x	y
0.13	0.56
0.56	0.11
0.11	0.40
0.40	0.63

The `get_data()` function is defined as follows:

```
def get_data(data, window_size):
    X = []
    y = []
    i = 0
    while (i + window_size) <= len(data) - 1:
        X.append(data[i:i+window_size])
        y.append(data[i+window_size])
        i += 1
    assert len(X) == len(y)
    return X, y
```

We choose `window_size` as 7 and generate the input and output:

```
X, y = get_data(data, window_size = 7)
```

Consider the first 1000 points as the train set and the rest of the points in the dataset as the test set:

```
#train set
X_train = np.array(X[:1000])
y_train = np.array(y[:1000])

#test set
X_test = np.array(X[1000:])
y_test = np.array(y[1000:])
```

The shape of `X_train` is shown as follows:

```
X_train.shape

(1000,7,1)
```

What does the preceding shape mean? It implies that the `sample_size`, `time_steps`, and `features` functions and the LSTM network require input exactly as follows:

- 1000 sets the number of data points (`sample_size`)
- 7 specifies the window size (`time_steps`)
- 1 specifies the dimension of our dataset (`features`)

Defining the parameters

Define the network parameters as shown:

```
batch_size = 7
window_size = 7
hidden_layer = 256
learning_rate = 0.001
```

Define `placeholders` for our input and output:

```
input = tf.placeholder(tf.float32, [batch_size, window_size, 1])
target = tf.placeholder(tf.float32, [batch_size, 1])
```

Let's now define all the weights we will use in our LSTM cell.

The weights of the input gate are defined as follows:

```
U_i = tf.Variable(tf.truncated_normal([1, hidden_layer], stddev=0.05))
W_i = tf.Variable(tf.truncated_normal([hidden_layer, hidden_layer],
stddev=0.05))
b_i = tf.Variable(tf.zeros([hidden_layer]))
```

The weights of the forget gate are defined as follows:

```
U_f = tf.Variable(tf.truncated_normal([1, hidden_layer], stddev=0.05))
W_f = tf.Variable(tf.truncated_normal([hidden_layer, hidden_layer],
stddev=0.05))
b_f = tf.Variable(tf.zeros([hidden_layer]))
```

The weights of the output gate are defined as given:

```
U_o = tf.Variable(tf.truncated_normal([1, hidden_layer], stddev=0.05))
W_o = tf.Variable(tf.truncated_normal([hidden_layer, hidden_layer],
stddev=0.05))
b_o = tf.Variable(tf.zeros([hidden_layer]))
```

The weights of the candidate state are defined as follows:

```
U_g = tf.Variable(tf.truncated_normal([1, hidden_layer], stddev=0.05))
W_g = tf.Variable(tf.truncated_normal([hidden_layer, hidden_layer],
stddev=0.05))
b_g = tf.Variable(tf.zeros([hidden_layer]))
```

The output layer weight is given as follows:

```
V = tf.Variable(tf.truncated_normal([hidden_layer, 1], stddev=0.05))
b_v = tf.Variable(tf.zeros([1]))
```

Define the LSTM cell

Now, we define the function called `LSTM_cell`, which returns the cell state and hidden state as an output. Recall the steps we saw in the forward propagation of LSTM, which is implemented as shown in the following code. `LSTM_cell` takes the input, previous hidden state, and previous cell state as inputs, and returns the current cell state and current hidden state as outputs:

```
def LSTM_cell(input, prev_hidden_state, prev_cell_state):
    it = tf.sigmoid(tf.matmul(input, U_i) + tf.matmul(prev_hidden_state,
W_i) + b_i)

    ft = tf.sigmoid(tf.matmul(input, U_f) + tf.matmul(prev_hidden_state,
W_f) + b_f)

    ot = tf.sigmoid(tf.matmul(input, U_o) + tf.matmul(prev_hidden_state,
W_o) + b_o)

    gt = tf.tanh(tf.matmul(input, U_g) + tf.matmul(prev_hidden_state, W_g)
+ b_g)

    ct = (prev_cell_state * ft) + (it * gt)

    ht = ot * tf.tanh(ct)

    return ct, ht
```

Defining forward propagation

Now we will perform forward propagation and predict the output, $\hat{y}_t$, and initialize a list called `y_hat` for storing the output:

```
y_hat = []
```

For each iteration, we compute the output and store it in the `y_hat` list:

```
for i in range(batch_size):
```

We initialize the hidden state and cell state:

```
        hidden_state = np.zeros([1, hidden_layer], dtype=np.float32)
        cell_state = np.zeros([1, hidden_layer], dtype=np.float32)
```

We perform the forward propagation and compute the hidden state and cell state of the LSTM cell for each time step:

```
for t in range(window_size):
    cell_state, hidden_state = LSTM_cell(tf.reshape(input[i][t], (-1,
1)), hidden_state, cell_state)
```

We know that output $\hat{y}_t$ can be computed as follows:

$$\hat{y}_t = Vh_t + b_v$$

Compute `y_hat`, and append it to the `y_hat` list:

```
y_hat.append(tf.matmul(hidden_state, V) + b_v)
```

Defining backpropagation

After performing forward propagation and predicting the output, we compute the loss. We use mean squared error as our loss function, and the total loss is the sum of losses across all of the time steps:

```
losses = []

for i in range(len(y_hat)):
    losses.append(tf.losses.mean_squared_error(tf.reshape(target[i], (-1,
1)), y_hat[i]))

loss = tf.reduce_mean(losses)
```

To avoid the exploding gradient problem, we perform gradient clipping:

```
gradients = tf.gradients(loss, tf.trainable_variables())
clipped, _ = tf.clip_by_global_norm(gradients, 4.0)
```

We use the Adam optimizer and minimize our loss function:

```
optimizer =
tf.train.AdamOptimizer(learning_rate).apply_gradients(zip(gradients,
tf.trainable_variables()))
```

Training the LSTM model

Start the TensorFlow session and initialize all the variables:

```
session = tf.Session()
session.run(tf.global_variables_initializer())
```

Set the number of epochs:

```
epochs = 100
```

Then, for each iteration, perform the following:

```
for i in range(epochs):

    train_predictions = []
    index = 0
    epoch_loss = []
```

Then sample a batch of data and train the network:

```
while(index + batch_size) <= len(X_train):

    X_batch = X_train[index:index+batch_size]
    y_batch = y_train[index:index+batch_size]
    #predict the price and compute the loss
    predicted, loss_val, _ = session.run([y_hat, loss, optimizer],
feed_dict={input:X_batch, target:y_batch})
    #store the loss in the epoch_loss list
    epoch_loss.append(loss_val)

    #store the predictions in the train_predictions list
    train_predictions.append(predicted)
    index += batch_size
```

Print the loss on every 10 iterations:

```
if (i % 10)== 0:
    print 'Epoch {}, Loss: {} '.format(i,np.mean(epoch_loss))
```

As you may see in the following output, the loss decreases over the epochs:

```
Epoch 0, Loss: 0.0402321927249
Epoch 10, Loss: 0.0244581680745
Epoch 20, Loss: 0.0177710317075
Epoch 30, Loss: 0.0117778982967
Epoch 40, Loss: 0.00901956297457
Epoch 50, Loss: 0.0112476013601
Epoch 60, Loss: 0.00944950990379
```

```
Epoch 70, Loss: 0.00822851061821
Epoch 80, Loss: 0.00766260037199
Epoch 90, Loss: 0.00710930628702
```

Making predictions using the LSTM model

Now we will start making predictions on the test set:

```
predicted_output = []
i = 0
while i+batch_size <= len(X_test):
    output = session.run([y_hat],feed_dict={input:X_test[i:i+batch_size]})
    i += batch_size
    predicted_output.append(output)
```

Print the predicted output:

```
predicted_output[0]
```

We will get the result as shown:

```
[[array([[-0.60426176]], dtype=float32),
  array([[-0.60155034]], dtype=float32),
  array([[-0.60079575]], dtype=float32),
  array([[-0.599668]], dtype=float32),
  array([[-0.5991149]], dtype=float32),
  array([[-0.6008351]], dtype=float32),
  array([[-0.5970466]], dtype=float32)]]
```

As you can see, the values of test predictions are in a nested list, so we will flatten them:

```
predicted_values_test = []
for i in range(len(predicted_output)):
    for j in range(len(predicted_output[i][0])):
        predicted_values_test.append(predicted_output[i][0][j])
```

Now, if we print the predicted values, they are no longer in a nested list:

```
predicted_values_test[0]
```

```
array([[-0.60426176]], dtype=float32)
```

As we took the first 1000 points as a training set, we make predictions for time steps greater than 1000:

```
predictions = []
for i in range(1280):
```

```
if i >= 1000:
   predictions.append(predicted_values_test[i-1019])
else:
   predictions.append(None)
```

We plot and see how well the predicted value matches the actual value:

```
plt.figure(figsize=(16, 7))
plt.plot(data, label='Actual')
plt.plot(predictions, label='Predicted')
plt.legend()
plt.xlabel('Days')
plt.ylabel('Price')
plt.grid()
plt.show()
```

As you can see in the following plot, the actual value is shown in red and the predicted value is shown in blue. As we are making predictions for time steps greater than 1000, you can see after time step **1000**, the red and blue lines into each other, which implies that our model has correctly predicted the actual values:

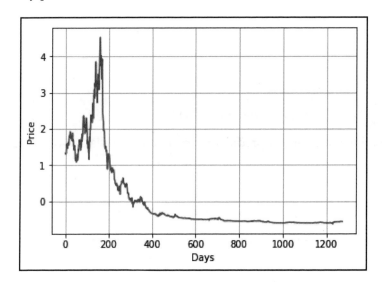

Gated recurrent units

So far, we have learned about how the LSTM cell uses different gates and how it solves the vanishing gradient problem of the RNN. But, as you may have noticed, the LSTM cell has too many parameters due to the presence of many gates and states.

Thus, while backpropagating the LSTM network, we need to update a lot of parameters in every iteration. This increases our training time. So, we introduce the **Gated Recurrent Units** (**GRU**) cell, which acts as a simplified version of the LSTM cell. Unlike the LSTM cell, the GRU cell has only two gates and one hidden state.

An RNN with a GRU cell is shown in the following diagram:

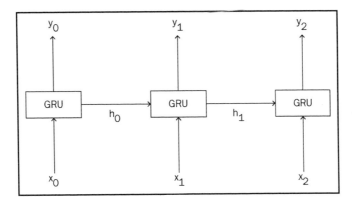

Understanding the GRU cell

As shown in the following diagram, a GRU cell has only two gates, called the update gate and the reset gate, and one hidden state:

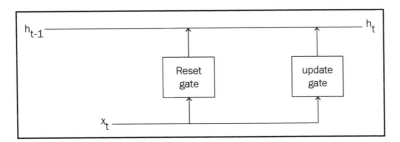

Let's delve deeper and see how these gates are used and how the hidden state is computed.

Update gate

The update gate helps to decide what information from the previous time step, h_{t-1}, can be taken forward to the next time step, h_t. It is basically a combination of an input gate and a forget gate, which we learned about in LSTM cells. Similar to the gates about the LSTM cell, the update gate is also regulated by the sigmoid function.

The update gate, z, at time step t is expressed as follows:

$$z_t = \sigma(U_z x_t + W_z h_{t-1} + b_z)$$

Here, the following applies:

- U_z is the input-to-hidden weights of the update gate
- W_z is the hidden-to-hidden weights of the update gate
- b_z is the bias of the update gate

The following diagram shows the update gate. As you can see, input x_t is multiplied with U_z, and the previous hidden state, h_{t-1}, 0 and 1:

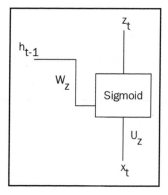

Reset gate

The reset gate helps to decide how to add the new information to the memory, that is, how much of the past information it can forget. The reset gate, r, at time step t is expressed as follows:

$$r_t = \sigma(U_r x_t + W_r h_{t-1} + b_r)$$

Here, the following applies:

- U_r is the input-to-hidden weights of the reset gate
- W_r is the hidden-to-hidden weights of the reset gate
- b_r is the bias of the reset gate

The reset gate is shown in the following diagram:

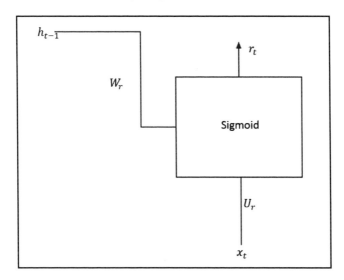

Updating hidden state

We just learned how the update and reset gates work, but how do these gates help in updating the hidden state? That is, how do you add new information to the hidden state and how do you remove unwanted information from the hidden state with the help of the reset and update gates?

First, we will see how to add new information to the hidden state.

We create a new state called the **content state**, c_t, for holding the information. We know that the reset gate is used to remove information that is not required. So, using the reset gate, we create a content state, c_t, that holds only the required information.

The content state, c, at time step t is expressed as follows:

$$c_t = tanh(U_c x_t + W_c(h_{t-1}) \cdot r_t)$$

The following diagram shows how the content state is created with the reset gate:

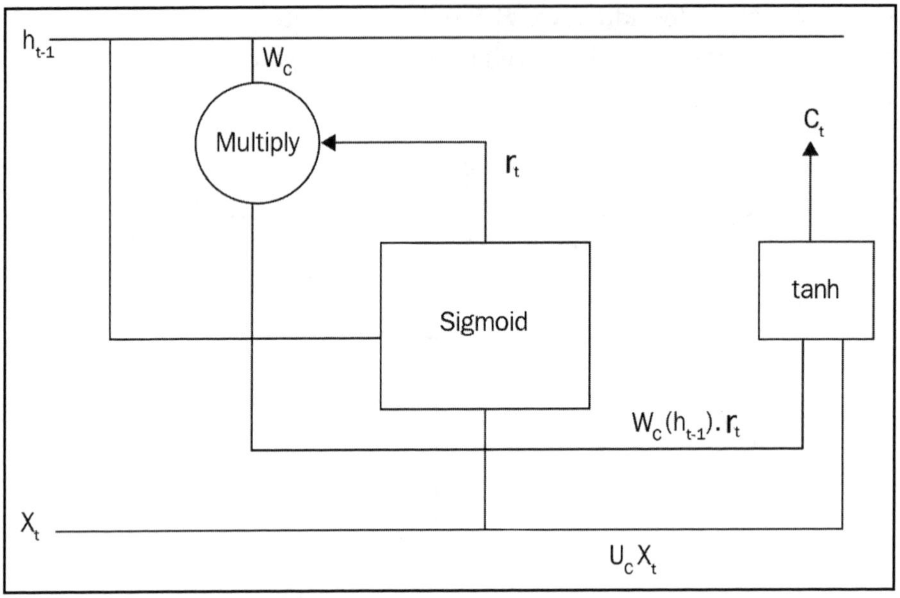

Now we will see how to remove information from the hidden state.

We learned that the update gate, z_t, helps to decide what information from the previous time step, h_{t-1}, can be taken forward to the next time step, h_t. Multiplying z_t and h_{t-1} gives us only the relevant information from the previous step. Instead of having a new gate, we just a take a complement of z_t, that is $(1 - z_t)$, and multiply them with c_t.

The hidden state is then updated as follows:

$$h_t = (1 - z_t) \cdot c_t + z_t \cdot h_{t-1}$$

Once the hidden state is computed, we can apply the softmax function and compute the output as follows:

$$\hat{y}_t = softmax(Vh_t)$$

Forward propagation in a GRU cell

Putting it all together, we learned in the previous section that the complete forward propagation steps in the GRU cell can be given as follows:

- **Update Gate:**

$$z_t = \sigma(U_z x_t + W_z h_{t-1}) \tag{5}$$

- **Reset Gate:**

$$r_t = \sigma(U_r x_t + W_r h_{t-1}) \tag{6}$$

- **Content State:**

$$c_t = tanh(U_c x_t + W_c(h_{t-1}) \cdot r_t) \tag{7}$$

- **Hidden State:**

$$h_t = (1 - z_t) \cdot c_t + z_t \cdot h_{t-1} \tag{8}$$

- **Output:**

$$y_t = softmax(V h_t) \tag{9}$$

Backpropagation in a GRU cell

The total loss, L, is the sum of losses at all time steps, and can be given as follows:

$$L = \sum_{j=0}^{T} L_j$$

To minimize the loss using gradient descent, we find the derivative of loss with respect to all of the weights used in the GRU cell as follows:

- We have three input-to-hidden layer weights, U_z, U_r, U_c, which are the input-to-hidden layer weights of the update gate, reset gate, and content state, respectively
- We have three hidden-to-hidden layer weights, W_z, W_r, W_c, which are the hidden-to-hidden layer weights of the update gate, reset gate, and content state respectively
- We have one hidden-to-output layer weight, V

We find the optimal values for all these weights through gradient descent and update the weights according to the weight update rule.

Gradient with respect to gates

As we saw when we discussed LSTM cells, calculating gradients of loss with respect to all the weights requires the gradients of all the gates and content state. So, first, we will see how to calculate them.

In the upcoming calculations, we will be using the gradients of loss with respect to the hidden state, h_t, which is $\frac{\partial L}{\partial h_t}$ at multiple places, so we will see how to calculate that. Computing the gradients of loss with respect to the hidden state, h_t, is exactly the same as we saw in the LSTM cell, and can be given as follows:

$$\boxed{\frac{\partial L}{\partial h_t} = (\hat{y}_t - y_t)V}$$

First, let's see how to calculate the **gradients of loss with respect to the content state,** c_t.

To calculate the gradients of loss with respect to content state, look at the equations of forward propagation and find out which equation has the c_t term. In the hidden state equation, that is, equation *(8)*, we have the c_t term, which is given as follows:

$$h_t = (1 - z_t) \cdot c_t + z_t \cdot h_{t-1}$$

So, by the chain rule, we can write the following:

$$\frac{\partial L}{\partial c_t} = \frac{\partial L}{\partial h_t} \frac{\partial h_t}{\partial c_t}$$
$$= \frac{\partial L}{\partial h_t}(1 - z_t)tanh'(c_t)$$

$$\frac{\partial L}{\partial c_t} = \frac{\partial L}{\partial h_t}(1 - z_t)(1 - tanh^2(c_t)) \qquad (10)$$

Let's see how to calculate the **gradient of loss with respect to the reset gate** r_t.

We have the r_t term in the content state equation and it can be given as shown:

$$c_t = tanh(U_c x_t + W_c(h_{t-1}) \cdot r_t)$$

Thus, by the chain rule, we can write the following:

$$\frac{\partial L}{\partial r_t} = \frac{\partial L}{\partial c_t}\frac{\partial c_t}{\partial r_t}$$

$$\frac{\partial L}{\partial r_t} = \frac{\partial L}{\partial c_t}W_c(h_{t-1}) \tag{11}$$

Finally, we see the **gradient of loss with respect to the update gate**, z_t.

We have the z_t term in the hidden state, h_t, equation, which can be given as follows:

$$h_t = (1 - z_t) \cdot c_t + z_t \cdot h_{t-1}$$

Thus, by the chain rule, we can write the following:

$$\frac{\partial L}{\partial z_t} = \frac{\partial L}{\partial h_t}\frac{\partial h_t}{\partial z_t}$$
$$= \frac{\partial L}{\partial h_t}(-c_t + h_{t-1})$$

$$\frac{\partial L}{\partial z_t} = \frac{\partial L}{\partial h_t}(h_{t-1} - c_t) \tag{12}$$

Now that we have calculated the gradients of loss with respect to all the gates and the content state, we will see how to calculate gradients of loss with respect to all the weights in our GRU cell.

Gradients with respect to weights

Now, we will see how to calculate gradients with respect to all the weights used in the GRU cell.

Gradients with respect to V

Since the final equation of GRU, that is, $\hat{y}_t = softmax(Vh_t)$, is the same as with the RNN, calculating the gradients of loss with respect to hidden-to-output layer weight V is exactly the same as what we computed in the RNN. Thus, we can directly write the following:

$$\boxed{\frac{\partial L}{\partial V} = \sum_{j=1}^{T}(\hat{y}_j - y_j) \otimes h_j}$$

Gradients with respect to W

Now, we will see how to calculate the gradients of loss with respect to hidden-to-hidden layer weights, W, for all the gates and the content state.

Let's calculate the **gradients of loss with respect to** W_r.

Recall the equation of the reset gate, which is given as follows:

$$r_t = \sigma(U_r x_t + W_r h_{t-1})$$

Using the chain rule, we can write the following:

$$\frac{\partial L}{\partial W_r} = \sum_{j=0}^{T-1} \frac{\partial L_j}{\partial r_j} \frac{\partial r_j}{\partial W_r}$$

Let's calculate each of the terms in the preceding equation. The first term, $\frac{\partial L_j}{\partial r_j}$, we already calculated in equation *(11)*. The second term is calculated as follows:

$$\frac{\partial r_j}{\partial W_r} = \sigma'(r_j) \otimes h_{j-1}$$
$$= r_j(1 - r_j) \otimes h_{j-1}$$

Thus, our final equation for calculating the gradient of loss with respect to W_r becomes the following:

$$\boxed{\frac{\partial L}{\partial W_r} = \sum_{j=0}^{T-1} \frac{\partial L_j}{\partial r_j} \cdot r_j(1 - r_j) \otimes h_{j-1}}$$

Now, let's move on to finding the **gradients of loss with respect to** W_z.

Recall the equation of the update gate, which is given as follows:

$$z_t = \sigma(U_z x_t + W_z h_{t-1})$$

Using the chain rule, we can write the following:

$$\frac{\partial L}{\partial W_z} = \sum_{j=0}^{T-1} \frac{\partial L_j}{\partial z_j} \frac{\partial z_j}{\partial W_z}$$

We have already computed the first term in equation *(12)*. The second term is computed as follows:

$$\frac{\partial z_j}{\partial W_z} = \sigma'(z_j) \otimes h_{j-1}$$

$$= z_j(1 - z_j) \otimes h_{j-1}$$

Thus, our final equation for calculating the gradient of loss with respect to W_z becomes the following:

$$\boxed{\frac{\partial L}{\partial W_z} = \sum_{j=0}^{T-1} \frac{\partial L_j}{\partial z_j} \cdot z_j(1 - z_j) \otimes h_{j-1}}$$

Now, we will find the **gradients of loss with respect to** W_c.

Recall the content state equation:

$$c_t = tanh(U_c x_t + W_c(h_{t-1}) \cdot r_t)$$

Using the chain rule, we can write the following:

$$\frac{\partial L}{\partial W_c} = \sum_{j=0}^{T-1} \frac{\partial L_j}{\partial c_j} \frac{\partial c_j}{\partial W_c}$$

Refer to equation *(10)* for the first term. The second term is given as follows:

$$\frac{\partial c_j}{\partial W_c} = tanh'(c_j) \otimes r_j h_{j-1}$$

$$= (1 - c_j^2) \otimes r_j h_{j-1}$$

Thus, our final equation for calculating the gradient of loss with respect to W_c becomes the following:

$$\boxed{\frac{\partial L}{\partial W_c} = \sum_{j=0}^{T-1} \frac{\partial L_j}{\partial c_j} \cdot (1 - c_j^2) \otimes r_j h_{j-1}}$$

Gradients with respect to U

Now we will see how to calculate the gradients of loss with respect to the input to the hidden weights, U, for all the gates and the content state. Computing gradients with respect to U is exactly the same as for those we computed with respect to W, except that the last term will be x_j instead of h_{j-1}, similar to what we learned when we covered the LSTM cell.

We can write the **gradients of loss with respect to** U_r as:

$$\frac{\partial L}{\partial U_r} = \sum_{j=0}^{T-1} \frac{\partial L_j}{\partial r_j} \cdot r_j(1 - r_j) \otimes x_j$$

The **gradients of loss with respect to** U_z are represented as follows:

$$\frac{\partial L}{\partial U_z} = \sum_{j=0}^{T-1} \frac{\partial L_j}{\partial z_j} \cdot z_j(1 - z_j) \otimes x_j$$

The **gradients of loss with respect to** U_c are represented as follows:

$$\frac{\partial L}{\partial U_c} = \sum_{j=0}^{T-1} \frac{\partial L_j}{\partial c_j} \cdot (1 - c_j^2) \otimes x_j$$

Implementing a GRU cell in TensorFlow

Now, we will see how to implement a GRU cell in TensorFlow. Instead of looking at the code, we will only see how to implement forward propagation of GRUs in TensorFlow.

Defining the weights

First, let's define all the weights. The weights of the update gate are defined as follows:

```
Uz = tf.get_variable("Uz", [vocab_size, hidden_size], initializer=init)
Wz = tf.get_variable("Wz", [hidden_size, hidden_size], initializer=init)
bz = tf.get_variable("bz", [hidden_size], initializer=init)
```

The weights of the reset gate are defined as shown:

```
Ur = tf.get_variable("Ur", [vocab_size, hidden_size], initializer=init)
Wr = tf.get_variable("Wr", [hidden_size, hidden_size], initializer=init)
br = tf.get_variable("br", [hidden_size], initializer=init)
```

The weights of the content state are defined as follows:

```
Uc = tf.get_variable("Uc", [vocab_size, hidden_size], initializer=init)
Wc = tf.get_variable("Wc", [hidden_size, hidden_size], initializer=init)
bc = tf.get_variable("bc", [hidden_size], initializer=init)
```

Weights of the output layer are defined as follows:

```
V = tf.get_variable("V", [hidden_size, vocab_size], initializer=init)
by = tf.get_variable("by", [vocab_size], initializer=init)
```

Defining forward propagation

Define the update gate as given in equation *(5)*:

```
zt = tf.sigmoid(tf.matmul(x_t, Uz) + tf.matmul(h_t, Wz) + bz)
```

Define the reset gate as given in equation *(6)*:

```
rt = tf.sigmoid(tf.matmul(x_t, Ur) + tf.matmul(h_t, Wr) + br)
```

Define the content state as given in equation *(7)*:

```
ct = tf.tanh(tf.matmul(x_t, Uc) + tf.matmul(tf.multiply(rt, h_t), Wc) + bc)
```

Define the hidden state as given in equation *(8)*:

```
h_t = tf.multiply((1 - zt), ct) + tf.multiply(zt, h_t)
```

Compute the output as given in equation *(9)*:

```
y_hat_t = tf.matmul(h_t, V) + by
```

Bidirectional RNN

In a bidirectional RNN, we have two different layers of hidden units. Both of these layers connect from the input layer to the output layer. In one layer, the hidden states are shared from left to right, and in the other layer, they are shared from right to left.

But what does this mean? To put it simply, one hidden layer moves forward through time from the start of the sequence, while the other hidden layer moves backward through time from the end of the sequence.

As shown in the following diagram, we have two hidden layers: a forward hidden layer and a backward hidden layer, which are described as follows:

- In the forward hidden layer, hidden state values are shared from past time steps, that is, h_0 is shared to h_1, h_1 is shared to h_2, and so on
- In the backward hidden layer, hidden start values are shared from future time steps, that is, z_3 to z_2, z_2 to z_1, and so on

The forward hidden layer and the backward hidden layer are represented as shown in the following diagram:

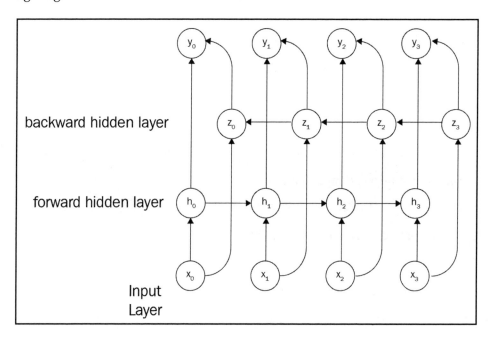

What is the use of bidirectional RNNs? In certain cases, reading the input sequence from both sides is very useful. So, a bidirectional RNN consists of two RNNs, one reading the sentence forward and the other reading the sentence backward.

For instance, consider the following sentence:

Archie lived for 13 years in _____. So he is good at speaking Chinese.

If we use an RNN to predict the blank in the preceding sentence, it would be ambiguous. As we know, an RNN can make predictions based on only the set of words it has seen so far. In the preceding sentence, to predict the blank, the RNN has seen only the words *Archie, lived, for, 13, years,* and *in,* but these words alone do not provide much context and do not give any clarity to predict the correct word. It just says *Archie lived for 13 years in.* With this information alone, we cannot predict the next word correctly.

But if we read the words following the blank as well, which are *So, he, is, good, at, speaking,* and *Chinese,* then we can say that Archie lived for 13 years in *China,* since it is given that he is good at speaking Chinese. So, in this circumstance, if we use a bidirectional RNN to predict the blank, it will predict correctly, since it reads the sentence in both forward and backward directions before making predictions.

Bidirectional RNNs have been used in various applications, such as **part-of-speech** (**POS**) tagging, in which it is vital to know the word before and after the target word, language translation, predicting protein structure, dependency parsing, and more. However, a bidirectional RNN is not suitable for online settings where we don't know the future.

The forward propagation steps in bidirectional RNNs are given as follows:

- Forward hidden layer:

$$h_t = \sigma(U_h x_t + W_h h_{t-1})$$

- Backward hidden layer:

$$z_t = \sigma(U_z x_t + W_z z_{t+1})$$

- Output:

$$\hat{y}_t = \text{softmax}(V_h h_t + V_z z_t)$$

Implementing a bidirectional RNN is simple with TensorFlow. Assuming we use the LSTM cell in the bidirectional RNN, we can do the following:

1. Import rnn from TensorFlow contrib as shown:

    ```
    from tensorflow.contrib import rnn
    ```

2. Define forward and backward hidden layers:

    ```
    forward_hidden_layer = rnn.BasicLSTMCell(num_hidden,
    forget_bias=1.0)

    backward_hidden_layer = rnn.BasicLSTMCell(num_hidden,
    forget_bias=1.0)
    ```

3. Define the bidirectional RNN with rnn.static_bidirectional_rnn:

    ```
    outputs, forward_states, backward_states =
    rnn.static_bidirectional_rnn(forward_hidden_layer,
    backward_hidden_layer, input)
    ```

Going deep with deep RNN

We know that a deep neural network is a network that has many hidden layers. Similarly, a deep RNN has more than one hidden layer, but how are the hidden states computed when we have more than one hidden layer? We know that an RNN computes the hidden state by taking inputs and the previous hidden state, but how are the hidden states in the later layers computed?

For instance, let's see how h_1^2 in hidden layer 2 is computed. It takes the previous hidden state, h_0^2, and the previous layer's output, h_1^1, as inputs to compute h_1^2.

Thus, when we have an RNN with more than one hidden layer, hidden layers at the later layers will be computed by taking the previous hidden state and the previous layer's output as input, as shown in the following diagram:

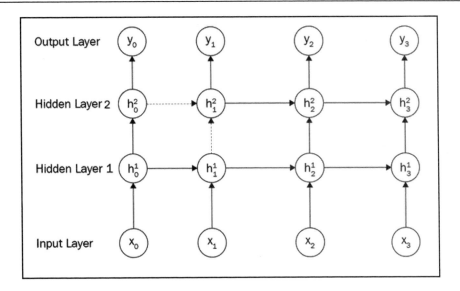

Language translation using the seq2seq model

The **sequence-to-sequence model (seq2seq)** is basically the many-to-many architecture of an RNN. It has been used for various applications because it can map an arbitrary-length input sequence to an arbitrary-length output sequence. Some of the applications of the seq2seq model include language translation, music generation, speech generation, and chatbots.

In most real-world scenarios, input and output sequences vary in length. For instance, let's take the language translation task, during which we need to convert a sentence from a source language to a target language. Let's assume we are converting from English (source) to French (target).

Consider our input sentence is *what are you doing?* Then, it would be mapped to *que faites vous?* As we can observe, the input sequence consists of four words, whereas the output sequence consists of three words. The seq2seq model handles this varying length of input and output sequence and maps the source to the target. Thus, they are widely used in applications for which the input and output sequences vary in length.

The architecture of the seq2seq model is very simple. It comprises two vital components, namely an encoder and a decoder. Let's consider the same language translation task. First, we feed the input sentence to an encoder.

The encoder learns the representation (embeddings) of the input sentence, but what is a representation? Representation, or embedding, is basically a vector comprising the meaning of the sentence. It is also called the **thought vector** or **context vector**. Once the encoder learns the embedding, it sends the embedding to the decoder. The decoder takes this embedding (thought vector) as input and tries to construct a target sentence. So, the decoder tries to generate the French translation for the English sentence.

As you can see in the following diagram, the encoder takes the input English sentence, learns the embeddings, and feeds the embeddings to the decoder, and then the decoder generates the translated French sentence using those embeddings:

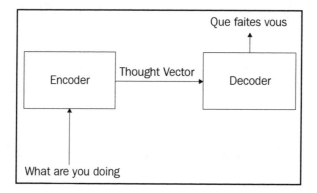

But how does this really work? How does the encoder understand the sentence? How does the decoder translate the sentence using the encoder's embeddings? Let's delve deeper and see how this works.

Encoder

An encoder is basically an RNN with LSTM or GRU cells. It can also be a bidirectional RNN. We feed the input sentence to an encoder and, instead of taking the output, we take the hidden state from the final time step as the embeddings. Let's better understand encoders with an example.

Consider we are using an RNN with a GRU cell and the input sentence is *what are you doing*. Let's represent the hidden state of the encoder with e:

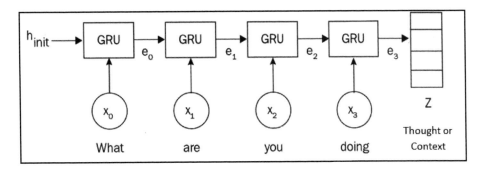

The preceding diagram shows how the encoder computes the thought vectors; this is explained as follows:

- In the first time step, $t = 0$. To a GRU cell, we pass the input, x_0, which is the first word in the input sentence, *what*, and also the initial hidden state, e_{init}, which is randomly initialized. With these inputs, the GRU cell computes the first hidden state, e_0, as follows:

$$e_0 = \text{GRU}(x_0, e_{init})$$

- In the next time step, $t = 1$, we pass the input, x_1, which is the next word in the input sentence, *are*, to the encoder. Along with this, we also pass the previous hidden state, e_0, and compute the hidden state, e_1:

$$e_1 = \text{GRU}(x_1, e_0)$$

- In the next time step, $t = 2$, we pass the input, x_2, which is the next word, *you*, to the encoder. Along with this, we also pass the previous hidden state, e_1, and compute the hidden state, e_2, as follows:

$$e_2 = \text{GRU}(x_2, e_1)$$

- In the final time step, $t = 3$, we feed the input, x_3, which is the last word in the input sentence, *doing*. Along with this, we also pass the previous hidden state, e_2, and compute the hidden state, e_3:

$$e_3 = \text{GRU}(x_3, e_2)$$

Thus, e_3 is our final hidden state. We learned that the RNN captures the context of all the words it has seen so far in its hidden state. Since e_3 is the final hidden state, it holds the context of all the words that the network has seen, which will be all the words in our input sentence, that is, *what, are, you,* and *doing.*

Since the final hidden state, e_3, holds the context of all the words in our input sentence, it holds the context of the input sentence, and this essentially forms our embedding, z, which is otherwise called a thought or context vector, as follows:

$$z = e_3$$

We feed the context vector, z, to the decoder to convert it to the target sentence.

Thus, in the encoder, on every time step, t, we feed an input word and along with it we also feed the previous hidden state, e_{t-1}, and compute the current hidden state, e_t. The hidden state at the final step, e_{T-1}, holds the context of the input sentence and it will become the embedding, z, which will be sent to the decoder to convert it to the target sentence.

Decoder

Now, we will learn how the decoder generates the target sentence by using the thought vector, z, generated by the encoder. A decoder is an RNN with LSTM or GRU cells. The goal of our decoder is to generate the target sentence for the given input (source) sentence.

We know that we start off an RNN by initializing its initial hidden state with random values, but for the decoder's RNN, we initialize the hidden state with the thought vector, z, generated by the encoder, instead of initializing them with random values. The decoder network is shown in the following diagram:

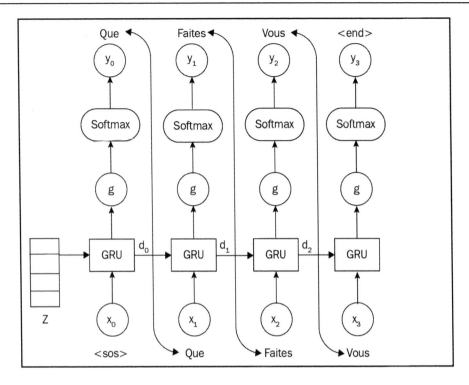

But, what should be the input to the decoder? We simply pass **<sos>** as an input to the decoder, which indicates the start of the sentence. So, once the decode receives **<sos>**, it tries to predict the actual starting word of the target sentence. Let's represent the decoder hidden state by d.

At the first time step, $t = 0$, we feed the first input, which is **<sos>**, to the decoder, and along with it, we pass the thought vector as the initial hidden state as follows:

$$d_0 = \text{GRU}(z, x_{sos})$$

Okay. What are we really doing here? We need to predict the output sequence, which is a French equivalent for our input English sentence. There are a lot of French words in our vocabulary. How does the decoder decide which word to output? That is, how does it decide the first word in our output sequence?

We feed the decoder hidden state, d_0, to $g(\cdot)$, which returns the score for all the words in our vocabulary to be the first output word. That is, the output word at a time step, $t = 0$ is computed as follows:

$$s_0 = g(d_0)$$

Instead of having raw scores, we convert them into probabilities. Since we learned that the softmax function squashes values between 0 to 1, we use the softmax function for converting the score, s, into a probability, p:

$$p_o = \text{softmax}(s_0)$$

Thus, we have probabilities for all the French words in our vocabulary to be the first output word. We select the word that has the highest probability as the first output word using the argmax function:

$$y_0 = \text{argmax}(p_0)$$

So, we have predicted that the first output word, y_0, is *Que*, as shown in the preceding diagram.

On the next time step, $t = 1$, we feed the output word predicted at the previous time step, y_0, as input to the decoder. Along with it, we also pass the previous hidden state, d_0:

$$d_1 = \text{GRU}(d_0, y_0)$$

Then, we compute the score for all the words in our vocabulary to be the next output word, that is, the output word at time step $t = 1$:

$$s_1 = g(d_1)$$

Then, we convert the scores to probabilities using the softmax function:

$$p_1 = \text{softmax}(s_1)$$

Next, we select the word that has the highest probability as the output word, y_1, at a time step, $t = 1$:

$$y_1 = \text{argmax}(p_1)$$

Thus, we initialize the decoder's initial hidden state with z, and, on every time step, t, we feed the predicted output word from the previous time step, y_{t-1}, and the previous hidden state, d_{t-1}, as an input to the decoder, d_t, at the current time step, and predict the current output, y_t.

But when does the decoder stop? Because our output sequence has to stop somewhere, we cannot keep on feeding the predicted output word from the previous time step as an input to the next time step. When the decoder predicts the output word as **<sos>**, this implies the end of the sentence. Then, the decoder learns that an input source sentence is converted to a meaningful target sentence and stops predicting the next word.

Thus, this is how the seq2seq model converts the source sentence to the target sentence.

Attention is all we need

We just learned how the seq2seq model works and how it translates a sentence from the source language to the target language. We learned that a context vector is basically a hidden state vector from the final time step of an encoder, which captures the meaning of the input sentence, and it is used by the decoder to generate the target sentence.

But when the input sentence is long, the context vector does not capture the meaning of the whole sentence, since it is just the hidden state from the final time step. So, instead of taking the last hidden state as a context vector and using it for the decoder, we take the sum of all the hidden states from the encoder and use it as a context vector.

Let's say the input sentence has 10 words; then we would have 10 hidden states. We take a sum of all these 10 hidden states and use it for the decoder to generate the target sentence. However, not all of these hidden states might be helpful in generating a target word at time step t. Some hidden states will be more useful than other hidden states. So, we need to know which hidden state is more important than another at time step t to predict the target word. To get this importance, we use the attention mechanism, which tells us which hidden state is more important to generate the target word at the time step t. Thus, attention mechanisms basically give the importance for each of the hidden states of the encoder to generate the target word at time step t.

How does an attention mechanism work? Let's say we have three hidden states of an encoder, e_0, e_1, and e_2, and a decoder hidden state, d_0, as shown in the following diagram:

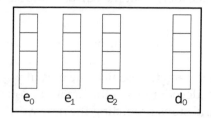

Now, we need to know the importance of all the hidden states of an encoder to generate a target word at time step t, So, we take each encoder hidden state, e, and decoder hidden state, d_0, and feed them to a function, $f(\cdot)$, which is called a **score function** or **alignment function**, and it returns the score for each of the encoder hidden states indicating their importance. But what is this score function? There are a number of choices for the score function, such as dot product, scaled dot product, cosine similarity, and more.

We use a simple dot product as the score function; that is, the dot product between the encoder hidden states and the decoder hidden states. For instance, to know the importance of e_0 in generating the target word, we simply compute the dot product between e_0 and d_0, which gives us a score indicating how similar e_0 and d_0 are.

Once we have the score, we convert them into probabilities using the softmax function as follows:

$$\alpha_t = \text{softmax}(score_t)$$

These probabilities, α_t, are called **attention weights**.

As you can see in the following diagram, we compute the similarity score between each of the encoder's hidden states with the decoder's hidden state using a function, $f(\cdot)$. Then, the similarity score is converted into probabilities using the softmax function, which are called attention weights:

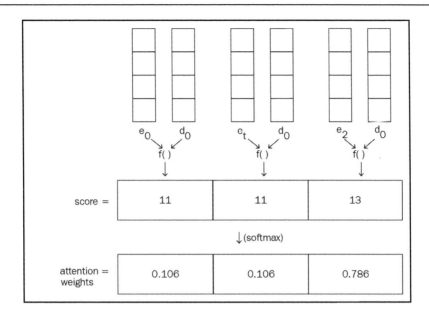

Thus, we have attention weights (probabilities) for each of the encoder's hidden states. Now, we multiply the attention weights with their corresponding encoder's hidden state, that is, $\alpha_t \cdot e_t$. As shown in the following diagram, the encoder's hidden state, e_0, is multiplied by **0.106**, e_1 is multiplied by **0.106**, and e_2 is multiplied by **0.786**:

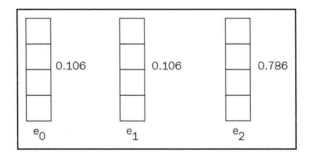

But, why do we have to multiply attention weights by the encoder's hidden state?

Multiplying the encoder's hidden states by their attention weights indicates that we are giving more importance to the hidden states that have more attention weights, and less importance to hidden states that have fewer attention weights. As shown in the preceding diagram, multiplying **0.786** with hidden state e_2 implies we are giving more importance to e_2 than the other two hidden states.

Thus, this is how the attention mechanism decides which hidden state is more important to generate the target word at time step t. After multiplying the encoder's hidden state by their attention weights, we simply sum them up, and this now forms our context/thought vector:

$$\text{context vector} = \sum_{t=0}^{N} \alpha_t e_t$$

As shown in the following diagram, the context vector is obtained by the sum of the encoder's hidden state multiplied by its respective attention weights:

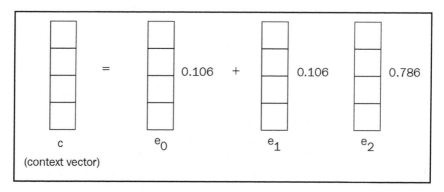

Thus, to generate a target word at time step t, the decoder uses context vector C at time step t. With the attention mechanism, instead of taking the last hidden state as a context vector and using it for the decoder, we take the sum of all the hidden states from the encoder and use it as a context vector.

Summary

In this chapter, we learned how the LSTM cell uses several gates to combat the vanishing gradient problem. Then, we saw how to use the LSTM cell to predict a Bitcoin's price in TensorFlow.

After looking at LSTM cells, we learned about the GRU cell, which is a simplified version of LSTM. We also learned about bidirectional RNNs, where we had two layers of hidden states with one layer moving forward through time from the start of the sequence, while another layer moved backward through time from the end of the sequence.

At the end of the chapter, we learned about the seq2seq model, which maps an input sequence of varying length to an output sequence of varying length. We also understood how the attention mechanism is used in the seq2seq model and how it focuses on important information.

In the next chapter, we will learn about convolutional neural networks and how they are used for recognizing images.

Questions

Let's put our newly gained knowledge to the test. Answer the following questions:

1. How does LSTM solve the vanishing gradient problem of RNN?
2. What are all the different gates and their functions in an LSTM cell?
3. What is the use of the cell state?
4. What is a GRU?
5. How do bidirectional RNNs work?
6. How do deep RNNs compute the hidden state?
7. What are encoders and decoders in the seq2seq architecture?
8. What is the use of the attention mechanism?

Further reading

Check out some of these cool projects on GitHub:

- Human activity recognition using LSTM: `https://github.com/guillaume-chevalier/LSTM-Human-Activity-Recognition`
- Building a chatbot using seq2seq: `https://github.com/tensorlayer/seq2seq-chatbot`
- Text summarization using a bidirectional GRU: `https://github.com/harpribot/deep-summarization`

6
Demystifying Convolutional Networks

Convolutional Neural Networks (CNNs) are one of the most commonly used deep learning algorithms. They are widely used for image-related tasks, such as image recognition, object detection, image segmentation, and more. The applications of CNNs are endless, ranging from powering vision in self-driving cars to the automatic tagging of friends in our Facebook pictures. Although CNNs are widely used for image datasets, they can also be applied to textual datasets.

In this chapter, we will look at CNNs in detail and get the hang of CNNs and how they work. First, we will learn about CNNs intuitively, and then we will deep-dive into the underlying math behind them. Following this, we will come to understand how to implement a CNN in TensorFlow step by step. Moving ahead, we will explore different types of CNN architectures such as LeNet, AlexNet, VGGNet, and GoogleNet. At the end of the chapter, we will study the shortcomings of CNNs and how these can be resolved using Capsule networks. Also, we will learn how to build Capsule networks using TensorFlow.

In this chapter, we will look at the following topics:

- What are CNNs?
- The math behind CNNs
- Implementing CNNs in TensorFlow
- Different CNN architectures
- Capsule networks
- Building Capsule networks in TensorFlow

What are CNNs?

A CNN, also known as a **ConvNet**, is one of the most widely used deep learning algorithms for computer vision tasks. Let's say we are performing an image-recognition task. Consider the following image. We want our CNN to recognize that it contains a horse:

How can we do that? When we feed the image to a computer, it basically converts it into a matrix of pixel values. The pixel values range from 0 to 255, and the dimensions of this matrix will be of [*image width* x *image height* x *number of channels*]. A grayscale image has one channel, and colored images have three channels **red, green, and blue (RGB)**.

Let's say we have a colored input image with a width of 11 and a height of 11, that is 11 x 11, then our matrix dimension would be of *[11 x 11 x 3]*. As you can see in *[11 x 11 x 3]*, 11 x 11 represents the image width and height and 3 represents the channel number, as we have a colored image. So, we will have a 3D matrix.

But it is hard to visualize a 3D matrix, so, for the sake of understanding, let's consider a grayscale image as our input. Since the grayscale image has only one channel, we will get a 2D matrix.

As shown in the following diagram, the input grayscale image will be converted into a matrix of pixel values ranging from 0 to 255, with the pixel values representing the intensity of pixels at that point:

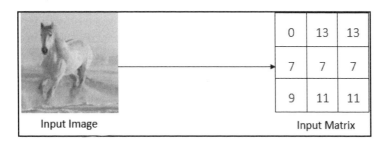

Input Image   Input Matrix

 The values given in the input matrix are just arbitrary values for our understanding.

Okay, now we have an input matrix of pixel values. What happens next? How does the CNN come to understand that the image contains a horse? CNNs consists of the following three important layers:

- The convolutional layer
- The pooling layer
- The fully connected layer

With the help of these three layers, the CNN recognizes that the image contains a horse. Now we will explore each of these layers in detail.

Convolutional layers

The convolutional layer is the first and core layer of the CNN. It is one of the building blocks of a CNN and is used for extracting important features from the image.

We have an image of a horse. What do you think are the features that will help us to understand that this is an image of a horse? We can say body structure, face, legs, tail, and so on. But how does the CNN understand these features? This is where we use a convolution operation that will extract all the important features from the image that characterize the horse. So, the convolution operation helps us to understand what the image is all about.

Okay, what exactly is this convolution operation? How it is performed? How does it extract the important features? Let's look at this in detail.

As we know, every input image is represented by a matrix of pixel values. Apart from the input matrix, we also have another matrix called the **filter matrix**. The filter matrix is also known as a **kernel**, or simply a **filter**, as shown in the following diagram:

0	13	13
7	7	7
9	11	11

Input Matrix

0	1
1	0

Filter

We take the filter matrix, slide it over the input matrix by one pixel, perform element-wise multiplication, sum up the results, and produce a single number. That's pretty confusing, isn't it? Let's understand this better with the aid of the following diagram:

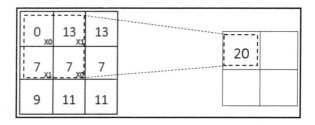

As you can see in the previous diagram, we took the filter matrix and placed it on top of the input matrix, performed element-wise multiplication, summed their results, and produced the single number. This is demonstrated as follows:

$$(0 * 0) + (13 * 1) + (7 * 1) + (7 * 0) = 20$$

Now, we slide the filter over the input matrix by one pixel and perform the same steps, as shown in the following diagram:

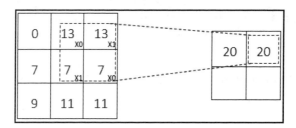

This is demonstrated as follows:

$$(13 * 0) + (13 * 1) + (7 * 1) + (7 * 0) = 20$$

Again, we slide the filter matrix by one pixel and perform the same operation, as shown in the following diagram:

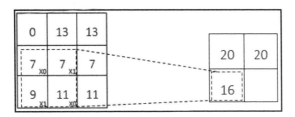

This is demonstrated as follows:

$$(7 * 0) + (7 * 1) + (9 * 1) + (11 * 0) = 16$$

Now, again, we slide the filter matrix over the input matrix by one pixel and perform the same operation, as shown in the following diagram:

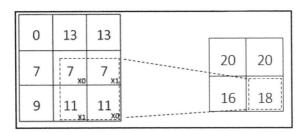

That is:

$$(7 * 0) + (7 * 1) + (11 * 1) + (11 * 0) = 18$$

Okay. What are we doing here? We are basically sliding the filter matrix over the entire input matrix by one pixel, performing element-wise multiplication and summing their results, which creates a new matrix called a **feature map** or **activation map**. This is called the **convolution operation**.

As we've learned, the convolution operation is used to extract features, and the new matrix, that is, the feature maps, represents the extracted features. If we plot the feature maps, then we can see the features extracted by the convolution operation.

The following diagram shows the actual image (the input image) and the convolved image (the feature map). We can see that our filter has detected the edges from the actual image as a feature:

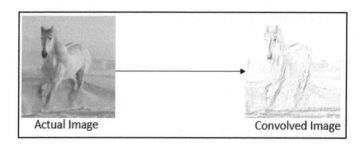

Actual Image   Convolved Image

Various filters are used for extracting different features from the image. For instance, if we use a sharpen filter, $\begin{bmatrix} 0 & -1 & 0 \\ -1 & 5 & -1 \\ 0 & 1 & 0 \end{bmatrix}$, then it will sharpen our image, as shown in the following figure:

Thus, we have learned that with filters, we can extract important features from the image using the convolution operation. So, instead of using one filter, we can use multiple filters for extracting different features from the image, and produce multiple feature maps. So, the depth of the feature map will be the number of filters. If we use seven filters to extract different features from the image, then the depth of our feature map will be seven:

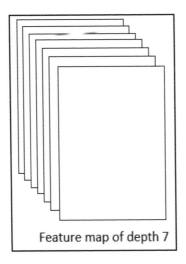

Feature map of depth 7

Okay, we have learned that different filters extract different features from the image. But the question is, how can we set the correct values for the filter matrix so that we can extract the important features from the image? Worry not! We just initialize the filter matrix randomly, and the optimal values of the filter matrix, with which we can extract the important features from the images, will be learned through backpropagation. However, we just need to specify the size of the filter and the number of filters we want to use.

Strides

We have just learned how a convolution operation works. We slide over the input matrix with the filter matrix by one pixel and perform the convolution operation. But we can't only slide over the input matrix by one pixel. We can also slide over the input matrix by any number of pixels.

The number of pixels we slide over the input matrix by the filter matrix is called a **stride**.

If we set the stride to 2, then we slide over the input matrix with the filter matrix by two pixels. The following diagram shows a convolution operation with a stride of 2:

17 x0	80 x1	14	63		17	80	14 x0	63 x1
13 x1	11 x0	43	79	**Stride of 2** →	13	11	43 x1	79 x0
27	33	7	4		27	33	7	4
255	69	77	63		255	69	77	63

But how do we choose the stride number? We just learned that a stride is the number of pixels along that we move our filter matrix. So, when the stride is set to a small number, we can encode a more detailed representation of the image than when the stride is set to a large number. However, a stride with a high value takes less time to compute than one with a low value.

Padding

With the convolution operation, we are sliding over the input matrix with a filter matrix. But in some cases, the filter does not perfectly fit the input matrix. What do we mean by that? For example, let's say we are performing a convolution operation with a stride of 2. There exists a situation where, when we move our filter matrix by two pixels, it reaches the border and the filter matrix does not fit the input matrix. That is, some part of our filter matrix is outside the input matrix, as shown in the following diagram:

17	80	14	63 x0	x1
13	11	43	79 x1	x0
27	33	7	4	
255	89	77	63	

In this case, we perform padding. We can simply pad the input matrix with zeros so that the filter can fit the input matrix, as shown in the following diagram. Padding with zeros on the input matrix is called **same padding** or **zero padding**:

17	80	14	63	0
13	11	43	79	0
27	33	7	4	
255	89	77	63	

Instead of padding them with zeros, we can also simply discard the region of the input matrix where the filter doesn't fit in. This is called **valid padding**:

17	80	14	63	
13	11	43	79	
27	33	7	4	
255	89	77	63	

Pooling layers

Okay. Now, we are done with the convolution operation. As a result of the convolution operation, we have some feature maps. But the feature maps are too large in dimension. In order to reduce the dimensions of feature maps, we perform a pooling operation. This reduces the dimensions of the feature maps and keeps only the necessary details so that the amount of computation can be reduced.

For example, to recognize a horse from the image, we need to extract and keep only the features of the horse; we can simply discard unwanted features, such as the background of the image and more. A pooling operation is also called a **downsampling** or **subsampling** operation, and it makes the CNN translation invariant. Thus, the pooling layer reduces spatial dimensions by keeping only the important features.

> The pooling operation will not change the depth of the feature maps; it will only affect the height and width.

There are different types of pooling operations, including max pooling, average pooling, and sum pooling.

In max pooling, we slide over the filter on the input matrix and simply take the maximum value from the filter window, as shown in the following diagram:

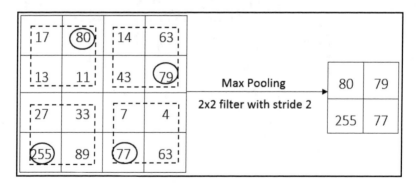

As the name suggests, in average pooling, we take the average value of the input matrix within the filter window, and in sum pooling, we sum all the values of the input matrix within the filter window, as shown in the following diagram:

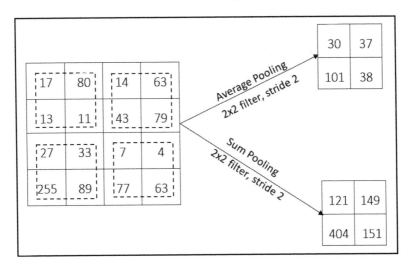

Max pooling is one of the most commonly used pooling operations.

Fully connected layers

So far, we've learned how convolutional and pooling layers work. A CNN can have multiple convolutional layers and pooling layers. However, these layers will only extract features from the input image and produce the feature map; that is, they are just the feature extractors.

Given any image, convolutional layers extract features from the image and produce a feature map. Now, we need to classify these extracted features. So, we need an algorithm that can classify these extracted features and tell us whether the extracted features are the features of a horse, or something else. In order to make this classification, we use a feedforward neural network. We flatten the feature map and convert it into a vector, and feed it as an input to the feedforward network. The feedforward network takes this flattened feature map as an input, applies an activation function, such as sigmoid, and returns the output, stating whether the image contains a horse or not; this is called a fully connected layer and is shown in the following diagram:

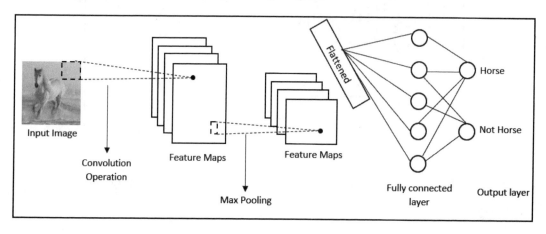

The architecture of CNNs

The architecture of a CNN is shown in the following diagram:

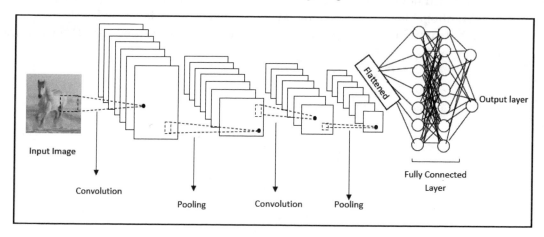

As you will notice, first we feed the input image to the convolutional layer, where we apply the convolution operation to extract important features from the image and create the feature maps. We then pass the feature maps to the pooling layer, where the dimensions of the feature maps will be reduced. As shown in the previous diagram, we can have multiple convolutional and pooling layers, and we should also note that the pooling layer does not necessarily have to be there after every convolutional layer; there can be many convolutional layers followed by a pooling layer.

So, after the convolutional and pooling layers, we flatten the resultant feature maps and feed it to a fully connected layer, which is basically a feedforward neural network that classifies the given input image based on the feature maps.

The math behind CNNs

So far, we have intuitively understood how a CNN works. But how exactly does a CNN learn? How does it find the optimal values for the filter using backpropagation? To answer this question, we will explore mathematically how the CNN works. Unlike in the Chapter 5, *Improvements to the RNN*, the math behind a CNN is pretty simple and very interesting.

Forward propagation

Let's begin with the forward propagation. We have already seen how forward propagation works and how a CNN classifies the given input image. Let's frame this mathematically. Let's consider an input matrix, X, and filter, W, with values shown as follows:

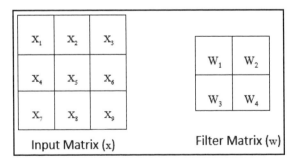

First, let's familiarize ourselves with the notations. Whenever we write x_{ij}, it implies the element in the ith row and the jth column of the input matrix. The same applies to the filter and output matrix; that is, w_{ij} and o_{ij} represent the ith row and the jth column value in the filter and output matrix, respectively. In the previous figure, $x_{11} = x_1$, that is, x_1 is the element in the first row and first column of the input matrix.

As shown in the following diagram, we take the filter, slide it over the input matrix, perform a convolution operation, and produce the output matrix (the feature map) just as we learned in the previous section:

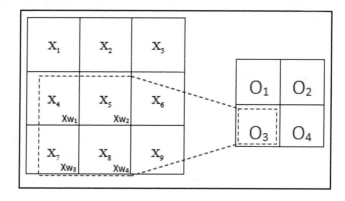

Thus, all the values in the output matrix (feature map) are computed as follows:

$$o_1 = x_1 w_1 + x_2 w_2 + x_4 w_3 + x_5 w_4$$

$$o_2 = x_2 w_1 + x_3 w_2 + x_5 w_3 + x_6 w_4$$

$$o_3 = x_4 w_1 + x_5 w_2 + x_7 w_3 + x_8 w_4$$

$$o_4 = x_5 w_1 + x_6 w_2 + x_8 w_3 + x_9 w_4$$

Okay, so we know this is how a convolution operation is performed and how the output is computed. Can we represent this in a simple equation? Let's say we have an input image, X, with a width of W and a height of H, and the filter of size $P \times Q$, then the convolution operation can be represented as follows:

$$o_{ij} = \sum_{m=0}^{p-1} \sum_{n=0}^{q-1} W_{m,n} \cdot X_{i+m, j+n} \qquad (1)$$

This equation basically represents how the output, o_{ij} (that is, the element in the ith row and the jth column of the output matrix), is computed using a convolution operation.

Once the convolution operation is performed, we feed the result, o_{ij}, to a feedforward network, f, and predict the output, $\hat{y}_i$:

$$\hat{y}_i = f(o_{ij})$$

Backward propagation

Once we have predicted the output, we compute the loss, L. We use the mean squared error as the loss function, that is, the mean of the squared difference between the actual output, y_i, and the predicted output, $\hat{y}_i$, which is given as follows:

$$L = \frac{1}{2} \sum_i (y_i - \hat{y}_i)^2$$

Now, we will see how can we use backpropagation to minimize the loss L. In order to minimize the loss, we need to find the optimal values for our filter W. Our filter matrix consists of four values, $w1$, $w2$, $w3$, and $w4$. To find the optimal filter matrix, we need to calculate the gradients of our loss function with respect to all these four values. How do we do that?

First, let's recollect the equations of the output matrix, as follows:

$$o_1 = x_1 w_1 + x_2 w_2 + x_4 w_3 + x_5 w_4$$

$$o_2 = x_2 w_1 + x_3 w_2 + x_5 w_3 + x_6 w_4$$

$$o_3 = x_4 w_1 + x_5 w_2 + x_7 w_3 + x_8 w_4$$

$$o_4 = x_5 w_1 + x_6 w_2 + x_8 w_3 + x_9 w_4$$

Don't get intimidated by the upcoming equations; they are actually pretty simple.

First, let's calculate gradients with respect to w_1. As you can see, w_1 appears in all the output equations; we calculate the partial derivatives of the loss with respect to w_1 as follows:

$$\frac{\partial L}{\partial w_1} = \frac{\partial L}{\partial o_1}\frac{\partial o_1}{\partial w_1} + \frac{\partial L}{\partial o_2}\frac{\partial o_2}{\partial w_1} + \frac{\partial L}{\partial o_3}\frac{\partial o_3}{\partial w_1} + \frac{\partial L}{\partial o_4}\frac{\partial o_4}{\partial w_1}$$

$$\frac{\partial L}{\partial w_1} = \frac{\partial L}{\partial o_1}x_1 + \frac{\partial L}{\partial o_2}x_2 + \frac{\partial L}{\partial o_3}x_4 + \frac{\partial L}{\partial o_4}x_5$$

Similarly, we calculate the partial derivative of the loss with respect to the w_2 weight as follows:

$$\frac{\partial L}{\partial w_2} = \frac{\partial L}{\partial o_1}\frac{\partial o_1}{\partial w_2} + \frac{\partial L}{\partial o_2}\frac{\partial o_2}{\partial w_2} + \frac{\partial L}{\partial o_3}\frac{\partial o_3}{\partial w_2} + \frac{\partial L}{\partial o_4}\frac{\partial o_4}{\partial w_2}$$

$$\frac{\partial L}{\partial w_2} = \frac{\partial L}{\partial o_1}x_2 + \frac{\partial L}{\partial o_2}x_3 + \frac{\partial L}{\partial o_3}x_5 + \frac{\partial L}{\partial o_4}x_6$$

The gradients of loss with respect to the w_3 weights, are calculated as follows:

$$\frac{\partial L}{\partial w_3} = \frac{\partial L}{\partial o_1}\frac{\partial o_1}{\partial w_3} + \frac{\partial L}{\partial o_2}\frac{\partial o_2}{\partial w_3} + \frac{\partial L}{\partial o_3}\frac{\partial o_3}{\partial w_3} + \frac{\partial L}{\partial o_4}\frac{\partial o_4}{\partial w_3}$$

$$\frac{\partial L}{\partial w_3} = \frac{\partial L}{\partial o_1}x_4 + \frac{\partial L}{\partial o_2}x_5 + \frac{\partial L}{\partial o_3}x_7 + \frac{\partial L}{\partial o_4}x_8$$

The gradients of loss with respect to the w_4 weights, are given as follows:

$$\frac{\partial L}{\partial w_4} = \frac{\partial L}{\partial o_1}\frac{\partial o_1}{\partial w_4} + \frac{\partial L}{\partial o_2}\frac{\partial o_2}{\partial w_4} + \frac{\partial L}{\partial o_3}\frac{\partial o_3}{\partial w_4} + \frac{\partial L}{\partial o_4}\frac{\partial o_4}{\partial w_4}$$

$$\frac{\partial L}{\partial w_4} = \frac{\partial L}{\partial o_1}x_5 + \frac{\partial L}{\partial o_2}x_6 + \frac{\partial L}{\partial o_3}x_8 + \frac{\partial L}{\partial o_4}x_9$$

So, in a nutshell, our final equations for the gradients of loss with respect to all the weights are as follows:

$$\frac{\partial L}{\partial w_1} = \frac{\partial L}{\partial o_1}x_1 + \frac{\partial L}{\partial o_2}x_2 + \frac{\partial L}{\partial o_3}x_4 + \frac{\partial L}{\partial o_4}x_5$$

$$\frac{\partial L}{\partial w_2} = \frac{\partial L}{\partial o_1}x_2 + \frac{\partial L}{\partial o_2}x_3 + \frac{\partial L}{\partial o_3}x_5 + \frac{\partial L}{\partial o_4}x_6$$

$$\frac{\partial L}{\partial w_3} = \frac{\partial L}{\partial o_1}x_4 + \frac{\partial L}{\partial o_2}x_5 + \frac{\partial L}{\partial o_3}x_7 + \frac{\partial L}{\partial o_4}x_8$$

$$\frac{\partial L}{\partial w_4} = \frac{\partial L}{\partial o_1}x_5 + \frac{\partial L}{\partial o_2}x_6 + \frac{\partial L}{\partial o_3}x_8 + \frac{\partial L}{\partial o_4}x_9$$

It turns out that computing the derivatives of loss with respect to the filter matrix is very simple—it is just another convolution operation. If we look at the preceding equations closely, we will notice they look like the result of a convolution operation between the input matrix and the gradient of the loss with respect to the output as a filter matrix, as depicted in the following diagram:

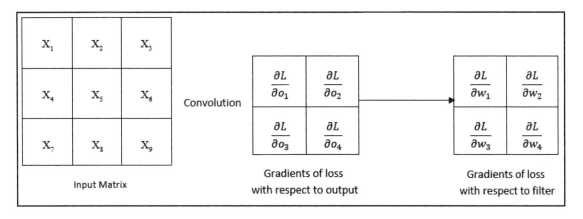

For example, let's see how the gradients of loss with respect to weight w_3 are computed by the convolution operation between the input matrix and the gradients of loss with respect to the output as a filter matrix, as shown in the following diagram:

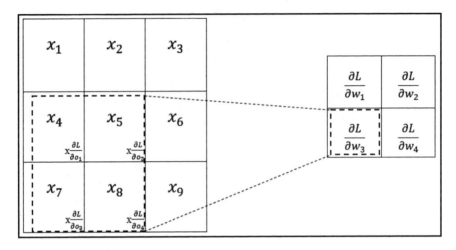

Thus, we can write the following:

$$\frac{\partial L}{\partial w_3} = \frac{\partial L}{\partial o_1}x_4 + \frac{\partial L}{\partial o_2}x_5 + \frac{\partial L}{\partial o_3}x_7 + \frac{\partial L}{\partial o_4}x_8$$

So, we understand that computing the gradients of loss with respect to the filter (that is, weights) is just the convolution operation between the input matrix and the gradient of loss with respect to the output as a filter matrix.

Apart from calculating the gradients of loss with respect to the filter, we also need to calculate the gradients of loss with respect to an input. But why do we do that? Because it is used for calculating the gradients of the filters present in the previous layer.

Our input matrix consists of nine values, from x_1 to x_9, so we need to calculate the gradients of loss with respect to all these nine values. Let's recollect how the output matrix is computed:

$$o_1 = x_1 w_1 + x_2 w_2 + x_4 w_3 + x_5 w_4$$

$$o_2 = x_2 w_1 + x_3 w_2 + x_5 w_3 + x_6 w_4$$

$$o_3 = x_4 w_1 + x_5 w_2 + x_7 w_3 + x_8 w_4$$

$$o_4 = x_5 w_1 + x_6 w_2 + x_8 w_3 + x_9 w_4$$

As you can see, x_1 is present only in o_1, so we can calculate the gradients of loss with respect to o_1 alone, as other terms would be zero:

$$\frac{\partial L}{\partial x_1} = \frac{\partial L}{\partial o_1}\frac{\partial o_1}{\partial x_1}$$

$$\frac{\partial L}{\partial x_1} = \frac{\partial L}{\partial o_1}w_1$$

Now, let's calculate the gradients with respect to x_2; as x_2 is present in only o_1 and o_2, we calculate the gradients with respect to o_1 and o_2 alone:

$$\frac{\partial L}{\partial x_2} = \frac{\partial L}{\partial o_1}\frac{\partial o_1}{\partial x_2} + \frac{\partial L}{\partial o_2}\frac{\partial o_2}{\partial x_2}$$

$$\frac{\partial L}{\partial x_2} = \frac{\partial L}{\partial o_1}w_1 + \frac{\partial L}{\partial o_2}w_2$$

In a very similar way, we calculate the gradients of loss with respect to all the inputs as follows:

$$\frac{\partial L}{\partial x_3} = \frac{\partial L}{\partial o_3}w_3$$

$$\frac{\partial L}{\partial x_4} = \frac{\partial L}{\partial o_1}w_1 + \frac{\partial L}{\partial o_4}w_4$$

$$\frac{\partial L}{\partial x_5} = \frac{\partial L}{\partial o_1}w_1 + \frac{\partial L}{\partial o_2}w_2 + \frac{\partial L}{\partial o_3}w_3 + \frac{\partial L}{\partial o_4}w_4$$

$$\frac{\partial L}{\partial x_6} = \frac{\partial L}{\partial o_2}w_2 + \frac{\partial L}{\partial o_4}w_4$$

$$\frac{\partial L}{\partial x_7} = \frac{\partial L}{\partial o_3}w_3$$

$$\frac{\partial L}{\partial x_8} = \frac{\partial L}{\partial o_3}w_3 + \frac{\partial L}{\partial o_4}w_4$$

$$\frac{\partial L}{\partial x_9} = \frac{\partial L}{\partial o_4}w_4$$

Just as we represented the gradients of the loss with respect to the weights using the convolution operation, can we also do the same here? It turns out that the answer is yes. We can actually represent the preceding equations, that is, the gradients of loss with respect to the inputs, using a convolution operation between the filter matrix as an input matrix and the gradients of loss with respect to the output matrix as a filter matrix. But the trick is that, instead of using the filter matrix directly, we rotate them 180 degrees and, also, instead of performing convolution, we perform full convolution. We are doing this so that we can derive the previous equations using a convolution operation.

The following shows what the kernel rotated by 180 degrees looks like:

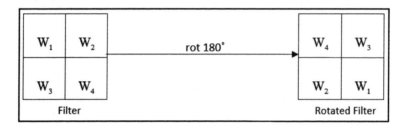

Okay, so what is full convolution? In the same way as a convolution operation, in full convolution, we use a filter and slide it over the input matrix, but the way we slide the filter is different from the convolution operation we looked at before. The following figure shows how full convolution operations work. As we can see, the shaded matrix represents the filter matrix and the unshaded one represents the input matrix; we can see how the filter slides over the input matrix step by step, as shown in this diagram:

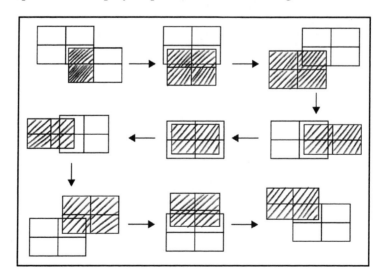

So, we can say that the gradient of loss with respect to the input matrix can be calculated using a full convolution operation between a filter rotated by 180 degrees as the input matrix and the gradient of the loss with respect to the output as a filter matrix:

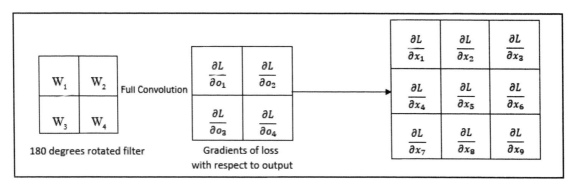

For example, as shown in the following figure, we will notice how the gradients of loss with respect to the input, w_1, is computed by the full convolution operation between the filter matrix rotated by 180 degrees, and the gradients of loss with respect to an output matrix as a filter matrix:

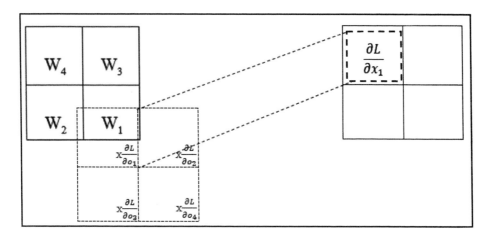

This is demonstrated as follows:

$$\frac{\partial L}{\partial x_1} = \frac{\partial L}{\partial o_1} w_1$$

Thus, we understand that computing the gradients of loss with respect to the input is just the full convolution operation. So, we can say that backpropagation in CNN is just another convolution operation.

Implementing a CNN in TensorFlow

Now we will learn how to build a CNN using TensorFlow. We will use the MNIST handwritten digits dataset and understand how a CNN recognizes handwritten digits, and we will also visualize how the convolutional layers extract important features from the image.

First, let's load the required libraries:

```
import warnings
warnings.filterwarnings('ignore')

import numpy as np
import tensorflow as tf
from tensorflow.examples.tutorials.mnist import input_data
tf.logging.set_verbosity(tf.logging.ERROR)

import matplotlib.pyplot as plt
%matplotlib inline
```

Load the MNIST dataset:

```
mnist = input_data.read_data_sets('data/mnist', one_hot=True)
```

Defining helper functions

Now we define the functions for initializing weights and bias, and for performing the convolution and pooling operations.

Initialize the weights by drawing from a truncated normal distribution. Remember, the weights are actually the filter matrix that we use while performing the convolution operation:

```
def initialize_weights(shape):
    return tf.Variable(tf.truncated_normal(shape, stddev=0.1))
```

Initialize the bias with a constant value of, say, 0.1:

```
def initialize_bias(shape):
    return tf.Variable(tf.constant(0.1, shape=shape))
```

We define a function called convolution using `tf.nn.conv2d()`, which actually performs the convolution operation; that is, the element-wise multiplication of the input matrix (x) by the filter (W) with a the stride of 1 and the same padding. We set `strides = [1,1,1,1]`. The first and last values of strides are set to 1, which implies that we don't want to move between training samples and different channels. The second and third values of `strides` are also set to 1, which implies that we move the filter by 1 pixel in both the height and width direction:

```
def convolution(x, W):
    rcturn tf.nn.conv2d(x, W, strides=[1,1,1,1], padding='SAME')
```

We define a function called `max_pooling`, using `tf.nn.max_pool()` to perform the pooling operation. We perform max pooling with a `stride` of 2 and the same `padding` and `ksize` implies our pooling window shape:

```
def max_pooling(x):
    return tf.nn.max_pool(x, ksize=[1,2,2,1], strides=[1,2,2,1],
    padding='SAME')
```

Define the placeholders for the input and output.

The `placeholder` for the input image is defined as follows:

```
X_ = tf.placeholder(tf.float32, [None, 784])
```

The `placeholder` for a reshaped input image is defined as follows:

```
X = tf.reshape(X_, [-1, 28, 28, 1])
```

The `placeholder` for the output label is defined as follows:

```
y = tf.placeholder(tf.float32, [None, 10])
```

Defining the convolutional network

Our network architecture consists of two convolutional layers. Each convolutional layer is followed by one pooling layer, and we use a fully connected layer that is followed by an output layer; that is, `conv1->pooling->conv2->pooling2->fully connected layer-> output layer`.

First, we define the first convolutional layer and pooling layer.

The weights are actually the filters in the convolutional layers. So, the weight matrix will be initialized as [`filter_shape[0]`, `filter_shape[1]`, `number_of_input_channel`, `filter_size`].

We use a 5 x 5 filter. Since we use grayscale images, the number of input channels will be 1 and we set the filter size as 32. So, the weight matrix of the first convolution layer will be [5,5,1,32]:

```
W1 = initialize_weights([5,5,1,32])
```

The shape of the bias is just the filter size, which is 32:

```
b1 = initialize_bias([32])
```

Perform the first convolution operation with ReLU activations followed by max pooling:

```
conv1 = tf.nn.relu(convolution(X, W1) + b1)
pool1 = max_pooling(conv1)
```

Next, we define the second convolution layer and pooling layer.

As the second convolutional layer takes the input from the first convolutional layer, which has 32-channel output, the number of input channel, to the second convolutional layer becomes 32 and we use the 5 x 5 filter with a filter size of 64. Thus, the weight matrix for the second convolutional layer becomes [5,5,32,64]:

```
W2 = initialize_weights([5,5,32,64])
```

The shape of the bias is just the filter size, which is 64:

```
b2 = initialize_bias([64])
```

Perform the second convolution operation with ReLU activations, followed by max pooling:

```
conv2 = tf.nn.relu(convolution(pool1, W2) + b2)
pool2 = max_pooling(conv2)
```

After two convolution and pooling layers, we need to flatten the output before feeding it to the fully connected layer. So, we flatten the result of the second pooling layer and feed it to the fully connected layer.

Flatten the result of the second pooling layer:

```
flattened = tf.reshape(pool2, [-1, 7*7*64])
```

Now we define the weights and bias for the fully connected layer. We know that we set the shape of the weight matrix as [number of neurons in the current layer, number of neurons layer in the next layer]. This is because the shape of the input image becomes 7x7x64 after flattening and we use 1024 neurons in the hidden layer. The shape of the weights becomes [7x7x64, 1024]:

```
W_fc = initialize_weights([7*7*64, 1024])
b_fc = initialize_bias([1024])
```

Here is a fully connected layer with ReLU activations:

```
fc_output = tf.nn.relu(tf.matmul(flattened, W_fc) + b_fc)
```

Define the output layer. We have 1024 neurons in the current layer, and since we need to predict 10 classes, we have 10 neurons in the next layer, thus the shape of the weight matrix becomes [1024 x 10]:

```
W_out = initialize_weights([1024, 10])
b_out = initialize_bias([10])
```

Compute the output with softmax activations:

```
YHat = tf.nn.softmax(tf.matmul(fc_output, W_out) + b_out)
```

Computing loss

Compute the loss using cross entropy. We know that the cross-entropy loss is given as follows:

$$\text{cross entropy} = -\sum_i y log(\hat{y}_i)$$

Here, y is the actual label and $\hat{y}$ is the predicted label. Thus, the cross-entropy loss is implemented as follows:

```
cross_entropy = -tf.reduce_sum(y*tf.log(YHat))
```

Minimize the loss using the Adam optimizer:

```
optimizer = tf.train.AdamOptimizer(1e-4).minimize(cross_entropy)
```

Calculate the accuracy:

```
predicted_digit = tf.argmax(y_hat, 1)
actual_digit = tf.argmax(y, 1)

correct_pred = tf.equal(predicted_digit,actual_digit)
accuracy = tf.reduce_mean(tf.cast(correct_pred, tf.float32))
```

Starting the training

Start a TensorFlow `Session` and initialize all the variables:

```
sess = tf.Session()
sess.run(tf.global_variables_initializer())
```

Train the model for 1000 epochs. Print the results for every 100 epochs:

```
for epoch in range(1000):
    #select some batch of data points according to the batch size (100)
    X_batch, y_batch = mnist.train.next_batch(batch_size=100)

    #train the network
    loss, acc, _ = sess.run([cross_entropy, accuracy, optimizer],
feed_dict={X_: X_batch, y: y_batch})

    #print the loss on every 100th epoch
    if epoch%100 == 0:
        print('Epoch: {}, Loss:{} Accuracy: {}'.format(epoch,loss,acc))
```

You will notice that the loss decreases and the accuracy increases over epochs:

```
Epoch: 0, Loss:631.2734375 Accuracy: 0.129999995232
Epoch: 100, Loss:28.9199733734 Accuracy: 0.930000007153
Epoch: 200, Loss:18.2174377441 Accuracy: 0.920000016689
Epoch: 300, Loss:21.740688324 Accuracy: 0.930000007153
```

Visualizing extracted features

Now that we have trained our CNN model, we can see what features our CNN has extracted to recognize the image. As we learned, each convolutional layer extracts important features from the image. We will see what features our first convolutional layer has extracted to recognize the handwritten digits.

First, let's select one image from the training set, say, digit 1:

```
plt.imshow(mnist.train.images[7].reshape([28, 28]))
```

The input image is shown here:

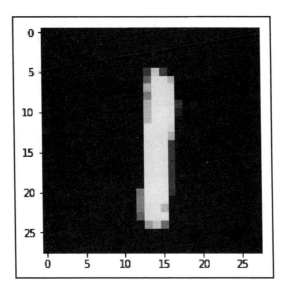

Feed this image to the first convolutional layer, that is, conv1, and get the feature maps:

```
image = mnist.train.images[7].reshape([-1, 784])
feature_map = sess.run([conv1], feed_dict={X_: image})[0]
```

Plot the feature map:

```
for i in range(32):
    feature = feature_map[:,:,:,i].reshape([28, 28])
    plt.subplot(4,8, i + 1)
    plt.imshow(feature)
    plt.axis('off')
plt.show()
```

As you can see in the following plot, the first convolutional layer has learned to extract edges from the given image:

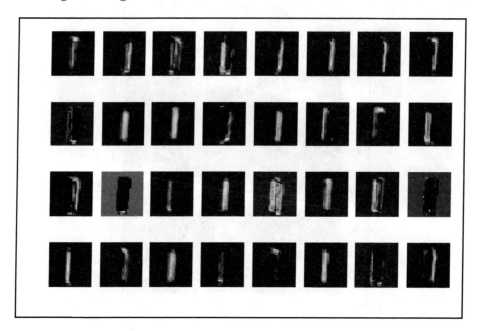

Thus, this is how the CNN uses multiple convolutional layers to extract important features from the image and feed these extracted features to a fully connected layer to classify the image. Now that we have learned how CNNs works, in the next section, we will learn about several interesting CNN architectures.

CNN architectures

In this section, we will explore different interesting types of CNN architecture. When we say different types of CNN architecture, we basically mean how convolutional and pooling layers are stacked on each other. Additionally, we will learn how many numbers of convolutional, pooling, and fully connected layers are used, what the number of filters and filter sizes are, and more.

LeNet architecture

The LeNet architecture is one of the classic architectures of a CNN. As shown in the following diagram, the architecture is very simple, and it consists of only seven layers. Out of these seven layers, there are three convolutional layers, two pooling layers, one fully connected layer, and one output layer. It uses a 5 x 5 convolution with a stride of 1, and uses average pooling. What is 5 x 5 convolution? It implies we are performing a convolution operation with a 5 x 5 filter.

As shown in the following diagram, LeNet consists of three convolutional layers (C1, C3, C5), two pooling layers (S2, S4), one fully connected layer (F6), and one output layer (OUTPUT), and each convolutional layer is followed by a pooling layer:

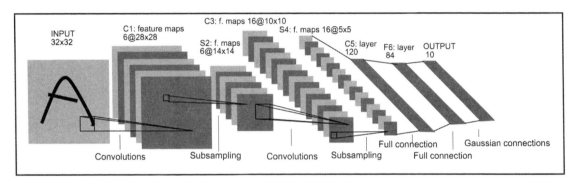

Understanding AlexNet

AlexNet is a classic and powerful deep learning architecture. It won the ILSVRC 2012 by significantly reducing the error rate from 26% to 15.3%. ILSVRC stands for ImageNet Large Scale Visual Recognition Competition, which is one of the biggest competitions focused on computer vision tasks, such as image classification, localization, object detection, and more. ImageNet is a huge dataset containing over 15 million labeled, high-resolution images, with over 22,000 categories. Every year, researchers compete to win the competition using innovative architecture.

AlexNet was designed by pioneering scientists, including Alex Krizhevsky, Geoffrey Hinton, and Ilya Sutskever. It consists of five convolutional layers and three fully connected layers, as shown in the following diagram. It uses the ReLU activation function instead of the tanh function, and ReLU is applied after every layer. It uses dropout to handle overfitting, and dropout is performed before the first and second fully connected layers. It uses data augmentation techniques, such as image translation, and is trained using batch stochastic gradient descent on two GTX 580 GPUs for 5 to 6 days:

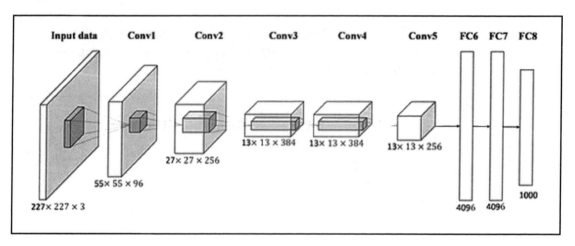

Architecture of VGGNet

VGGNet is one of the most popularly used CNN architectures. It was invented by the **Visual Geometry Group** (**VGG**) at the University of Oxford. It started to get very popular when it became the first runner-up of ILSVRC 2014.

It is basically a deep convolutional network and is widely used for object-detection tasks. The weights and structure of the network are made available to the public by the Oxford team, so we can use these weights directly to carry out several computer vision tasks. It is also widely used as a good baseline feature extractor for images.

The architecture of the VGG network is very simple. It consists of convolutional layers followed by a pooling layer. It uses 3 x 3 convolution and 2 x 2 pooling throughout the network. It is referred to as VGG-*n*, where *n* corresponds to a number of layers, excluding the pooling and softmax layer. The following figure shows the architecture of the VGG-16 network:

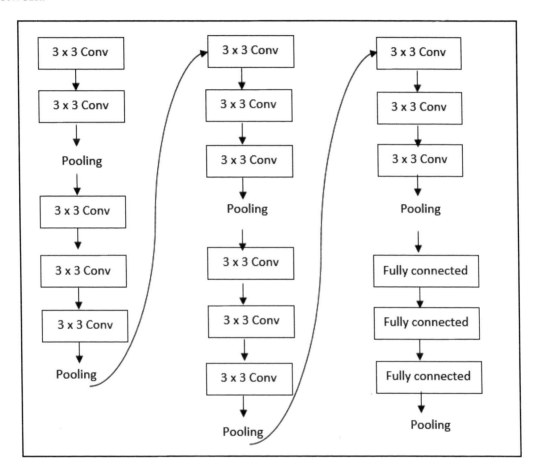

As you can see in the following figure, the architecture of AlexNet is characterized by a pyramidal shape, as the initial layers are wide and the later layers are narrow. You will notice it consists of multiple convolutional layers followed by a pooling layer. Since the pooling layer reduces the spatial dimension, it narrows the network as we go deeper into the network:

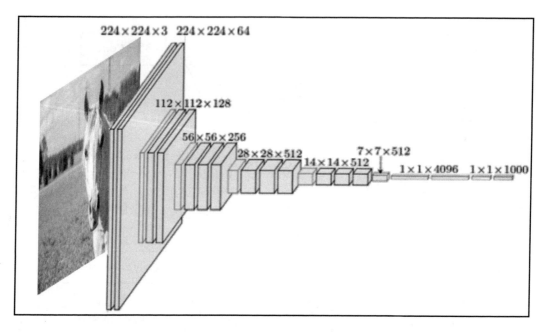

The one shortcoming of VGGNet is that it is computationally expensive, and it has over 160 million parameters.

GoogleNet

GoogleNet, also known as **inception net**, was the winner of ILSVRC 2014. It consists of various versions, and each version is an improved version of the previous one. We will explore each version one by one.

Inception v1

Inception v1 is the first version of the network. An object in an image appears in different sizes and in a different positions. For example, look at the first image; as you can see, the parrot, when viewed closer, takes up the whole portion of the image but in the second image, when the parrot is viewed from a distance, it takes up a smaller region of the image:

Thus, we can say objects (in the given image, it's a parrot) can appear on any region of the image. It might be small or big. It might take up a whole region of the image, or just a very small portion. Our network has to exactly identify the object. But what's the problem here? Remember how we learned that we use a filter to extract features from the image? Now, because our object of interest varies in size and location in each image, choosing the right filter size is difficult.

We can use a filter of a large size when the object size is large, but a large filter size is not suitable when we have to detect an object that is in a small corner of an image. Since we use a fixed receptive field that is a fixed filter size, it is difficult to recognize objects in the images whose position varies greatly. We can use deep networks, but they are more vulnerable to overfitting.

To overcome this, instead of using a single filter of the same size, the inception network uses multiple filters of varying sizes on the same input. An inception network consists of nine inception blocks stacked over one another. A single inception block is shown in the following figure. As you will notice, we perform convolution operations on a given image with three different filters of varying size, that is, 1 x 1, 3 x 3, and 5 x 5. Once the convolution operation is performed by all these different filters, we concatenate the results and feed it to the next inception block:

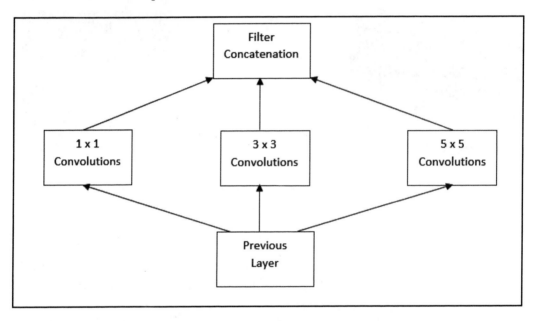

As we are concatenating output from multiple filters, the depth of the concatenated result will increase. Although we use padding that only matches the shape of the input and output to be the same but we will still have different depths. Since the result of one inception block is the feed to another, the depth keeps on increasing. So, to avoid the increase in the depth, we just add a 1 x 1 convolution before the 3 x 3 and 5 x 5 convolution, as shown in the following figure. We also perform a max pooling operation, and a 1 x 1 convolution is added after the max pooling operation as well:

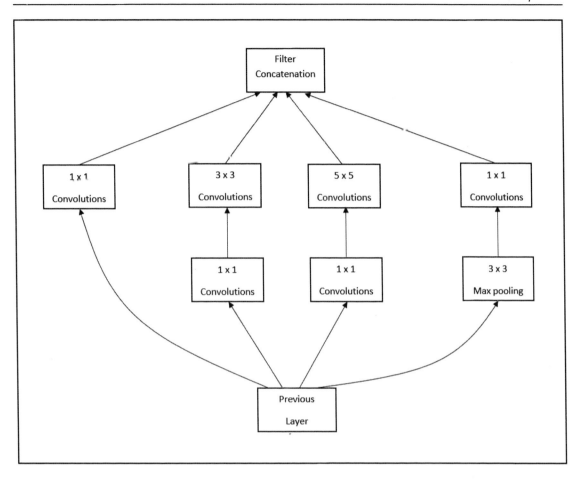

Each inception block extracts some features and feeds them to the next inception block. Let's say we are trying to recognize a picture of a parrot. The inception block in the first few layers detects basic features, and the later inception blocks detect high-level features. As we saw, in a convolutional network, inception blocks will only extract features, and don't perform any classification. So, we feed the features extracted by the inception block to a classifier, which will predict whether the image contains a parrot or not.

As the inception network is deep, with nine inception blocks, it is susceptible to the vanishing-gradient problem. To avoid this, we introduce classifiers between the inception blocks. Since each inception block learns the meaningful feature of the image, we try to perform classification and compute loss from the intermediate layers as well. As shown in the following figure, we have nine inception blocks. We take the result of the third inception block, I_3, and feed it to an intermediate classifier, and also the result of the sixth inception block, I_6, to another intermediate classifier. There is also yet another classifier at the end of the final inception blocks. This classifier basically consists of average pooling, 1 x 1 convolutions, and a linear layer with softmax activations:

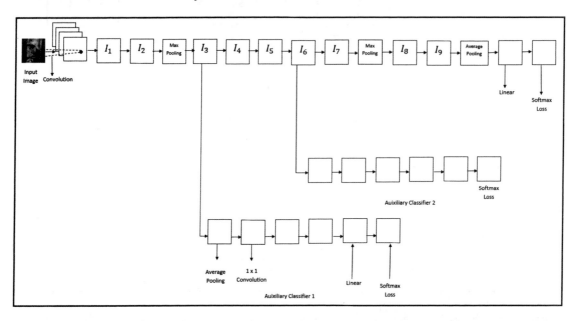

The intermediate classifiers are actually called auxiliary classifiers. So, the final loss of the inception network is the weighted sum of the auxiliary classifier's loss and the loss of the final classifier (real loss), as follows:

$$Loss = Real\ Loss + 0.3 * Auxiliary\ Loss1 + 0.3 * Auxiliary\ Loss2$$

Inception v2 and v3

Inception v2 and v3 are introduced in the paper, *Going Deeper with Convolutions* by Christian Szegedy as mentioned in *Further reading* section. The authors suggest the use of factorized convolution, that is, we can break down a convolutional layer with a larger filter size into a stack of convolutional layers with a smaller filter size. So, in the inception block, a convolutional layer with a 5 x 5 filter can be broken down into two convolutional layers with 3 x 3 filters, as shown in the following diagram. Having a factorized convolution increases performance and speed:

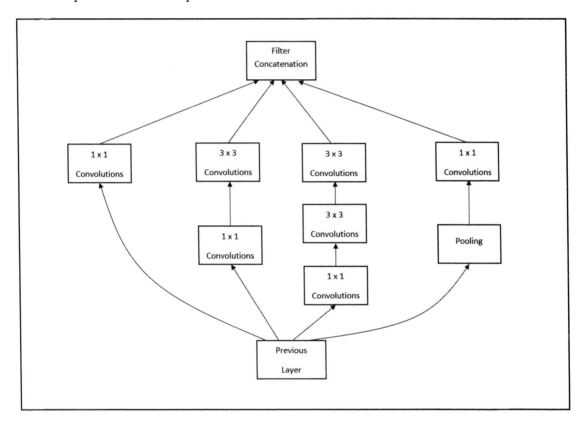

The authors also suggest breaking down a convolutional layer of filter size *n* x *n* into a stack of convolutional layers with filter sizes *1* x *n* and *n* x *1*. For example, in the previous figure, we have *3* x *3* convolution, which is now broken down into *1* x *3* convolution, followed by *3* x *1* convolution, as shown in the following diagram:

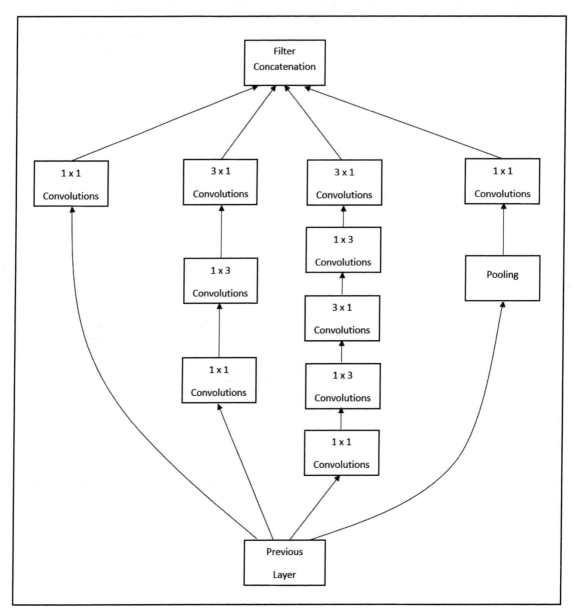

As you will notice in the previous diagram, we are basically expanding our network in a deeper fashion, which will lead us to lose information. So, instead of making it deeper, we make our network wider, shown as follows:

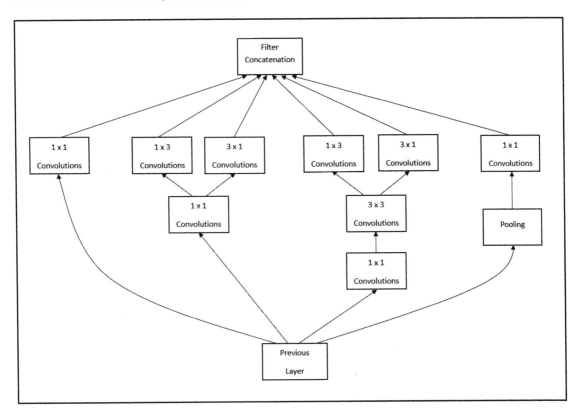

In inception net v3, we use factorized 7 x 7 convolutions with RMSProp optimizers. Also, we apply batch normalization in the auxiliary classifiers.

Capsule networks

Capsule networks (CapsNets) were introduced by Geoffrey Hinton to overcome the limitations of convolutional networks.

Hinton stated the following:

> *"The pooling operation used in convolutional neural networks is a big mistake and the fact that it works so well is a disaster."*

But what is wrong with the pooling operation? Remember when we used the pooling operation to reduce the dimension and to remove unwanted information? The pooling operation makes our CNN representation invariant to small translations in the input.

This translation invariance property of a CNN is not always beneficial, and can be prone to misclassifications. For example, let's say we need to recognize whether an image has a face; the CNN will look for whether the image has eyes, a nose, a mouth, and ears. It does not care about which location they are in. If it finds all such features, then it classifies it as a face.

Consider two images, as shown in the following figure. The first image is the actual face, and in the second image, the eyes are placed on the left side, one above the another, and the ears and mouth are placed on the right. But the CNN will still classify both the images as a face as both images have all the features of a face, that is, ears, eyes, a mouth, and a nose. The CNN thinks that both images consist of a face. It does not learn the spatial relationship between each feature; that the eyes should be placed at the top and should be followed by a nose, and so on. All it checks for is the existence of the features that make up the face.

This problem will become worse when we have a deep network, as in the deep network, the features will become abstract, and it will also shrink in size due to the several pooling operations:

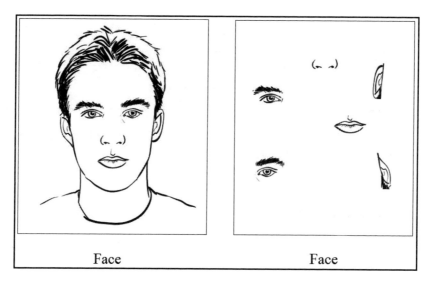

Face   Face

To overcome this, Hinton introduced a new network called the Capsule network, which consists of capsules instead of neurons. Like a CNN, the Capsule network checks for the presence of certain features to classify the image, but apart from detecting the features, it will also check the spatial relationship between them. That is, it learns the hierarchy of the features. Taking our example of recognizing a face, the Capsule network will learn that the eyes should be at the top and the nose should be in the middle, followed by a mouth and so on. If the image does not follow this relationship, then the Capsule network will not classify it as a face:

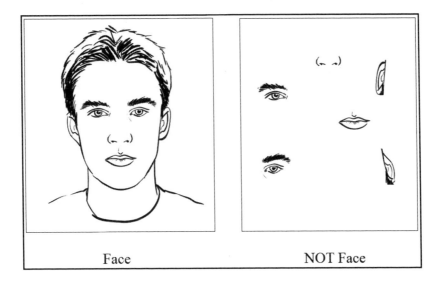

Face   NOT Face

A Capsule network consists of several capsules connected together. But, wait. What is a capsule?

A capsule is a group of neurons that learn to detect a particular feature in the image; say, eyes. Unlike neurons, which return a scalar, capsules return a vector. The length of the vector tells us whether a particular feature exists in a given location, and the elements of the vector represent the properties of the features, such as, position, angle, and so on.

Let's say we have a vector, v, as follows:

$$v = [0.3, 1.2]$$

The length of the vector can be calculated as follows:

$$length = \sqrt{(0.3)^2 + (1.2)^2} = 1.53$$

We have learned that the length of the vector represents the probability of the existence of the features. But the preceding length does not represent a probability, as it exceeds 1. So, we convert this value into a probability using a function called the squash function. The squash function has an advantage. Along with calculating probability, it also preserves the direction of the vector:

$$probability = squash(length(v))$$

Just like a CNN, capsules in the earlier layers detect basic features including eyes, a nose, and so on, and the capsules in the higher layers detect high-level features, such as the overall face. Thus, capsules in the higher layers take input from the capsules in the lower layers. In order for the capsules in the higher layers to detect a face, they not only check for the presence of features such as a nose and eyes, but also check their spatial relationships.

Now that we have a basic understanding of what a capsule is, we will go into this in more detail and see how exactly a Capsule network works.

Understanding Capsule networks

Let's say we have two layers, l and $l+1$. l will be the lower layer and it has i capsules, and $l+1$ will be the higher layer and it has j capsules. Capsules from the lower layer send their outputs to capsules in the higher layer. u_i will be the activations of the capsules from the lower layer, l. v_j will be the activations of the capsules from the higher layer, $l+1$.

The following figure represents a capsule, j, and as you can observe, it takes the outputs of the previous capsules, u_i, as inputs and computes its output, v_j:

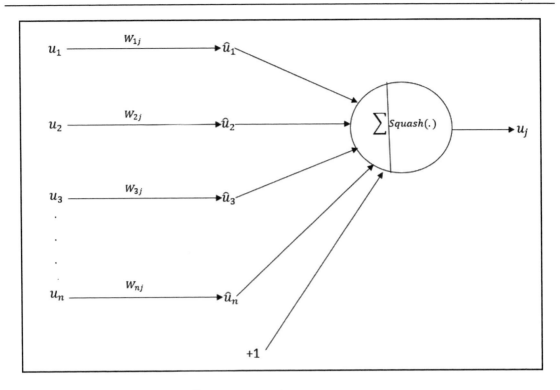

We will move on to learn how v_j is computed.

Computing prediction vectors

In the previous figure, u_1, u_2, and, u_3 represent the output vectors from the previous capsule. First, we multiply these vectors by the weight matrix and compute a prediction vector:

$$\hat{u}_{j|i} = W_{ij}u_i$$

Okay, so what exactly are we doing here, and what are prediction vectors? Let's consider a simple example. Say that capsule j is trying to predict whether an image has a face. We have learned that capsules in the earlier layers detect basic features and send their results to the capsules in the higher layer. So, the capsules in the earlier layer, u_1, u_2, and, u_3 detect basic low features, such as eyes, a nose, and a mouth, and send their results to the capsules in the high-level layer, that is, capsule j, which detects the face.

Thus, capsule j takes the previous capsules, u_1, u_2, and, u_3, as inputs and multiplies them by a weights matrix, W.

The weights matrix W represents the spatial and other relationship between low-level features and high-level features. For instance, the weight W_{1j} tells us that eyes should be on the top. W_{2j} tells us that a nose should be in the middle. W_{3j} tells us that a mouth should be on the bottom. Note that the weight matrix not only captures the position (that is, the spatial relationship), but also other relationships.

So, by multiplying the inputs by weights, we can predict the position of the face:

- $\hat{u}_{j|1} = W_{1j}u_1$ implies the predicted position of the face based on the eyes
- $\hat{u}_{j|2} = W_{2j}u_2$ implies the predicted position of the face based on the nose
- $\hat{u}_{j|3} = W_{1j}u_3$ implies the predicted position of the face based on the mouth

When all the predicted positions of the face are the same, that is, in agreement with each other, then we can say that the image contains a face. We learn these weights using backward propagation.

Coupling coefficients

Next, we multiply the prediction vectors $\hat{u}_{j|i}$ by the coupling coefficients c_{ij}. The coupling coefficients exist between any two capsules. We know that capsules from the lower layer send their output to the capsules in the higher layer. The coupling coefficient helps the capsule in the lower layer to understand which capsule in the higher layer it has to send its output to.

For instance, let's consider the same example, where we are trying to predict whether an image consists of a face. c_{ij} represents the agreement between i and j.

c_{1j} represents the agreement between an eye and a face. Since we know that the eye is on the face, the c_{1j} value will be increased. We know that the prediction vector $\hat{u}_{j|1}$ implies the predicted position of the face based on the eyes. Multiplying $\hat{u}_{j|1}$ by c_{1j} implies that we are increasing the importance of the eyes, as the value of c_{1j} is high.

c_{2j} represents the agreement between nose and face. Since we know that the nose is on the face, the c_{2j} value will be increased. We know that the prediction vector $\hat{u}_{j|2}$ implies the predicted position of the face based on the nose. Multiplying $\hat{u}_{j|2}$ by c_{2j} implies that we are increasing the importance of the nose, as the value of c_{2j} is high.

Let's consider another low-level feature, say, u_4, which detects a finger. Now, c_{4j} represents the agreement between a finger and a face, which will be low. Multiplying $\hat{u}_{j|4}$ by c_{4j} implies that we are decreasing the importance of the finger, as the value of c_{4j} is low.

But how are these coupling coefficients learned? Unlike weights, the coupling coefficients are learned in the forward propagation itself, and they are learned using an algorithm called dynamic routing, which we will discuss later in an upcoming section.

After multiplying $\hat{u}_{j|i}$ by c_{ij}, we sum them up, as follows:

$$\vec{s}_j = \hat{u}_{j|1}c_{1j} + \hat{u}_{j|2}c_{2j} + \hat{u}_{j|3}c_{3j}$$

Thus, we can write our equation as follows:

$$\vec{s_j} = \sum_i \hat{u}_{j|i}c_{ij}$$

Squashing function

We started off saying that capsule j tries to detect the face in the image. So, we need to convert $\vec{s_j}$ into probabilities to get the probability of the existence of a face in the image.

Apart from calculating probabilities, we also need to preserve the direction of the vectors, so we use an activation function called the squash function. It is given as follows:

$$\vec{v}_j = \frac{\|\vec{s}_j\|^2}{1 + \|\vec{s}_j\|^2} \frac{\vec{s}_j}{\|\vec{s}_j\|}$$

Now, $\vec{v}_j$ (also referred to as the activity vector) gives us the probability of the existence of a face in a given image.

Dynamic routing algorithm

Now, we will see how the dynamic routing algorithm computes the coupling coefficients. Let's introduce a new variable called b_{ij}, which is just a temporary variable, and is the same as the coupling coefficients c_{ij}. First, we initialize b_{ij} to 0. It implies coupling coefficients between the capsules i in the lower layer l and capsules j in the higher layer $l+1$ are set to 0.

Let $\vec{b}_i$ be the vector representation of b_{ij}. Given the prediction vectors $\hat{u}_{j|i}$, for some n number of iterations, we do the following:

1. For all the capsules i in the layer l, compute the following:

$$c_i = \text{softmax}(\vec{b_i})$$

2. For all the capsules j in the layer $l+1$, compute the following:

$$\vec{s_j} = \sum_i \hat{u}_{j|i} c_{ij}$$

$$\vec{v}_j = \frac{\|\vec{s}_j\|^2}{1 + \|\vec{s}_j\|^2} \frac{\vec{s}_j}{\|\vec{s}_j\|}$$

3. For all capsules i in l and for all capsules in j, compute b_{ij} as follows:

$$b_{ij} = b_{ij} + \hat{u}_{j|i} \cdot \vec{v}_j$$

The previous equation has to be noted carefully. It is where we update our coupling coefficient. The dot product $\hat{u}_{j|i} \cdot \vec{v}_j$ implies the dot product between the prediction vectors $\hat{u}_{j|i}$ of the capsule in the lower layer and the output vector $\vec{v}_j$ of the capsule in the higher layer. If the dot product is high, b_{ij} will increase the respective coupling coefficient c_{ij}, which makes the $\hat{u}_{j|i} \cdot \vec{v}_j$ stronger.

Architecture of the Capsule network

Let's suppose our network is trying to predict handwritten digits. We know that capsules in the earlier layers detect basic features, and those in the later layers detect the digit. So, let's call the capsules in the earlier layers **primary capsules** and those in the later layers **digit capsules**.

The architecture of a Capsule network is shown here:

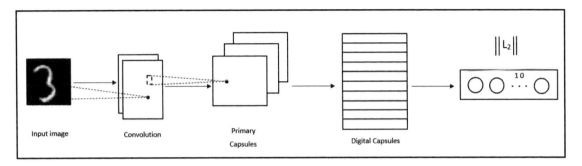

In the preceding diagram, we can observe the following:

1. First, we take the input image and feed it to a standard convolution layer, and we call the result convolutional inputs.
2. Then, we feed the convolutional inputs to the primary capsules layer and get the primary capsules.
3. Next, we compute digit capsules with primary capsules as input using the dynamic-routing algorithm.
4. The digit capsules consist of 10 rows, and each of the rows represents the probability of the predicted digit. That is, row 1 represents the probability of the input digit to be 0, row 2 represents the probability of the digit 1, and so on.
5. Since the input image is digit 3 in the preceding image, row 4, which represents the probability of digit 3, will be high in the digit capsules.

The loss function

Now we will explore the loss function of the Capsule network. The loss function is the weighted sum of two loss functions called margin loss and reconstruction loss.

Margin loss

We learned that the capsule returns a vector and the length of a vector represents the probability of the existence of the features. Say our network is trying to recognize the handwritten digits in an image. To detect multiple digits in a given image, we use margin loss, L_k, for each digit capsule, k, as follows:

$$L_k = T_k \max(0, m^+ - \|v_k\|)^2 + \lambda(1 - T_k) \max(0, \|v_k\| - m^-)^2$$

Here, the following is the case:

- $T_k = 1$, if the digit of a class k is present
- m is the margin, and m^+ is set to 0.9 and m^- is set to 0.1
- λ prevents the initial learning from shrinking the lengths of the vectors of all the digit capsules and is usually set to 0.5

The total margin loss is the sum of the loss of all classes, k, as follows:

$$\text{Margin Loss} = \sum_k L_k$$

Reconstruction loss

In order to make sure that the network has learned the important features in the capsules, we use reconstruction loss. This means that we use a three-layer network called a decoder network, which tries to reconstruct the original image from the digit capsules:

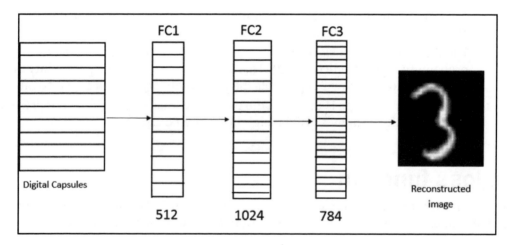

Reconstruction loss is given as the squared difference between the reconstructed and original image, as follows:

$$\text{Reconstruction Loss} = (\text{Reconstructed Image} - \text{Original Image})^2$$

The final loss is given as follows:

$$\text{Loss} = \text{Margin Loss} + \text{alpha} * \text{Reconstruction Loss}$$

Here, alpha is a regularization term, because we don't want the reconstruction loss to have more priority than the margin loss. So, alpha is multiplied by the reconstruction loss to scale down its importance, and is usually set to 0.0005.

Building Capsule networks in TensorFlow

Now we will learn how to implement Capsule networks in TensorFlow. We will use our favorite MNIST dataset to learn how a Capsule network recognizes the handwritten image.

Import the required libraries:

```
import warnings
warnings.filterwarnings('ignore')

import numpy as np
import tensorflow as tf

from tensorflow.examples.tutorials.mnist import input_data
tf.logging.set_verbosity(tf.logging.ERROR)
```

Load the MNIST dataset:

```
mnist = input_data.read_data_sets("data/mnist",one_hot=True)
```

Defining the squash function

We learned that the squash function converts the length of the vector into probability, and it is given as follows:

$$\vec{v}_j = \frac{\|\vec{s}_j\|^2}{1 + \|\vec{s}_j\|^2} \frac{\vec{s}_j}{\|\vec{s}_j\|}$$

The squash function can be defined as follows:

```
def squash(sj):
    sj_norm = tf.reduce_sum(tf.square(sj), -2, keep_dims=True)
    scalar_factor = sj_norm / (1 + sj_norm) / tf.sqrt(sj_norm + epsilon)

    vj = scalar_factor * sj

    return vj
```

Defining a dynamic routing algorithm

Now we will look at how the dynamic routing algorithm is implemented. We use variable names of the same notations that we learned in the dynamic routing algorithm, so that we can easily follow the steps. We will look at each line in our function step by step. You can also check the complete code on GitHub, at http://bit.ly/2HQqDEZ.

First, define the function called `dynamic_routing`, which takes the previous capsules, `ui`, coupling coefficients, `bij`, and number of routing iterations, `num_routing` as inputs as follows:

```
def dynamic_routing(ui, bij, num_routing=10):
```

Initialize the `wij` weights by drawing from a random normal distribution, and initialize `biases` with a constant value:

```
wij = tf.get_variable('Weight', shape=(1, 1152, 160, 8, 1),
dtype=tf.float32,

                    initializer=tf.random_normal_initializer(0.01))

biases = tf.get_variable('bias', shape=(1, 1, 10, 16, 1))
```

Define the primary capsules `ui` (`tf.tile` replicates the tensor *n* times):

```
ui = tf.tile(ui, [1, 1, 160, 1, 1])
```

Compute the prediction vector, $\hat{u}_{j|i} = W_{ij}u_i$, as follows:

```
u_hat = tf.reduce_sum(wij * ui, axis=3, keep_dims=True)
```

Reshape the prediction vector:

```
u_hat = tf.reshape(u_hat, shape=[-1, 1152, 10, 16, 1])
```

Stop gradient computation in the prediction vector:

```
u_hat_stopped = tf.stop_gradient(u_hat, name='stop_gradient')
```

Perform dynamic routing for a number of routing iterations, as follows:

```
for r in range(num_routing):

    with tf.variable_scope('iter_' + str(r)):
        #step 1
        cij = tf.nn.softmax(bij, dim=2)
        #step 2
```

```
        if r == num_routing - 1:

            sj = tf.multiply(cij, u_hat)

            sj = tf.reduce_sum(sj, axis=1, keep_dims=True) + biases

            vj = squash(sj)

        elif r < num_routing - 1:

            sj = tf.multiply(cij, u_hat_stopped)

            sj = tf.reduce_sum(sj, axis=1, keep_dims=True) + biases

            vj = squash(sj)

            vj_tiled = tf.tile(vj, [1, 1152, 1, 1, 1])

            coupling_coeff = tf.reduce_sum(u_hat_stopped * vj_tiled,
    axis=3, keep_dims=True)

            #step 3
            bij += coupling_coeff
    return vj
```

Computing primary and digit capsules

Now we will compute the primary capsules, which extract the basic features, and the digit capsules, which recognizes the digits.

Start the TensorFlow `Graph`:

```
graph = tf.Graph()
with graph.as_default() as g:
```

Define the placeholders for input and output:

```
x = tf.placeholder(tf.float32, [batch_size, 784])
y = tf.placeholder(tf.float32, [batch_size,10])
x_image = tf.reshape(x, [-1,28,28,1])
```

Perform the convolution operation and get the convolutional input:

```
with tf.name_scope('convolutional_input'):
    input_data = tf.contrib.layers.conv2d(inputs=x_image,
num_outputs=256, kernel_size=9, padding='valid')
```

Compute the primary capsules that extract the basic features, such as edges. First, compute the capsules using the convolution operation as follows:

```
capsules = []

for i in range(8):

with tf.name_scope('capsules_' + str(i)):

#convolution operation
output = tf.contrib.layers.conv2d(inputs=input_data,
num_outputs=32,kernel_size=9, stride=2, padding='valid')

#reshape the output
output = tf.reshape(output, [batch_size, -1, 1, 1])

#store the output which is capsule in the capsules list
capsules.append(output)
```

Concatenate all the capsules and form the primary capsules, squash the primary capsules, and get the probability as follows:

```
primary_capsule = tf.concat(capsules, axis=2)
```

Apply the `squash` function to the primary capsules and get the probability:

```
primary_capsule = squash(primary_capsule)
```

Compute the digit capsules using a dynamic-routing algorithm as follows:

```
    with tf.name_scope('dynamic_routing'):
        #reshape the primary capsule
        outputs = tf.reshape(primary_capsule, shape=(batch_size, -1, 1,
primary_capsule.shape[-2].value, 1))

        #initialize bij with 0s
        bij = tf.constant(np.zeros([1, primary_capsule.shape[1].value, 10,
1, 1], dtype=np.float32))


        #compute the digit capsules using dynamic routing algorithm which
takes
        #the reshaped primary capsules and bij as inputs and returns the
activity vector
        digit_capsules = dynamic_routing(outputs, bij)
  digit_capsules = tf.squeeze(digit_capsules, axis=1)
```

Masking the digit capsule

Why do we need to mask the digit capsule? We learned that in order to make sure that the network has learned the important features, we use a three-layer network called a decoder network, which tries to reconstruct the original image from the digit capsules. If the decoder is able to reconstruct the image successfully from the digit capsules, then it means the network has learned the important features of the image; otherwise, the network has not learned the correct features of the image.

The digit capsules contain the activity vector for all the digits. But the decoder wants to reconstruct only the given input digit (the input image). So, we mask out the activity vector of all the digits, except for the correct digit. Then we use this masked digit capsule to reconstruct the given input image:

```
with graph.as_default() as g:
    with tf.variable_scope('Masking'):
        #select the activity vector of given input image using the actual
label y and mask out others
        masked_v = tf.multiply(tf.squeeze(digit_capsules), tf.reshape(y,
(-1, 10, 1)))
```

Defining the decoder

Define the decoder network for reconstructing the image. It consists of three fully connected networks, as follows:

```
with tf.name_scope('Decoder'):

    #masked digit capsule
    v_j = tf.reshape(masked_v, shape=(batch_size, -1))

    #first fully connected layer
    fc1 = tf.contrib.layers.fully_connected(v_j, num_outputs=512)

    #second fully connected layer
    fc2 = tf.contrib.layers.fully_connected(fc1, num_outputs=1024)

    #reconstructed image
    reconstructed_image = tf.contrib.layers.fully_connected(fc2,
num_outputs=784, activation_fn=tf.sigmoid)
```

Computing the accuracy of the model

Now we compute the accuracy of our model:

```
with graph.as_default() as g:
    with tf.variable_scope('accuracy'):
```

Compute the length of each activity vector in the digit capsule:

```
        v_length = tf.sqrt(tf.reduce_sum(tf.square(digit_capsules), axis=2,
    keep_dims=True) + epsilon)
```

Apply `softmax` to the length and get the probabilities:

```
        softmax_v = tf.nn.softmax(v_length, dim=1)
```

Select the index that had the highest probability; this will give us the predicted digit:

```
        argmax_idx = tf.to_int32(tf.argmax(softmax_v, axis=1))
        predicted_digit = tf.reshape(argmax_idx, shape=(batch_size, ))
```

Compute the `accuracy`:

```
        actual_digit = tf.to_int32(tf.argmax(y, axis=1))

        correct_pred = tf.equal(predicted_digit,actual_digit)
        accuracy = tf.reduce_mean(tf.cast(correct_pred, tf.float32))
```

Calculating loss

As we know, we compute two types of loss—margin loss and reconstruction loss.

Margin loss

We know that margin loss is given as follows:

$$L_k = T_k \max(0, m^+ - \|v_k\|)^2 + \lambda(1 - T_k) \max(0, \|v_k\| - m^-)^2$$

Compute the maximum value in the left and maximum value in the right:

```
max_left = tf.square(tf.maximum(0.,0.9 - v_length))
max_right = tf.square(tf.maximum(0., v_length - 0.1))
```

Set T_k to y:

```
T_k = y

lambda_ = 0.5
L_k = T_k * max_left + lambda_ * (1 - T_k) * max_right
```

The total margin loss is computed as follows:

```
margin_loss = tf.reduce_mean(tf.reduce_sum(L_k, axis=1))
```

Reconstruction loss

Reshape and get the original image by using the following code:

```
original_image = tf.reshape(x, shape=(batch_size, -1))
```

Compute the mean of the squared difference between the reconstructed and the original image:

```
squared = tf.square(reconstructed_image - original_image)
```

Compute the reconstruction loss:

```
reconstruction_loss = tf.reduce_mean(squared)
```

Total loss

Define the total loss, which is the weighted sum of the margin loss and the reconstructed loss:

```
alpha = 0.0005
total_loss = margin_loss + alpha * reconstruction_loss
```

Optimize the loss using the Adam optimizer:

```
optimizer = tf.train.AdamOptimizer(0.0001)
train_op = optimizer.minimize(total_loss)
```

Training the Capsule network

Set the number of epochs and number of steps:

```
num_epochs = 100
num_steps = int(len(mnist.train.images)/batch_size)
```

Now start the TensorFlow `Session` and perform training:

```
with tf.Session(graph=graph) as sess:

    init_op = tf.global_variables_initializer()
    sess.run(init_op)


    for epoch in range(num_epochs):
        for iteration in range(num_steps):
            batch_data, batch_labels = mnist.train.next_batch(batch_size)
            feed_dict = {x : batch_data, y : batch_labels}
            _, loss, acc = sess.run([train_op, total_loss, accuracy],
feed_dict=feed_dict)

            if iteration%10 == 0:
                print('Epoch: {}, iteration:{}, Loss:{} Accuracy:
{}'.format(epoch,iteration,loss,acc))
```

You can see how the loss decreases over various iterations:

```
Epoch: 0, iteration:0, Loss:0.55281829834 Accuracy: 0.0399999991059
Epoch: 0, iteration:10, Loss:0.541650533676 Accuracy: 0.20000000298
Epoch: 0, iteration:20, Loss:0.233602654934 Accuracy: 0.40000007153
```

Thus, we have learned how Capsule networks work step by step, and how to build a Capsule network in TensorFlow.

Summary

We started off the chapter by understanding CNNs. We learned about the different layers of a CNN, such as convolution and pooling; where the important features from the image will be extracted and are fed to the fully collected layer; and where the extracted feature will be classified. We also visualized the features extracted from the convolutional layer using TensorFlow by classifying handwritten digits.

Later, we learned about several architectures of CNN, including LeNet, AlexNet, VGGNet, and GoogleNet. At the end of the chapter, we studied Capsule networks, which overcome the shortcomings of a convolutional network. We learned that Capsule networks use a dynamic routing algorithm for classifying the image.

In the next chapter, we will study the various algorithms used for learning text representations.

Questions

Let's try answering the following questions to assess our knowledge of CNNs:

1. What are the different layers of a CNN?
2. Define stride.
3. Why is padding required?
4. Define pooling. What are the different types of pooling operations?
5. Explain the architecture of VGGNet.
6. What is factorized convolution in the inception network?
7. How do Capsule networks differ from CNNs?
8. Define the squash function.

Further reading

Refer to the following for more information:

- *Very Deep Convolutional Networks for Large-Scale Image Recognition* by Karen Simonyan and Andrew Zisserman, available at `https://arxiv.org/pdf/1409.1556.pdf`
- A paper on inception net, *Going Deeper with Convolutions* by Christian Szegedy et al., available at `https://www.cv-foundation.org/openaccess/content_cvpr_2015/papers/Szegedy_Going_Deeper_With_2015_CVPR_paper.pdf`
- *Dynamic Routing Between Capsules* by Sara Sabour, Nicholas Frosst, and Geoffrey E. Hinton, available at `https://arxiv.org/pdf/1710.09829.pdf`

Learning Text Representations

7

Neural networks require inputs only in numbers. So when we have textual data, we convert them into numeric or vector representation and feed it to the network. There are various methods for converting the input text to numeric form. Some of the popular methods include **term frequency-inverse document frequency (tf-idf)**, **bag of words (BOW)**, and so on. However, these methods do not capture the semantics of the word. This means that these methods will not understand the meaning of the words.

In this chapter, we will learn about an algorithm called **word2vec** which converts the textual input to a meaningful vector. They learn the semantic vector representation for each word in the given input text. We will start off the chapter by understanding about word2vec model and two different types of word2vec model called **continuous bag-of-words (CBOW)** and skip-gram model. Next, we will learn how to build word2vec model using gensim library and how to visualize high dimensional word embeddings in tensorboard.

Going ahead, we will learn about **doc2vec** model which is used for learning the representations for a document. We will understand two different methods in doc2vec called **Paragraph Vector - Distributed Memory Model (PV-DM)** and **Paragraph Vector - Distributed Bag of Words (PV-DBOW)**. We will also see how to perform document classification using doc2vec. At the end of the chapter, we will learn about skip-thoughts algorithms and quick thoughts algorithm which is used for learning the sentence representations.

In this chapter, we will understand the following topics:

- The word2vec model
- Building a word2vec model using gensim
- Visualizing word embeddings in TensorBoard
- Doc2vec model
- Finding similar documents using doc2vec
- Skip-thoughts
- Quick-thoughts

Understanding the word2vec model

Word2vec is one of the most popular and widely used models for generating the word embeddings. What are word embeddings though? Word embeddings are the vector representations of words in a vector space. The embedding generated by the word2vec model captures the syntactic and semantic meanings of a word. Having a meaningful vector representation of a word helps the neural network to understand the word better.

For instance, let's consider the following text: *Archie used to live in New York, he then moved to Santa Clara. He loves apples and strawberries.*

Word2vec model generates the vector representation for each of the words in the preceding text. If we project and visualize the vectors in embedding space, we can see how all the similar words are plotted close together. As you can see in the following figure, words *apples* and *strawberries* are plotted close together, and *New York* and *Santa Clara* are plotted close together. They are plotted close together because the word2vec model has learned that *apples* and *strawberries* are similar entities that is, fruits and *New York* and *Santa Clara* are similar entities, that is *cities*, and so their vectors (embeddings) are similar to each other, and which is why the distance between them is less:

Thus, with word2vec model, we can learn the meaningful vector representation of a word which helps the neural networks to understand what the word is about. Having a good representation of a word would be useful in various tasks. Since our network can understand the contextual and syntactic meaning of words, this will branch out to various use cases such as text summarization, sentiment analysis, text generation, and more.

Okay. But how do the word2vec model learn the word embeddings? There are two types of word2vec models for learning the embeddings of a word:

1. CBOW model
2. Skip-gram model

We will go into detail and learn how each of these models learns the vector representations of a word.

Understanding the CBOW model

Let's say we have a neural network with an input layer, a hidden layer, and an output layer. The goal of the network is to predict a word given its surrounding words. The word that we are trying to predict is called the **target word** and the words surrounding the target word are called the **context words**.

How many context words do we use to predict the target word? We use a window of size n to choose the context word. If the window size is 2, then we use two words before and two words after the target word as the context words.

Let's consider the sentence, *The Sun rises in the east* with the word *rises* as the target word. If we set the window size as 2, then we take the words *the* and *sun,* which are the two words before, and *in* and *the* which are the two words after the target word *rises* as context words, as shown in the following figure:

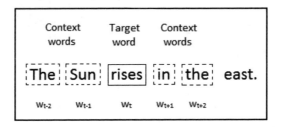

So the input to the network is context words and the output is a target word. How do we feed these inputs to the network? The neural network accepts only numeric input so we cannot feed the raw context words directly as an input to the network. Hence, we convert all the words in the given sentence into a numeric form using the one-hot encoding technique, as shown in the following figure:

the	=	[1 0 0 0 0]
sun	=	[0 1 0 0 0]
rises	=	[0 0 1 0 0]
in	=	[0 0 0 1 0]
east	=	[0 0 0 0 1]

The architecture of the CBOW model is shown in the following figure. As you can see, we feed the context words, *the, sun, in,* and *the,* as inputs to the network and it predicts the target word *rises* as an output:

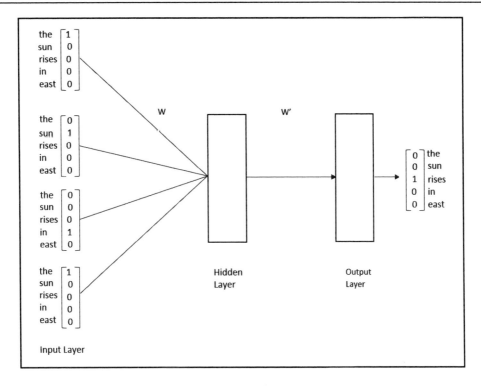

In the initial iteration, the network cannot predict the target word correctly. But over a series of iterations, it learns to predict the correct target word using gradient descent. With gradient descent, we update the weights of the network and find the optimal weights with which we can predict the correct target word.

As we have one input, one hidden, and one output layer, as shown in the preceding figure, we will have two weights:

- Input layer to hidden layer weight, W
- Hidden layer to output layer weight, W'

During the training process, the network will try to find the optimal values for these two sets of weights so that it can predict the correct target word.

It turns out that the optimal weights between the input to a hidden layer W forms the vector representation of words. They basically constitute the semantic meaning of the words. So, after training, we simply remove the output layer and take the weights between the input and hidden layers and assign them to the corresponding words.

After training, if we look at the W matrix, it represents the embeddings for each of the words. So, the embedding for the word *sun* is [0.0, 0.3,0.3,0.6,0.1]:

$$W = \begin{array}{c} \text{the} \\ \text{sun} \\ \text{rises} \\ \text{in} \\ \text{east} \end{array} \begin{bmatrix} 0.01 & 0.02 & 0.1 & 0.5 & 0.37 \\ 0.0 & 0.3 & 0.3 & 0.6 & 0.1 \\ 0.4 & 0.34 & 0.11 & 0.61 & 0.43 \\ 0.1 & 0.11 & 0.1 & 0.17 & 0.369 \\ 0.33 & 0.4 & 0.3 & 0.17 & 0.1 \end{bmatrix}$$

Thus, the CBOW model learns to predict the target word with the given context words. It learns to predict the correct target word using gradient descent. During training, it updates the weights of the network through gradient descent and finds the optimal weights with which we can predict the correct target word. The optimal weights between the input and hidden layers form the vector representations of a word. So, after training, we simply take the weights between the input and hidden layers and assign them as a vector to the corresponding words.

Now that we have an intuitive understanding of the CBOW model, we will go into detail and learn mathematically how exactly the word embeddings are computed.

We learned that weights between the input and the hidden layers basically form the vector representation of the words. But how exactly does the CBOW model predicts the target word? How does it learn the optimal weights using backpropagation? Let's look at this in the next section.

CBOW with a single context word

We learned that, in the CBOW model, we try to predict the target word given the context words, so it takes some C number of context words as an input and returns one target word as an output. In CBOW model with a single context word, we will have only one context word, that is, $C = 1$. So, the network takes only one context word as an input and returns one target word as an output.

Before going ahead, first, let's familiarize ourselves with the notations. All the unique words we have in our corpus is called the **vocabulary**. Considering the example we saw in the *Understanding the CBOW model* section, we have five unique words in the sentence—*the, sun, rises, in,* and *east*. These five words are our vocabulary.

Let V denote the size of the vocabulary (that is, number of words) and N denotes the number of neurons in the hidden layer. We learned that we have one input, one hidden, and one output layer:

- The input layer is represented by $X = \{x_1, x_2, x_3 \ldots x_k \ldots x_V\}$. When we say x_k, it represents the kth input word in the vocabulary.
- The hidden layer is represented by $h = \{h_1, h_2, h_3 \ldots h_i \ldots h_N\}$. When we say h_i, it represents the ith neuron in the hidden layer.
- Output layer is represented by $y = \{y_1, y_2, y_3 \ldots y_j \ldots y_V\}$. When we say y_j it represents the jth output word in the vocabulary.

The dimension of input to hidden layer weight W is $V \times N$ (which is the *size of our vocabulary x the number of neurons in the hidden layer*) and the dimension of hidden to output layer weight, W' is $N \times V$ (that is, the *number of neurons in the hidden layer x the size of the vocabulary*). The representation of the elements of the matrix is as follows:

- W_{ki} represents an element in the matrix between node x_k of the input layer and node h_i of the hidden layer
- W_{ij} represents an element in the matrix between node h_i of the hidden layer and node y_j of the output layer

The following figure will help us to attain clarity on the notations:

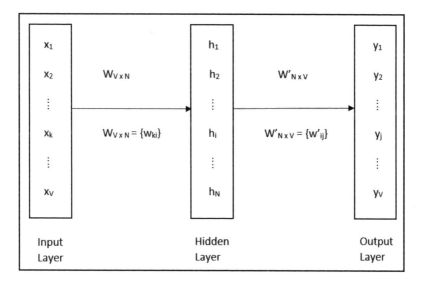

Forward propagation

In order to predict target words given a context word, we need to perform forward propagation.

First, we multiply the input X with the input to hidden layer weight W:

$$h = XW^T$$

We know each of the input words is one-hot encoded, so when we multiply X with W, we basically obtain the kth row of W to h. So, we can directly write as follows:

$$h = W_{(k,.)}$$

$W_{(k,.)}$ basically implies the vector representation of the input word. Let's denote the vector representation for the input word w_I by Z_{w_I}. So, the previous equation can be written as follows:

$$h = Z_{w_I} \tag{1}$$

Now we are in the hidden layer h and we have another set of weight, which is hidden to output layer weight, w'. We know that we have V number of words in our vocabulary and we need to compute the probability for each of the words in our vocabulary to be the target word.

Let u_j denote the score for the jth word in our vocabulary to be a target word. The score u_j is computed by multiplying the value of hidden layer h and the hidden to output layer weight w'. Since we are calculating the score for a word j, we multiply the hidden layer h with the jth column of the matrix, W'_{ij}:

$$u_j = W'^T_{ij} . h$$

The jth column of the weight matrix W'_{ij} basically denotes the vector representation of the word j. Let's denote the vector representation of the jth word by $Z'_{w_j^T}$. So, the preceding equation can be written as follows:

$$u_j = Z'_{w_j^T} . h \tag{2}$$

Substituting equation *(1)* in equation *(2)*, we can write the following:

$$u_j = Z'_{w_j}{}^T \cdot Z_{w_i}$$

Can you infer what the preceding equation is trying to say? We are basically computing the dot product between the input context word representation, Z_{w_i}, and the representation of ith word in our vocabulary, $Z'_{w_j}{}^T$.

Computing the dot product between any two vectors helps us to understand how similar they are. Hence, computing the dot product between Z_{w_i} and $Z'_{w_j}{}^T$ tells us how similar the jth word in our vocabulary is to the input context word. Thus, when the score for a jth word in the vocabulary, u_j is high, then it implies that the word j is similar to the given input word and it is the target word. Similarly, when the score for a jth word in the vocabulary, u_j, is low, then it implies that the word j is not similar to the given input word and it is not the target word.

Thus, u_j basically gives us the score for a word j to be the target word. But instead of having u_j as a raw score, we convert them into probabilities. We know that the softmax function squashes values between 0 to 1, so we can use the softmax function for converting u_j into the probability.

We can write our output as follows:

$$\boxed{y_j = \frac{exp(u_j)}{\sum_{j'=1}^{V} exp(u'_j)}} \quad (3)$$

Here, y_j tells us the probability for a word j to the target word given an input context word. We compute the probability for all the words in our vocabulary and select the word that has a high probability as the target word.

Okay, what is our objective function? that is, how do we compute the loss?

Our goal is to find the correct target word. Let y_j^* denote the probability of the correct target word. So, we need to maximize this probability:

$$max \ y_j^*$$

Instead of maximizing the raw probabilities, we maximize the log probabilities:

$$max \quad log(y_j)^*$$

But why do we want to maximize the log probability instead of the raw probability? Because machines have limitations in representing a floating point of a fraction and when we multiply many probabilities, it will lead to a value that is infinitely small. So, to avoid that, we use log probabilities and it will ensure numerical stability.

Now we have a maximization objective, we need to convert this to a minimization objective so that we can apply our favorite gradient descent algorithm for minimizing the objective function. How can we change our maximization objective to the minimization objective? We can do that by simply adding the negative sign. So our objective function becomes the following:

$$min \quad - log(y_j*)$$

The loss function can be given as follows:

$$L = -log(y_j*) \tag{4}$$

Substituting equation *(3)* in equation *(4)*, we get the following:

$$L = -log(\frac{exp(u_j^*)}{\sum_{j'=1}^{V} exp(u'_j)})$$

According to the logarithm quotient rule, *log(a/b) = log(a) - log(b)*, we can rewrite the previous equation as follows:

$$L = -(log(\exp u_{j^*}) - log(\sum_{j'=1}^{V} \exp u'_j))$$

$$= -(log(\exp u_{j^*}) + log(\sum_{j'=1}^{V} \exp u'_j)$$

We know that *log* and *exp* cancel each other, so we can cancel the *log* and *exp* in the first term and our final loss function becomes the following:

$$L = -u_{j^*} + log(\sum_{j'=1}^{V} exp\, u'_j)$$

Backward propagation

We minimize the loss function using the gradient descent algorithm. So, we backpropagate the network, calculate the gradient of the loss function with respect to weights, and update the weights. We have two sets of weights, input to hidden layer weight W and hidden to output layer weights W'. We calculate gradients of loss with respect to both of these weights and update them according to the weight update rule:

$$W = W - \alpha \frac{\partial L}{\partial W}$$

$$W' = W' - \alpha \frac{\partial L}{\partial W'}$$

In order to better understand the backpropagation, let's recollect the steps involved in the forward propagation:

$$h = XW^T$$

$$u_j = W'^T_{ij} . h$$

$$L = -u_{j^*} + log(\sum_{j'=1}^{V} exp\, u'_j)$$

First, we compute the gradients of loss with respect to the hidden to output layer W'. We cannot calculate the gradient of loss L with respect to W' directly from L, as there is no W' term in the loss function L, so we apply the chain rule as follows:

$$\frac{\partial L}{\partial W'_{ij}} = \frac{\partial L}{\partial u_j} . \frac{\partial u_j}{\partial W'_{ij}}$$

Please refer to the equations of forward propagation to understand how derivatives are calculated.

The derivative of the first term is as follows:

$$\frac{\partial L}{\partial u_j} = e_j \tag{5}$$

Here, e_j is the error term, which is the difference between the actual word and predicted word.

Now, we will calculate the derivative of the second term.

Since we know $u_j = W_{ij}'^T . h$:

$$\frac{\partial u_j}{\partial W_{ij}'} = h$$

Thus, the **gradient of loss L with respect to W'** is given as:

$$\boxed{\frac{\partial L}{\partial W_{ij}'} = e_j . h}$$

Now, we compute the gradient with respect to the input to hidden layer weight W. We cannot calculate the derivative directly from L, as there is no W term in L, so we apply the chain rule as follows:

$$\frac{\partial L}{\partial W_{ki}} = \frac{\partial L}{\partial h_i} . \frac{\partial h_i}{\partial W_{ki}}$$

In order to compute the derivative of the first term in the preceding equation, we again apply the chain rule, as we cannot compute the derivative of L with respect to h_i directly from L:

$$\frac{\partial L}{\partial h_i} = \sum_{j=1}^{V} \frac{\partial L}{\partial u_j} . \frac{\partial u_j}{\partial h_i}$$

From equation (5), we can write:

$$\frac{\partial L}{\partial h_i} = \sum_{j=1}^{V} e_j . \frac{\partial u_j}{\partial h_i}$$

Since we know $u_j = W_{ij}'^T . h$:

$$\frac{\partial L}{\partial h_i} = \sum_{j=1}^{V} e_j . W_{ij}'$$

Instead of having the sum, we can write:

$$\frac{\partial L}{\partial h_i} = LH^T$$

LH^T denotes the sum of the output vector of all words in the vocabulary, weighted by their prediction error.

Let's now calculate the derivative of the second term.

Since we know, $h = XW^T$:

$$\frac{\partial h_i}{\partial W_{ki}} = X$$

Thus, the **gradient of loss** L **with respect to** W is given as:

$$\boxed{\frac{\partial L}{\partial W_{ki}} = LH^T . X}$$

So, our weight update equation becomes the following:

$$W = W - \alpha LH^T . X$$

$$W' = W' - \alpha e_j . h$$

We update the weights of our network using the preceding equation and obtain an optimal weights during training. The optimal input to hidden layer weight, *W*, becomes the vector representation for the words in our vocabulary.

The Python code for `Single_context_CBOW` is as follows:

```python
def Single_context_CBOW(x, label, W1, W2, loss):

    #forward propagation
    h = np.dot(W1.T, x)
    u = np.dot(W2.T, h)
    y_pred = softmax(u)

    #error
    e = -label + y_pred

    #backward propagation
    dW2 = np.outer(h, e)
    dW1 = np.outer(x, np.dot(W2.T, e))

    #update weights
    W1 = W1 - lr * dW1
    W2 = W2 - lr * dW2

    #loss function
    loss += -float(u[label == 1]) + np.log(np.sum(np.exp(u)))

    return W1, W2, loss
```

CBOW with multiple context words

Now that we understood how the CBOW model works with a single word as a context, we will see how it will work when you have multiple words as context words. The architecture of CBOW with multiple input words as a context is shown in the following figure:

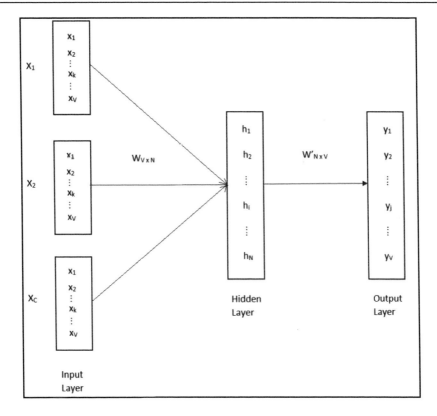

There is not much difference between the multiple words as a context and a single word as a context. The difference is that, with multiple contexts words as inputs, we take the average of all the input context words. That is, as a first step, we forward propagate the network and compute the value of h by multiplying input X and weights w, as we saw in the CBOW *with a single context word* section:

$$h = XW^T$$

But, here, since we have multiple context words, we will have multiple inputs (that is $X_1, X_2, \ldots X_C$), where c is the number of context words, and we simply take the average of them and multiply with the weight matrix, shown as follows:

$$h = \frac{(X_1 + X_2 + \ldots + X_C)}{C} W^T$$

$$h = \frac{1}{C}(X_1 W^T + X_2 W^T + \ldots X_C W^T)$$

Similar to what we learned in the *CBOW with single context word* section, $x_1 w^T$ represents the vector representation of the input context word w_1. $x_2 w^T$ represents the vector representation of the input word w_2, and so on.

We denote the representation of the input context word w_1 by z_{w_1}, the representation of the input context word w_2 by z_{w_2}, and so on. So, we can rewrite the preceding equation as:

$$h = \frac{1}{C}(Z_{w_1} + Z_{w_2} + \dots Z_{w_c}^T) \tag{6}$$

Here, C represents the number of context words.

Computing the value of u_j is the same as we saw in the previous section:

$$u_j = Z'_{w_j^T} \cdot h \tag{7}$$

Here, $Z'_{w_j^T}$ denotes the vector representation of the jth word in the vocabulary.

Substituting equation *(6)* in equation *(7)*, we write the following:

$$u_j = Z'_{w_j^T} \cdot \frac{1}{C}(Z_{w_1} + Z_{w_2} + \dots Z_{w_c}^T)$$

The preceding equation gives us the similarity between the jth word in the vocabulary and the average representations of given input context words.

The loss function is the same as we saw in the single word context and it is given as:

$$L = -u_{j^*} + log(\sum_{j'=1}^{V} \exp u'_j)$$

Now, there is a small difference in backpropagation. We know that in backpropagation we compute gradients and update our weights according to the weight update rule. Recall that, in the previous section, this is how we update the weights:

$$W = W - \alpha LH^T . X$$

$$W' = W' - \alpha e_j . h$$

Since, here, we have multiple context words as an input, we take an average of context words while computing W:

$$W = W - \alpha LH^T . \frac{(X_1 + X_2 + .. X_C)}{C}$$

Computing W' is the same as we saw in the previous section:

$$W' = W' - \alpha e_j . h$$

So, in a nutshell, in the multi-word context, we just take the average of multiple context input words and build the model as we did in the single word context of CBOW.

Understanding skip-gram model

Now, let's look at another interesting type of the word2vec model, called skip-gram. Skip-gram is just the reverse of the CBOW model,. That is in a skip-gram model, we try to predict the context words given the target word as an input. As shown in the following figure, we can notice that we have the target word as *rises* and we need to predict the context words *the, sun, in,* and *the*:

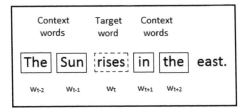

Similar to the CBOW model, we use the window size to determine how many context words we need to predict. The architecture of the skip-gram model is shown in the following figure.

As we can see that it takes the single target word as input and tries to predict the multiple context words:

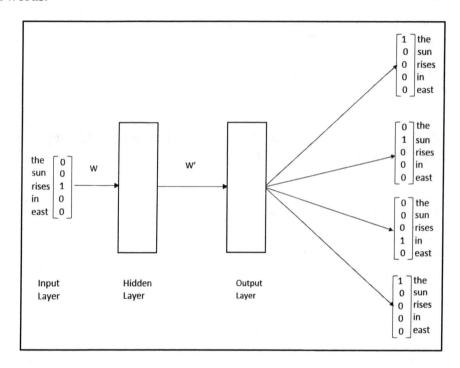

In the skip-gram model, we try to predict the context words based on the target word. So, it takes one target word as an input and returns C context words as output, as shown in the above figure. So, after training the skip-gram model to predict the context words, the weights between our input to hidden layer W becomes the vector representation of the words, just like we saw in the CBOW model.

Now that we have a basic understanding of the skip-gram model, let us dive into detail and learn how they work.

Forward propagation in skip-gram

First, we will understand how forward propagation works in the skip-gram model. Let's use the same notations we used in the CBOW model. The architecture of the skip-gram model is shown in the following figure. As you can see, we feed only one target word X as an input and it returns the C context words as an output Y:

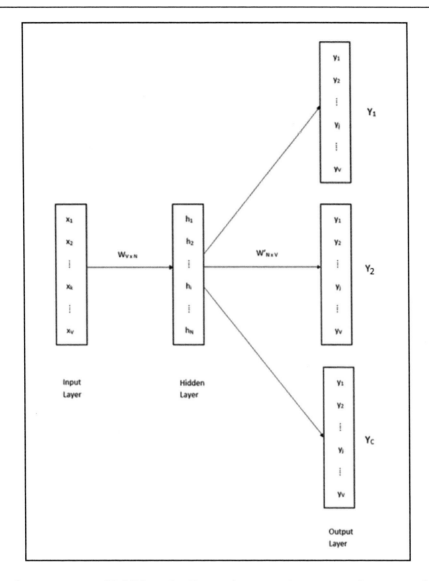

Similar to what we saw in CBOW, in the *Forward propagation* section, first we multiply our input X with the input to hidden layer weights W:

$$h = XW^T$$

We can directly rewrite the preceding equation as:

$$h = Z w_I$$

Here, Z_{w_I} implies the vector representation for the input word w_I.

Next, we compute u_j, which implies a similarity score between the word j^{th} word in our vocabulary and the input target word. Similar to what we saw in the CBOW model, u_j can be given as:

$$u_j = W_{ij}'^T . h$$

We can directly rewrite the above equation as:

$$u_j = Z'_{w_j^T} . h$$

Here, $Z'_{w_j^T}$ implies the vector representation of the word j.

But, unlike the CBOW model where we just predicted the one target word, here we are predicting the C number of context words. So, we can rewrite the above equation as:

$$u_{c,j} = Z'_{w_j^T} . h \quad for \ c = 1, 2, 3 .... C$$

Thus, $u_{c,j}$ implies the score for the jth word in the vocabulary to be the context word c. That is:

- $u_{1,j}$ implies the score for the word j to be the first context word
- $u_{2,j}$ implies the score for the word j to be the second context word
- $u_{3,j}$ implies the score for the word j to be the third context word

And since we want to convert our scores to probabilities, we apply the softmax function and compute $y_{c,j}$:

$$y_{c,j} = \frac{exp(u_{c,j})}{\sum_{j'=1}^{V} exp(u_{j'})} \tag{8}$$

Here, $y_{c,j}$ implies the probability of the jth word in the vocabulary to be the context word c.

Now, let us see how to compute the loss function. Let $y^*_{c,j}$ denote the probability of the correct context word. So, we need to maximize this probability:

$$max \quad y^*_{c,j}$$

Instead of maximizing raw probabilities, maximize the log probabilities:

$$max \quad log(y^*_{c,j})$$

Similar to what we saw in the CBOW model, we convert this into the minimization objective function by adding the negative sign:

$$min \quad - log(y^*_{c,j})$$

Substituting equation *(8)* in the preceding equation, we can write the following:

$$L = - \log \frac{\exp(u_{c,j^*})}{\sum_{j'=1}^{V} \exp(u_{j'})}$$

Since we have C context words, we take the product sum of the probabilities as:

$$L = - \log \prod_{c=1}^{C} \frac{\exp(u_{c,j^*})}{\sum_{j'=1}^{V} \exp(u_{j'})}$$

So, according to logarithm rules, we can rewrite the above equation and our final loss function becomes:

$$L = - \sum_{c=1}^{C} u_{c,j^*} + C . \log \sum_{j'=1}^{V} \exp(u_{j'})$$

Look at the loss function of the CBOW and skip-gram models. You'll notice that the only difference between the CBOW loss function and skip-gram loss function is the addition of the context word c.

Backward propagation

We minimize the loss function using the gradient descent algorithm. So, we backpropagate the network, calculate the gradient of the loss function with respect to weights, and update the weights according to the weight update rule.

First, we compute the gradient of loss with respect to hidden to output layer W'. We cannot calculate the derivative of loss with respect to W' directly from L as it has no W' term in it, so we apply the chain rule as shown below. It is basically the same as what we saw in the CBOW model, except that here we sum over all the context words:

$$\frac{\partial L}{\partial W'_{ij}} = \sum_{c=1}^{C} \frac{\partial L}{\partial u_{c,j}} \cdot \frac{\partial u_{c,j}}{\partial W'_{ij}}$$

First, let's compute the first term:

$$\frac{\partial L}{\partial u_j} = e_{c,j}$$

We know that $e_{c,j}$ is the error term, which is the difference between the actual word and the predicted word. For notation simplicity, we can write this sum over all the context words as:

$$EI_J = \sum_{c=1}^{C} e_{c,j}$$

So, we can say that:

$$\frac{\partial L}{\partial u_j} = EI_j$$

Now, let's compute the second term. Since we know $u_j = W'^{T}_{ij} \cdot h$, we can write:

$$\frac{\partial u_j}{\partial W'_{ij}} = h$$

Thus, the **gradient of loss L with respect to** W' is given as follows:

$$\boxed{\frac{\partial L}{\partial W'_{ij}} = EI_j \cdot h}$$

Now, we compute the gradient of loss with respect to the input to hidden layer weight W. It is simple and exactly same as we saw in the CBOW model:

$$\frac{\partial L}{\partial W_{ki}} = \frac{\partial L}{\partial h_i} \cdot \frac{\partial h_i}{\partial W_{ki}}$$

Thus, the **gradient of loss L with respect to** w is given as:

$$\boxed{\frac{\partial L}{\partial W_{ki}} = LH^T . X}$$

After computing the gradients, we update our weights W and W' as:

$$W' = W' - \alpha\, EI_j . h$$

$$W = W - \alpha\, LH^T . X$$

Thus, while training the network, we update the weights of our network using the preceding equation and obtain optimal weights. The optimal weight between the input to hidden layer, W becomes the vector representation for the words in our vocabulary.

Various training strategies

Now, we will look at different training strategies which can optimize and increase the efficiency of our word2vec model.

Hierarchical softmax

In both the CBOW and skip-gram models, we used the softmax function for computing the probability of the occurrence of a word. But computing the probability using the softmax function is computationally expensive. Say, we are building a CBOW model; we compute the probability of the jth word in our vocabulary to be the target word as:

$$y_j = \frac{exp(u_j)}{\sum_{j'=1}^{V} exp(u'_j)}$$

If you look at the preceding equation, we are basically driving the exponent of the u_j with the exponent of all the words u'_j in the vocabulary. Our complexity would be $O(V)$, where V is the vocabulary size. When we train the word2vec model with a vocabulary comprising millions of words, it is definitely going to be computationally expensive. So, to combat this problem, instead of using the softmax function, we use the hierarchical softmax function.

The hierarchical softmax function uses a Huffman binary search tree and significantly reduces the complexity to $O(log_2(V))$. As shown in the following diagram, in hierarchical softmax, we replace the output layer with a binary search tree:

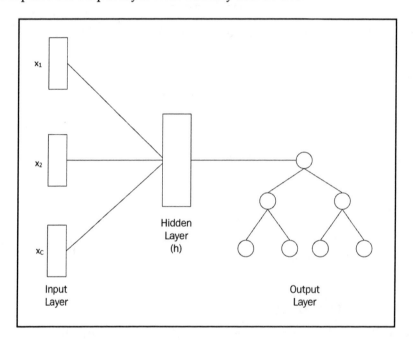

Each leaf node in the tree represents a word in the vocabulary and all the intermediate nodes represent the relative probability of their child node.

How do we compute the probability of a target word given a context word? We simply traverse the tree by making a decision whether to turn left or right. As shown in the following figure, the probability of the word *flew* to be the target word, given some context word c, is computed as a product of the probabilities along the path:

$$p(flew|c) = p_{n_0}(left|c) * p_{n_1}(right|c)$$

$$p(flew|c) = 0.6 * 0.8 = 0.48$$

The probability of the target word is shown as follows:

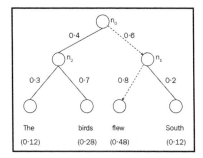

But how do we compute these probabilities? Each node n has an embedding associated with it (say, v'_n). To compute the probability for a node, we multiply the node's embedding v'_n with hidden layer value h and apply a sigmoid function. For instance, the probability of a node n to take a right, given a context word c, is computed as:

$$p(right|n, c) = \sigma(v'_n . h)$$

Once we computed the probability of taking right, we can easily compute the probability of taking left by simply subtracting the probability of taking right from 1:

$$p(left|n, c) = 1 - p(right|n, c)$$

If we sum the probability of all the leaf nodes, then it equals to 1, meaning that our tree is already normalized, and to find a probability of a word, we need to evaluate only log_2 nodes.

Negative sampling

Let's say we are building a CBOW model and we have a sentence *Birds are flying in the sky*. Let the context words be *birds, are, in,* and *the* and the target word be *flying*.

We need to update the weights of the network every time it predicts the incorrect target word. So, except for the word *flying*, if a different word is predicted as a target word, then we update the network.

But this is just a small set of vocabulary. Consider the case where we have millions of words in the vocabulary. In that case, we need to perform numerous weight updates until the network predict the correct target word. It is time-consuming and also not an efficient method. So, instead of doing this, we mark the correct target word as a positive class and sample a few words from the vocabulary and mark it as a negative class.

What we are essentially doing here is that we are converting our multinomial class problem to a binary classification problem (that is, instead of trying to predict the target word, the model classifies whether the given word is target word or not).

The probability that the word is chosen as a negative sample is given as:

$$p(w_i) = \frac{frequency(w_i)^{3/4}}{\sum_{j=0}^{n} frequency(w_j)^{3/4}}$$

Subsampling frequent words

In our corpus, there will be certain words that occur very frequently, such as *the, is*, and so on, and there are certain words that occur infrequently. To maintain a balance between these two, we use a subsampling technique. So, we remove the words that occur frequently more than a certain threshold with the probability p, and it can be represented as:

$$p(w_i) = 1 - \sqrt{\frac{t}{f(w_i)}}$$

Here, t is the threshold and $f(w_i)$ is the frequency of the word i.

Building the word2vec model using gensim

Now that we have understood how the word2vec model works, let's see how to build the word2vec model using the `gensim` library. Gensim is one of the popular scientific software packages widely used for building vector space models. It can be installed via `pip`. So, we can just type the following command in the terminal to install the `gensim` library:

```
pip install -U gensim
```

Now that we have installed gensim, we will see how to build the word2vec model using that. You can download the dataset used in this section along with complete code with step by step explanation from GitHub at `http://bit.ly/2Xjndj4`.

First, we will import the necessary libraries:

```
import warnings
warnings.filterwarnings(action='ignore')

#data processing
import pandas as pd
import re
from nltk.corpus import stopwords
stopWords = stopwords.words('english')

#modelling
from gensim.models import Word2Vec
from gensim.models import Phrases
from gensim.models.phrases import Phraser
```

Loading the dataset

Load the dataset:

```
data = pd.read_csv('data/text.csv',header=None)
```

Let's see what we got in our data:

```
data.head()
```

The preceding code generates the following output:

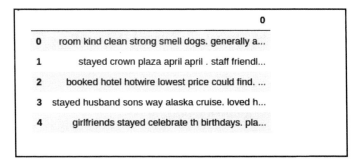

	0
0	room kind clean strong smell dogs. generally a...
1	stayed crown plaza april april . staff friendl...
2	booked hotel hotwire lowest price could find. ...
3	stayed husband sons way alaska cruise. loved h...
4	girlfriends stayed celebrate th birthdays. pla...

Preprocessing and preparing the dataset

Define a function for preprocessing the dataset:

```
def pre_process(text):
    # convert to lowercase
    text = str(text).lower()
    # remove all special characters and keep only alpha numeric characters
and spaces
    text = re.sub(r'[^A-Za-z0-9\s.]',r'',text)
    #remove new lines
    text = re.sub(r'\n',r' ',text)
    # remove stop words
    text = " ".join([word for word in text.split() if word not in
stopWords])
    return text
```

We can see how the preprocessed text looks like by running the following code:

```
pre_process(data[0][50])
```

We get the output as:

```
'agree fancy. everything needed. breakfast pool hot tub nice shuttle
airport later checkout time. noise issue tough sleep through. awhile forget
noisy door nearby noisy guests. complained management later email credit
compd us amount requested would return.'
```

Preprocess the whole dataset:

```
data[0] = data[0].map(lambda x: pre_process(x))
```

The genism library requires input in the form of a list of lists:

```
text = [ [word1, word2, word3], [word1, word2, word3] ]
```

We know that each row in our data contains a set of sentences. So, we split them by '.' and convert them into a list:

```
data[0][1].split('.')[:5]
```

The preceding code generates the following output:

```
['stayed crown plaza april april ',
 ' staff friendly attentive',
 ' elevators tiny ',
 ' food restaurant delicious priced little high side',
 ' course washington dc']
```

Thus, as shown, now, we have the data in a list. But we need to convert them into a list of lists. So, now again we split it by a space `' '`. That is, first, we split the data by `'.'` and then we split them by `' '` so that we can get our data in a list of lists:

```
corpus = []
for line in data[0][1].split('.'):
    words = [x for x in line.split()]
    corpus.append(words)
```

You can see that we have our inputs in the form of a list of lists:

```
corpus[:2]

[['stayed', 'crown', 'plaza', 'april', 'april'], ['staff', 'friendly',
'attentive']]
```

Convert the whole text in our dataset to a list of lists:

```
data = data[0].map(lambda x: x.split('.'))

corpus = []
for i in (range(len(data))):
    for line in data[i]:
        words = [x for x in line.split()]
        corpus.append(words)


print corpus[:2]
```

As shown, we successfully converted the whole text in our dataset into a list of lists:

```
[['room', 'kind', 'clean', 'strong', 'smell', 'dogs'],

['generally', 'average', 'ok', 'overnight', 'stay', 'youre', 'fussy']]
```

Now, the problem we have is that our corpus contains only unigrams and it will not give us results when we give a bigram as an input, for example, *san francisco*.

So we use gensim's `Phrases` functions, which collects all the words that occur together and adds an underscore between them. So, now *san francisco* becomes *san_francisco*.

We set the `min_count` parameter to 25, which implies that we ignore all the words and bigrams that appear less the `min_count`:

```
phrases = Phrases(sentences=corpus,min_count=25,threshold=50)
bigram = Phraser(phrases)

for index,sentence in enumerate(corpus):
    corpus[index] = bigram[sentence]
```

As you can see, now, an underscore has been added to the bigrams in our corpus:

```
corpus[111]

[u'connected', u'rivercenter', u'mall', u'downtown', u'san_antonio']
```

We check one more value from the corpus to see how an underscore is added for bigrams:

```
corpus[9]

[u'course', u'washington_dc']
```

Building the model

Now let's build our model. Let's define some of the important hyperparameters that our model needs:

- The `size` parameter represents the size of the vector, that is, dimensions of our vector, to represent a word. The size can be chosen according to our data size. If our data is very small, then we can set the size to a small value, but if we have a significantly large dataset, then we can set the size to 300. In our case, we set the size to 100.

- The `window_size` parameter represents the distance that should be considered between the target word and its neighboring word. Words exceeding the window size from the target word will not be considered for learning. Typically, a small window size is preferred.

- The `min_count` parameter represents the minimum frequency of words. If the particular word's occurrence is less than a `min_count`, then we can simply ignore that word.

- The `workers` parameter specifies the number of worker threads we need to train the model.

- Setting `sg=1` implies that we use the skip-gram model for training, but if it is set to `sg=0`, then it implies that we use CBOW model for training.

Define all the hyperparameters using following code:

```
size = 100
window_size = 2
epochs = 100
min_count = 2
workers = 4
sg = 1
```

Let's train the model using the `Word2Vec` function from gensim:

```
model = Word2Vec(corpus, sg=1,window=window_size,size=size,
min_count=min_count,workers=workers,iter=epochs)
```

Once, we trained the model successfully, we save them. Saving and loading the model is very simple; we can simply use the `save` and `load` functions, respectively:

```
model.save('model/word2vec.model')
```

We can also `load` the already saved `Word2Vec` model by using the following code:

```
model = Word2Vec.load('model/word2vec.model')
```

Evaluating the embeddings

Let's now evaluate what our model has learned and how well our model has understood the semantics of the text. The `genism` library provides the `most_similar` function, which gives us the top similar words related to the given word.

As you can see in the following code, given `san_diego` as an input, we are getting all the other related city names that are most similar:

```
model.most_similar('san_diego')

[(u'san_antonio', 0.8147615790367126),
 (u'indianapolis', 0.7657858729362488),
 (u'austin', 0.7620342969894409),
 (u'memphis', 0.7541092038154602),
 (u'phoenix', 0.7481759786605835),
 (u'seattle', 0.7471771240234375),
 (u'dallas', 0.7407466769218445),
 (u'san_francisco', 0.7373261451721191),
 (u'la', 0.7354192137718201),
 (u'boston', 0.7213659286499023)]
```

We can also apply arithmetic operations on our vectors to check how accurate our vectors are as follows:

```
model.most_similar(positive=['woman', 'king'], negative=['man'], topn=1)

[(u'queen', 0.7255150675773621)]
```

We can also find the words that do not match in the given set of words; for instance, in the following list called `text`, other than the word `holiday`, all others are city names. Since Word2Vec has understood this difference, it returns the word `holiday` as the one that does not match with the other words in the list as shown:

```
text = ['los_angeles','indianapolis', 'holiday', 'san_antonio','new_york']

model.doesnt_match(text)

'holiday'
```

Visualizing word embeddings in TensorBoard

In the previous section, we learned how to build word2vec model for generating word embeddings using gensim. Now, we will see how to visualize those embeddings using TensorBoard. Visualizing word embeddings help us to understand the projection space and also helps us to easily validate the embeddings. TensorBoard provides us a built-in visualizer called the **embedding projector** for interactively visualizing and analyzing the high-dimensional data like our word embeddings. We will learn how can we use the TensorBoard's projector for visualizing the word embeddings step by step.

Import the required libraries:

```
import warnings
warnings.filterwarnings(action='ignore')

import tensorflow as tf
from tensorflow.contrib.tensorboard.plugins import projector
tf.logging.set_verbosity(tf.logging.ERROR)

import numpy as np
import gensim
import os
```

Load the saved model:

```
file_name = "model/word2vec.model"
model = gensim.models.keyedvectors.KeyedVectors.load(file_name)
```

After loading the model, we will save the number of words in our model to the `max_size` variable:

```
max_size = len(model.wv.vocab)-1
```

We know that the dimension of word vectors will be $V \times N$. So, we initialize a matrix named `w2v` with the shape as our `max_size`, which is the vocabulary size, and the model's first layer size, which is the number of neurons in the hidden layer:

```
w2v = np.zeros((max_size,model.layer1_size))
```

Now, we create a new file called `metadata.tsv`, where we save all the words in our model and we store the embedding of each word in the `w2v` matrix:

```
if not os.path.exists('projections'):
    os.makedirs('projections')
with open("projections/metadata.tsv", 'w+') as file_metadata:
    for i, word in enumerate(model.wv.index2word[:max_size]):
        #store the embeddings of the word
        w2v[i] = model.wv[word]
        #write the word to a file
        file_metadata.write(word + '\n')
```

Next, we initialize the TensorFlow session:

```
sess = tf.InteractiveSession()
```

Initialize the TensorFlow variable called `embedding` that holds the word embeddings:

```
with tf.device("/cpu:0"):
    embedding = tf.Variable(w2v, trainable=False, name='embedding')
```

Initialize all the variables:

```
tf.global_variables_initializer().run()
```

Create an object to the `saver` class, which is actually used for saving and restoring variables to and from our checkpoints:

```
saver = tf.train.Saver()
```

Using `FileWriter`, we can save our summaries and events to our event file:

```
writer = tf.summary.FileWriter('projections', sess.graph)
```

Now, we initialize the projectors and add the `embeddings`:

```
config = projector.ProjectorConfig()
embed = config.embeddings.add()
```

Next, we specify our `tensor_name` as `embedding` and `metadata_path` to the `metadata.tsv` file, where we have the words:

```
embed.tensor_name = 'embedding'
embed.metadata_path = 'metadata.tsv'
```

And, finally, save the model:

```
projector.visualize_embeddings(writer, config)

saver.save(sess, 'projections/model.ckpt', global_step=max_size)
```

Now, open the terminal and type the following command to open the `tensorboard`:

```
tensorboard --logdir=projections --port=8000
```

Once the **TensorBoard** is opened, go to the **PROJECTOR** tab. We can see the output, as shown in the following screenshot. As you can notice, when we type the word `delighted`, we can see all the related words, such as `pleasant`, `surprise`, and many more similar words, adjacent to that:

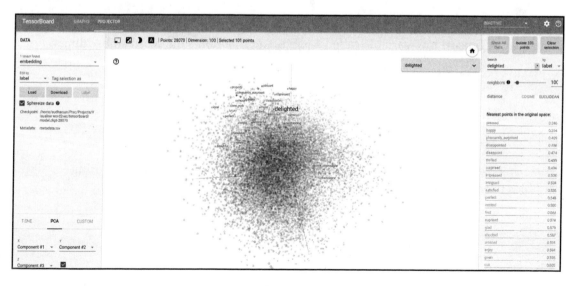

Doc2vec

So far, we have seen how to generate embeddings for a word. But how can we generate the embeddings for a document? A naive method would be to compute a word vector for each word in the document and take an average of it. Mikilow and Le introduced a new method for generating the embeddings for documents instead of just taking the average of word embeddings. They introduced two new methods, called PV-DM and PV-DBOW. Both of these methods just add a new vector, called **paragraph id**. Let's see how exactly these two methods work.

Paragraph Vector – Distributed Memory model

PV-DM is similar to the CBOW model, where we try to predict the target word given a context word. In PV-DM, along with word vectors, we introduce one more vector, called the paragraph vector. As the name suggests, the paragraph vector learns the vector representation of the whole paragraph and it captures the subject of the paragraph.

As shown in the following figure, each paragraph is mapped to a unique vector and each word is also mapped to a unique vector. So, in order to predict the target word, we combine the word vectors and paragraph vector by either concatenating or averaging them:

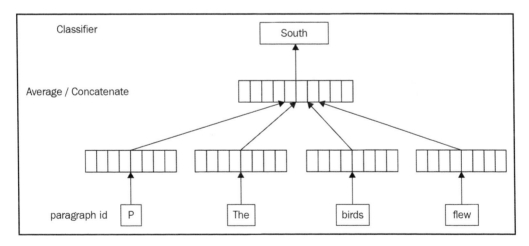

But having said all that, how is the paragraph vector useful in predicting the target word? What is really the use of having the paragraph vector? We know that we try to predict the target word based on the context words. Context words are of a fixed length and they are sampled within a sliding window from a paragraph.

Along with context words, we also make use of the paragraph vector for predicting the target word. Since the paragraph vector contains information about the subject of the paragraph, they contain meanings that the context words do not hold. That is, context word contains information about the particular words alone but the paragraph vector contains the information about the whole paragraph. So, we can think of the paragraph vector as a new word that is used along with context words for predicting the target word.

Paragraph vector is the same for all the context words sampled from the same paragraph and are not shared across paragraphs. Let's say that we have three paragraphs, *p1*, *p2*, and *p3*. If the context is sampled from a paragraph *p1*, then the *p1* vector is used to predict the target word. If a context is sampled from paragraph *p2*, then the *p2* vector is used. Thus, Paragraph vectors are not shared across paragraphs. However, word vectors are shared across all paragraphs. That is, the vector for the word *sun* is the same across all the paragraphs. We call our model as a distributed memory model of paragraph vectors, as our paragraph vectors serve as a memory that holds information that is missing from the current context words.

So, both of the paragraph vectors and word vectors are learned using stochastic gradient descent. On each iteration, we sample context words from a random paragraph, try to predict the target word, calculate the error, and update the parameters. After training, the paragraph vectors capture the embeddings of the paragraphs (documents).

Paragraph Vector – Distributed Bag of Words model

PV-DBOW is similar to the skip-gram model, where we try to predict the context words based on the target word:

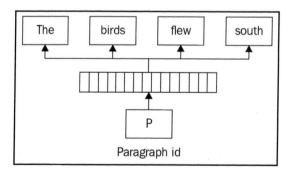

Unlike previous methods, here we do not try to predict the next words. Instead, we use a paragraph vector to classify the words in the document. But how do they work? We train the model to understand whether the word belongs to a paragraph or not. We sample some set of words and then feed it to a classifier, which tells us whether the words belong to a particular paragraph or not, and in such a way we learn the paragraph vector.

Finding similar documents using doc2vec

Now, we will see how to perform document classification using doc2vec. In this section, we will use the 20 `news_dataset`. It consists of 20,000 documents over 20 different news categories. We will use only four categories: `Electronics`, `Politics`, `Science`, and `Sports`. So, we have 1,000 documents under each of these four categories. We rename the documents with a prefix, `category_`. For example, all science documents are renamed as `Science_1`, `Science_2`, and so on. After renaming them, we combine all the documents and place them in a single folder. The combined data, along with complete code is available at available as a Jupyter Notebook on GitHub at `http://bit.ly/2KgBWYv`.

Now, we train our doc2vec model to classify and find similarities between these documents.

First, we import all the necessary libraries:

```
import warnings
warnings.filterwarnings('ignore')

import os
import gensim
from gensim.models.doc2vec import TaggedDocument

from nltk import RegexpTokenizer
from nltk.corpus import stopwords

tokenizer = RegexpTokenizer(r'\w+')
stopWords = set(stopwords.words('english'))
```

Now, we load all our documents and save the document names in the `docLabels` list and the document content in a list called `data`:

```
docLabels = []
docLabels = [f for f in os.listdir('data/news_dataset') if
f.endswith('.txt')]

data = []
for doc in docLabels:
      data.append(open('data/news_dataset/'+doc).read())
```

You can see in `docLabels` list we have all our documents' names:

```
docLabels[:5]

['Electronics_827.txt',
 'Electronics_848.txt',
 'Science829.txt',
 'Politics_38.txt',
 'Politics_688.txt']
```

Define a class called `DocIterator`, which acts as an iterator to run over all the documents:

```
class DocIterator(object):
    def __init__(self, doc_list, labels_list):
        self.labels_list = labels_list
        self.doc_list = doc_list

    def __iter__(self):
        for idx, doc in enumerate(self.doc_list):
            yield TaggedDocument(words=doc.split(), tags=
[self.labels_list[idx]])
```

Create an object called `it` to the `DocIterator` class:

```
it = DocIterator(data, docLabels)
```

Now, let's build the model. Let's first, define some of the important hyperparameters of the model:

- The `size` parameter represents our embedding size.
- The `alpha` parameter represents our learning rate.
- The `min_alpha` parameter implies that our learning rate, `alpha`, will decay to `min_alpha` during training.

- Setting `dm=1` implies that we use the distributed memory (PV-DM) model and if we set `dm=0`, it implies that we use the distributed bag of words (PV-DBOW) model for training.
- The `min_count` parameter represents the minimum frequency of words. If the particular word's occurrence is less than a `min_count`, than we can simply ignore that word.

These hyperparameters are defined as:

```
size = 100
alpha = 0.025
min_alpha = 0.025
dm = 1
min_count = 1
```

Now let's define the model using `gensim.models.Doc2ec()` class:

```
model = gensim.models.Doc2Vec(size=size, min_count=min_count, alpha=alpha,
min_alpha=min_alpha, dm=dm)
model.build_vocab(it)
```

Train the model:

```
for epoch in range(100):
    model.train(it,total_examples=120)
    model.alpha -= 0.002
    model.min_alpha = model.alpha
```

After training, we can save the model, using the `save` function:

```
model.save('model/doc2vec.model')
```

We can load the saved model, using the `load` function:

```
d2v_model = gensim.models.doc2vec.Doc2Vec.load('model/doc2vec.model')
```

Now, let's evaluate our model's performance. The following code shows that when we feed the `Sports_1.txt` document as an input, it will input all the related documents with the corresponding scores:

```
d2v_model.docvecs.most_similar('Sports_1.txt')

[('Sports_957.txt', 0.719024658203125),
 ('Sports_694.txt', 0.6904895305633545),
 ('Sports_836.txt', 0.6636477708816528),
 ('Sports_869.txt', 0.657712459564209),
 ('Sports_123.txt', 0.6526877880096436),
```

```
('Sports_4.txt', 0.6499642729759216),
('Sports_749.txt', 0.6472041606903076),
('Sports_369.txt', 0.64080250026321411),
('Sports_167.txt', 0.6392412781715393),
('Sports_104.txt', 0.6284008026123047)]
```

Understanding skip-thoughts algorithm

Skip-thoughts is one of the popular unsupervised learning algorithms for learning the sentence embedding. We can see skip-thoughts as an analogy to the skip-gram model. We learned that in the skip-gram model, we try to predict the context word given a target word, whereas in skip-thoughts, we try to predict the context sentence given a target sentence. In other words, we can say that skip-gram is used for learning word-level vectors and skip-thoughts is used for learning sentence-level vectors.

The algorithm of skip-thoughts is very simple. It consists of an encoder-decoder architecture. The role of the encoder is to map the sentence to a vector and the role of the decoder is to generate the surrounding sentences that is the previous and next sentence of the given input sentence. As shown in the following diagram, the skip-thoughts vector consists of one encoder and two decoders, called a previous decoder and next decoder:

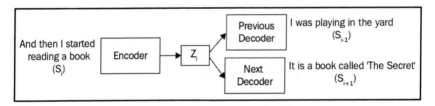

The working of an encoder and decoder is discussed next:

- **Encoder**: An encoder takes the words in a sentence sequentially and generates the embeddings. Let's say we have a list of sentences. $S = [s_1, s_2, s_3 \ldots s_n]$. w_i^t denotes the tth word in a sentence s_i and z_i^t denotes its word embeddings. So the hidden state of an encoder is interpreted as a sentence representation.

- **Decoder**: There are two decoders, called a previous decoder and next decoder. As the name suggests, the previous decoder is used to generate the previous sentence, and the next decoder is used to generate the next sentence. Let's say we have a sentence s_i and its embeddings are z_i. Both of the decoders take the embeddings z_i as an input and the previous decoder tries to generate the previous sentence, s_{i-1}, and the next decoder tries to generate the next sentence, s_{i+1}.

So, we train our model by minimizing the reconstruction error of both the previous and next decoders. Because when the decoders reconstruct/generate correct previous and next sentences, it means that we have a meaningful sentence embedding z_i. We send the reconstruction error to the encoder, so that encoder can optimize the embeddings and send better representations to the decoder. Once we have trained our model, we use our encoder to generate the embedding for a new sentence.

Quick-thoughts for sentence embeddings

Quick-thoughts is another interesting algorithm for learning the sentence embeddings. In skip-thoughts, we saw how we used the encoder-decoder architecture to learn the sentence embeddings. In quick-thoughts, we try to learn whether a given sentence is related to the candidate sentence. So, instead of using a decoder, we use a classifier to learn whether a given input sentence is related to the candidate sentence.

Let s be the input sentence and S_{cand} be the set of candidate sentences containing both valid context and invalid context sentences related to the given input sentence s. Let s_{cand} be any candidate sentence from the S_{cand}.

We use two encoding functions, f and g. The role of these two functions, f and g, is to learn the embeddings, that is, to learn the vector representations of a given sentence s and candidate sentence s_{cand}, respectively.

Once these two functions generate the embeddings, we use a classifier c, which returns the probability for each candidate sentence to be related to the given input sentence.

As shown in the following figure, the probability of the second candidate sentence s_{cand2} is high, as it is related to the given input sentence s:

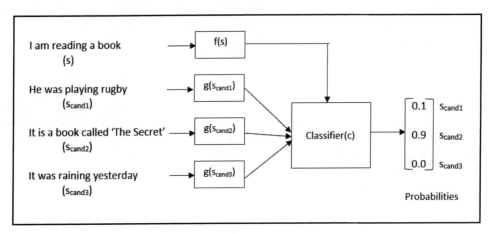

Thus, the probability that s_{cand} is a correct sentence ,that is, s_{cand} is related to the given input sentence s is computed as:

$$p(s_{cand}|s, S_{cand}) = \frac{exp[c(f(s), g(s_{cand}))]}{\sum_{s' \in S_{cand}} exp[c(f(s), g(s'))]}$$

Here, c is a classifier.

The goal of our classifier is to identify the valid context sentence related to a given input sentence s. So, our cost function is to maximize the probability of finding the correct context sentence for the given input sentence s. If it classifies the sentence correctly, then it means that our encoders learned the better representation of the sentence.

Summary

We started off the chapter by understanding word embeddings and we looked at two different types of Word2Vec model, called CBOW, where we try to predict the target word given the context word, and skip-gram, where we try to predict the context word given the target word.

Then, we learned about various training strategies in Word2Vec. We looked at hierarchical softmax, where we replace the output layer of the network with a Huffman binary tree and reduce the complexity to $\mathcal{O}(log_2(V))$. We also learned about negative sampling and subsampling frequent word methods. Then we understood how to build the Word2Vec model using a gensim library and how to project the high-dimensional word embeddings to visualize them in TensorBoard. Going forward, we studied how the doc2vec model works with two types of doc2vec models—PV-DM and PV-DBOW. Following this, we learned about the skip-thoughts model, where we learn the embedding of a sentence by predicting the previous and next sentences of the given sentence and we also explored the quick-thoughts model at the end of the chapter.

In the next chapter, we will learn about generative models and how generative models are used to generate images.

Questions

Let's evaluate our newly acquired knowledge by answering the following questions:

1. What is the difference between the skip-gram and CBOW models?
2. What is the loss function of the CBOW model?
3. What is the need for negative sampling?
4. Define PV-DM.
5. What is the role of the encoder and decoder in the skip-thoughts vector?
6. What are quick thoughts vector?

Further reading

Explore the following links to gain more insights into learning representation of text:

- *Distributed Representations of Words and Phrases and their Compositionality*, by Tomas Mikolov, et al., `https://papers.nips.cc/paper/5021-distributed-representations-of-words-and-phrases-and-their-compositionality.pdf`
- *Distributed Representations of Sentences and Documents*, by Quoc Le and Tomas Mikolov, `https://cs.stanford.edu/~quocle/paragraph_vector.pdf`
- *Skip-thought Vectors*, by Ryan Kiros, et al., `https://arxiv.org/pdf/1506.06726.pdf`
- *An Efficient Framework for Learning Sentence Representations*, by Lajanugen Logeswaran and Honglak Lee, `https://arxiv.org/pdf/1803.02893.pdf`

Section 3: Advanced Deep Learning Algorithms

3

In this section, we will explore advanced deep learning algorithms in detail and we will see how to implement them using TensorFlow. We will learn about **generative adversarial networks (GANs)** and autoencoders. We will explore their types and applications.

The following chapters are included in this section:

- Chapter 8, *Generating Images Using GANs*
- Chapter 9, *Learning More about GANs*
- Chapter 10, *Reconstructing Inputs Using Autoencoders*
- Chapter 11, *Exploring Few-Shot Learning Algorithms*

8
Generating Images Using GANs

So far, we have learned about the discriminative model, which learns to discriminate between the classes. That is, given an input, it tells us which class they belong to. For instance, to predict whether an email is a spam or ham, the model learns the decision boundary that best separates the two classes (spam and ham), and when a new email comes in they can tell us which class the new email belongs to.

In this chapter, we will learn about a generative model that learns the class distribution, that is, the characteristics of the classes rather than learning the decision boundary. As the name suggests, with the generative models, we can generate new data points similar to the data points present in the training set.

We will start off the chapter by understanding the difference between the discriminative and generative models in detail. Then, we will deep dive into one of the most popularly used generative algorithms, called **Generative Adversarial Networks** (**GANs**). We will understand how GANs work and how they are used to generate new data points. Going ahead, we will explore the architecture of GANs and we will learn about the loss function. Later, we will see how to implement GANs in TensorFlow to generate handwritten digits.

We shall also scrutinize the **Deep Convolutional Generative Adversarial Network** (**DCGAN**), which acts as a small extension to the vanilla GAN by using convolutional networks in their architecture. Going forward, we will explore **Least Squares GAN** (**LSGAN**), which adopts the least square loss for generating better and quality images.

At the end of the chapter, we will get the hang of **Wasserstein GAN (WGAN)** which uses the Wasserstein metric in the GAN's loss function for better results.

The chapter will cover the following topics:

- Differences between generative and discriminative models
- GANs
- Architecture of GANs
- Building GANs in TensorFlow
- Deep convolutional GANs
- Generating CIFAR images using DCGAN
- Least Squares GANs
- Wasserstein GANs

Differences between discriminative and generative models

Given some data points, the discriminative model learns to classify the data points into their respective classes by learning the decision boundary that separates the classes in an optimal way. The generative models can also classify given data points, but instead of learning the decision boundary, they learn the characteristics of each of the classes.

For instance, let's consider the image classification task for predicting whether a given image is an apple or an orange. As shown in the following figure, to classify between apple and orange, the discriminative model learns the optimal decision boundary that separates the apples and oranges classes, while generative models learn their distribution by learning the characteristics of the apple and orange classes:

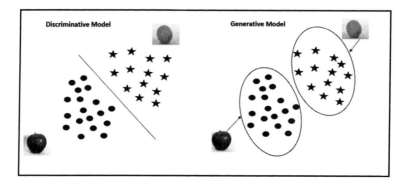

To put it simply, discriminative models learn to find the decision boundary that separates the classes in an optimal way, while the generative models learn about the characteristics of each class.

Discriminative models predict the labels conditioned on the input $p(y|x)$, whereas generative models learn the joint probability distribution $p(x,y)$. Examples of discriminative models include logistic regression, **Support Vector Machine (SVM)**, and so on, where we can directly estimate $p(y|x)$ from the training set. Examples of generative models include **Markov random fields** and **naive Bayes**, where first we estimate $p(x,y)$ to determine $p(y|x)$:

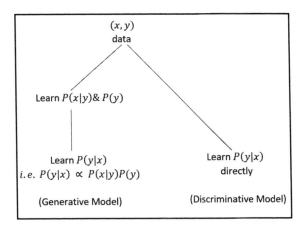

Say hello to GANs!

GAN was first introduced by Ian J Goodfellow, Jean Pouget-Abadie, Mehdi Mirza, Bing Xu, David Warde-Farley, Sherjil Ozair, Aaron Courville, and Yoshua Bengio in their paper, *Generative Adversarial Networks*, in 2014.

GANs are used extensively for generating new data points. They can be applied to any type of dataset, but they are popularly used for generating images. Some of the applications of GANs include generating realistic human faces, converting grayscale images to colored images, translating text descriptions into realistic images, and many more.

Yann LeCun said the following about GANs:

> *"The coolest idea in deep learning in the last 20 years."*

GANs have evolved so much in recent years that they can generate a very realistic image. The following figure shows the evolution of GANs in generating images over the course of five years:

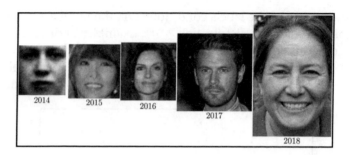

Excited about GANs already? Now, we will see how exactly they work. Before going ahead, let's consider a simple analogy. Let's say you are the police and your task is to find counterfeit money, and the role of the counterfeiter is to create fake money and cheat the police.

The counterfeiter constantly tries to create fake money in a way that is so realistic that it cannot be differentiated from the real money. But the police have to identify whether the money is real or fake. So, the counterfeiter and the police essentially play a two-player game where one tries to defeat the other. GANs work something like this. They consist of two important components:

- Generator
- Discriminator

You can perceive the generator as analogous to the counterfeiter, while the discriminator is analogous to the police. That is, the role of the generator is to create fake money, and the role of the discriminator is to identify whether the money is fake or real.

Without going into detail, first, we will get a basic understanding of GANs. Let's say we want our GAN to generate handwritten digits. How can we do that? First, we will take a dataset containing a collection of handwritten digits; say, the MNIST dataset. The generator learns the distribution of images in our dataset. Thus, it learns the distribution of handwritten digits in our training set. Once, it learns the distribution of the images in our dataset, and we feed a random noise to the generator, it will convert the random noise into a new handwritten digit similar to the one in our training set based on the learned distribution:

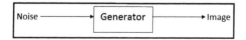

The goal of the discriminator is to perform a classification task. Given an image, it classifies it as real or fake; that is, whether the image is from the training set or the image is generated by the generator:

The generator component of GAN is basically a generative model, and the discriminator component is basically a discriminative model. Thus, the generator learns the distribution of the class and the discriminator learns the decision boundary of a class.

As shown in the following figure, we feed a random noise to the generator, and it then converts this random noise into a new image *similar* to the one we have in our training set, but not *exactly* the same as the images in the training set. The image generated by the generator is called a fake image, and the images in our training set are called real images. We feed both the real and fake images to the discriminator, which tells us the probability of them being real. It returns 0 if the image is fake and 1 if the image is real:

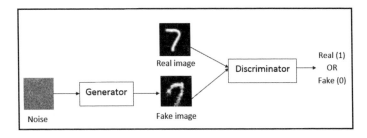

Now that we have a basic understanding of generators and discriminators, we will study each of the components in detail.

Breaking down the generator

The generator component of a GAN is a generative model. When we say the generative model, there are two types of generative models— an **implicit** and an **explicit** density model. The implicit density model does not use any explicit density function to learn the probability distribution, whereas the explicit density model, as the name suggests, uses an explicit density function. GANs falls into the first category. That is, they are an implicit density model. Let's study in detail and understand how GANs are an implicit density model.

Let's say we have a generator, G. It is basically a neural network parametrized by θ_g. The role of the generator network is to generate new images. How do they do that? What should be the input to the generator?

We sample a random noise, z, from a normal or uniform distribution, p_z. We feed this random noise, z, as an input to the generator and then it converts this noise to an image:

$$G(z; \theta_g)$$

Surprising, isn't it? How does the generator converts a random noise to a realistic image?

Let's say we have a dataset containing a collection of human faces and we want our generator to generate a new human face. First, the generator learns all the features of the face by learning the probability distribution of the images in our training set. Once the generator learns the correct probability distribution, it can generate totally new human faces.

But how does the generator learn the distribution of the training set? That is, how does the generator learn the distribution of images of human faces in the training set?

A generator is nothing but a neural network. So, what happens is that the neural network learns the distribution of the images in our training set implicitly; let's call this distribution a generator distribution, p_g. At the first iteration, the generator generates a really noisy image. But over a series of iterations, it learns the exact probability distribution of our training set and learns to generate a correct image by tuning its θ_g parameter.

 It is important to note that we are not using the uniform distribution p_z for learning the distribution of our training set. It is only used for sampling random noise, and we feed this random noise as an input to the generator. The generator network implicitly learns the distribution of our training set and we call this distribution a generator distribution, p_g and that is why we call our generator network an implicit density model.

Breaking down the discriminator

As the name suggests, the discriminator is a discriminative model. Let's say we have a discriminator, D. It is also a neural network and it is parametrized by θ_d.

The goal of the discriminator to discriminate between two classes. That is, given an image x, it has to identify whether the image is from a real distribution or a fake distribution (generator distribution). That is, discriminator has to identify whether the given input image is from the training set or the fake image generated by the generator:

$$D(x; \theta_d)$$

Let's call the distribution of our training set the real data distribution, which is represented by p_r. We know that the generator distribution is represented by p_g.

So, the discriminator D essentially tries to discriminate whether the image x is from p_r or p_g.

How do they learn though?

So far, we just studied the role of the generator and discriminator, but how do they learn exactly? How does the generator learn to generate new realistic images and how does the discriminator learn to discriminate between images correctly?

We know that the goal of the generator is to generate an image in such a way as to fool the discriminator into believing that the generated image is from a real distribution.

In the first iteration, the generator generates a noisy image. When we feed this image to the discriminator, discriminator can easily detect that the image is from a generator distribution. The generator takes this as a loss and tries to improve itself, as its goal is to fool the discriminator. That is, if the generator knows that the discriminator is easily detecting the generated image as a fake image, then it means that it is not generating an image similar to those in the training set. This implies that it has not learned the probability distribution of the training set yet.

So, the generator tunes its parameters in such a way as to learn the correct probability distribution of the training set. As we know that the generator is a neural network, we simply update the parameters of the network through backpropagation. Once it has learned the probability distribution of the real images, then it can generate images similar to the ones in the training set.

Okay, what about the discriminator? How does it learn? As we know, the role of the discriminator is to discriminate between real and fake images.

If the discriminator incorrectly classifies the generated image; that is, if the discriminator classifies the fake image as a real image, then it implies that the discriminator has not learned to differentiate between the real and fake image. So, we update the parameter of the discriminator network through backpropagation to make the discriminator learn to classify between real and fake images.

So, basically, the generator is trying to fool the discriminator by learning the real data distribution, p_r, and the discriminator is trying to find out whether the image is from a real or fake distribution. Now the question is, when do we stop training the network in light of the fact that both generator and discriminator are competing against each other?

Basically, the goal of the GAN is to generate images similar to the one in the training set. Say we want to generate a human face—we learn the distribution of images in the training set and generate new faces. So, for a generator, we need to find the optimal discriminator. What do we mean by that?

We know that a generator distribution is represented by p_g and the real data distribution is represented by p_r. If the generator learns the real data distribution perfectly, then p_g equals p_r, as shown in the following plot:

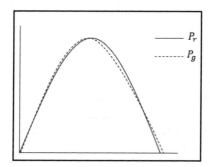

When $p_g = p_r$, then the discriminator cannot differentiate between whether the input image is from a real or a fake distribution, so it will just return 0.5 as a probability, as the discriminator will become confused between the two distributions when they are same.

So, for a generator, the optimal discriminator can be given as follows:

$$D(x) = \frac{p_r(x)}{p_r(x) + p_g(x)} = \frac{1}{2}$$

So, when the discriminator just returns the probability of 0.5 for any image, then we can say that the generator has learned the distribution of images in our training set and fooled the discriminator successfully.

Architecture of a GAN

The architecture of a GAN is shown in the following diagram:

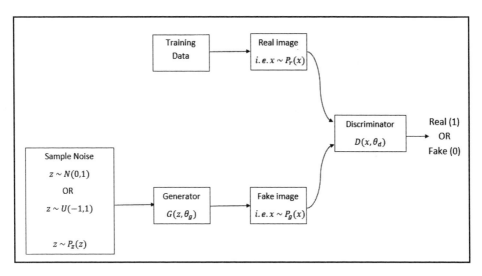

As shown in the preceding diagram, **Generator** G takes the random noise, z, as input by sampling from a uniform or normal distribution and generates a fake image by implicitly learning the distribution of the training set.

We sample an image, x, from the real data distribution, $x \sim p_r(x)$, and fake data distribution, $x \sim p_g(x)$, and feed it to the discriminator, D. We feed real and fake images to the discriminator and the discriminator performs a binary classification task. That is, it returns 0 when the image is fake and 1 when the image is real.

Demystifying the loss function

Now we will examine the loss function of GAN. Before going ahead, let's recap the notations:

- A noise that is fed as an input to the generator is represented by z
- The uniform or normal distribution from which the noise z is sampled is represented by p_z
- An input image is represented by x
- Real data distribution or distribution of our training set is represented by p_r
- Fake data distribution or distribution of the generator is represented by p_g

When we write $x \sim p_r(x)$, it implies that image x is sampled from the real distribution, p_r. Similarly, $x \sim p_g(x)$ denotes that image x is sampled from the generator distribution, p_g, and $z \sim p_z(z)$ implies that the generator input, z, is sampled from the uniform distribution, p_z.

We've learned that both the generator and discriminator are neural networks and both of them update their parameters through backpropagation. We now need to find the optimal generator parameter, θ_g, and the discriminator parameter, θ_d.

Discriminator loss

Now we will look at the loss function of the discriminator. We know that the goal of the discriminator is to classify whether the image is a real or a fake image. Let's denote discriminator by D.

The loss function of the discriminator is given as follows:

$$\max_d L(D, G) = \mathbb{E}_{x \sim p_r(x)}[\log D(x; \theta_d)] + \mathbb{E}_{z \sim p_z(z)}[\log(1 - D(G(z; \theta_g); \theta_d))]$$

What does this mean, though? Let's understand each of the terms one by one.

First term

Let's look at the first term:

$$\mathbb{E}_{x \sim p_r} \log(D(x))$$

Here, $x \sim p_r(x)$ implies that we are sampling input x from the real data distribution, p_r, so x is a real image.

$D(x)$ implies that we are feeding the input image x to the discriminator D, and the discriminator will return the probability of input image x to be a real image. As x is sampled from real data distribution p_r, we know that x is a real image. So, we need to maximize the probability of $D(x)$:

$$max\ D(x)$$

But instead of maximizing raw probabilities, we maximize log probabilities; as we learned in Chapter 7, *Learning Text Representations*, we can write the following:

$$max\ \log D(x)$$

So, our final equation becomes the following:

$$max \; \mathbb{E}_{x \sim p_r(x)}[\log D(x)]$$

$\mathbb{E}_{x \sim p_r(x)}[\log D(x)]$ implies the expectations of the log likelihood of input images sampled from the real data distribution being real.

Second term

Now, let's look at the second term:

$$\mathbb{E}_{z \sim p_z(z)}[\log(1 - D(G(z)))]$$

Here, $z \sim p_z(z)$ implies that we are sampling a random noise z from the uniform distribution p_z. $G(z)$ implies that the generator G takes the random noise z as an input and returns a fake image based on its implicitly learned distribution p_g.

$D(G(z))$ implies that we are feeding the fake image generated by the generator to the discriminator D and it will return the probability of the fake input image being a real image.

If we subtract $D(G(z))$ from 1, then it will return the probability of the fake input image being a fake image:

$$1 - D(G(z))$$

Since we know z is not a real image, the discriminator will maximize this probability. That is, the discriminator maximizes the probability of z being classified as a fake image, so we write:

$$max \; 1 - D(G(z))$$

Instead of maximizing raw probabilities, we maximize the log probability:

$$max \; log(1 - D(G(z)))$$

$\mathbb{E}_{z \sim p_z(z)}[\log(1 - D(G(z)))]$ implies the expectations of the log likelihood of the input images generated by the generator being fake.

Final term

So, combining these two terms, the loss function of the discriminator is given as follows:

$$\max_{d} L(D, G) = \mathbb{E}_{x \sim p_r(x)}[\log D(x; \theta_d)] + \mathbb{E}_{z \sim p_z(z)}[\log(1 - D(G(z; \theta_g); \theta_d))]$$

Here, θ_g and θ_d are the parameters of the generator and discriminator network respectively. So, the discriminator's goal is to find the right θ_d so that it can classify the image correctly.

Generator loss

The loss function of the generator is given as follows:

$$\min_{g} L(D, G) = \mathbb{E}_{z \sim p_z(z)}[\log(1 - D(G(z; \theta_g); \theta_d))]$$

We know that the goal of the generator is to fool the discriminator to classify the fake image as a real image.

In the *Discriminator loss* section, we saw that $\mathbb{E}_{z \sim p_z(z)}[\log(1 - D(G(z)))]$ implies the probability of classifying the fake input image as a fake image, and the discriminator maximizes the probabilities for correctly classifying the fake image as fake.

But the generator wants to minimize this probability. As the generator wants to fool the discriminator, it minimizes this probability of a fake input image being classified as fake by the discriminator. Thus, the loss function of the generator can be expressed as follows:

$$\min_{g} L(D, G) = \mathbb{E}_{z \sim p_z(z)}[\log(1 - D(G(z; \theta_g); \theta_d))]$$

Total loss

We just learned the loss function of the generator and the discriminator combining these two losses, and we write our final loss function as follows:

$$\boxed{\min_{G} \max_{D} L(D, G) = \mathbb{E}_{x \sim p_r(x)}[\log D(x)] + \mathbb{E}_{z \sim p_z(z)}[\log(1 - D(G(z)))]}$$

So, our objective function is basically a min-max objective function, that is, a maximization for the discriminator and minimization for the generator, and we find the optimal generator parameter, θ_g, and discriminator parameter, θ_d, through backpropagating the respective networks.

So, we perform gradient ascent; that is, maximization on the discriminator:

$$\nabla_{\theta_d} \frac{1}{m} \sum_{i=1}^{m} \left[\log D\left(x^{(i)}\right) + \log\left(1 - D\left(G\left(z^{(i)}\right)\right)\right) \right]$$

And, we perform gradient descent; that is, minimization on the generator:

$$\nabla_{\theta_g} \frac{1}{m} \sum_{i=1}^{m} \log\left(1 - D\left(G\left(z^{(i)}\right)\right)\right)$$

However, optimizing the preceding generator objective function does not work properly and causes a stability issue. So, we introduce a new form of loss called **heuristic loss**.

Heuristic loss

There is no change in the loss function of the discriminator. It can be directly written as follows:

$$\max_{d} L(D, G) = \mathbb{E}_{x \sim p_r(x)} [\log D(x; \theta_d)] + \mathbb{E}_{z \sim p_z(z)} [\log(1 - D(G(z; \theta_g); \theta_d))]$$

Now, let's look at the generator loss:

$$\min_{g} L(D, G) = \mathbb{E}_{z \sim p_z(z)} [\log(1 - D(G(z; \theta_g); \theta_d))]$$

Can we change the minimization objective in our generator loss function into a maximization objective just like the discriminator loss? How can we do that? We know that $1 - D(G(Z))$ returns the probability of the fake input image being fake, and the generator is minimizing this probability.

Instead of doing this, we can write $D(G(z))$. It implies the probability of the fake input image being real, and now the generator can maximize this probability. It implies a generator is maximizing the probability of the fake input image being classified as a real image. So, the loss function of our generator now becomes the following:

$$\max_{g} L(D, G) = \mathbb{E}_{z \sim p_z(z)} [\log(D(G(z; \theta_g); \theta_d))]$$

So now, we have both the loss function of our discriminator and generator in maximizing terms:

$$\max_{d} L(D,G) = \mathbb{E}_{x \sim p_r(x)}[\log D(x; \theta_d)] + \mathbb{E}_{z \sim p_z(z)}[\log(1 - D(G(z; \theta_g); \theta_d))]$$

$$\max_{g} L(D,G) = \mathbb{E}_{z \sim p_z(z)}[\log(D(G(z; \theta_g); \theta_d))]$$

But, instead of maximizing, if we can minimize the loss, then we can apply our favorite gradient descent algorithm. So, how can we convert our maximizing problem into a minimization problem? We can do that by simply adding a negative sign.

So, our final loss function for the discriminator is given as follows:

$$L^D = -\mathbb{E}_{x \sim p_r(x)}[\log D(x)] - \mathbb{E}_{z \sim p_z(z)}[\log(1 - D(G(z)))]$$

Also, the generator loss is given as follows:

$$L^G = -\mathbb{E}_{z \sim p_z(z)}[\log(D(G(z)))]$$

Generating images using GANs in TensorFlow

Let's strengthen our understanding of GANs by building them to generate handwritten digits in TensorFlow. You can also check the complete code used in this section here at http://bit.ly/2wwBvRU.

First, we will import all the necessary libraries:

```
import warnings
warnings.filterwarnings('ignore')

import numpy as np
import tensorflow as tf
from tensorflow.examples.tutorials.mnist import input_data
tf.logging.set_verbosity(tf.logging.ERROR)

import matplotlib.pyplot as plt
%matplotlib inline

tf.reset_default_graph()
```

Reading the dataset

Load the MNIST dataset:

```
data = input_data.read_data_sets("data/mnist",one_hot=True)
```

Let's plot one image:

```
plt.imshow(data.train.images[13].reshape(28,28),cmap="gray")
```

The input image looks as follows:

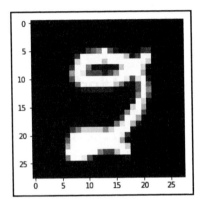

Defining the generator

Generator *G* takes the noise *z* as an input and returns an image. We define the generator as a feedforward network with three layers. Instead of coding the generator network from scratch, we can use tf.layers.dense(), which can be used to create a dense layer. It takes three parameters: inputs, the number of units, and the activation function:

```
def generator(z,reuse=None):
    with tf.variable_scope('generator',reuse=reuse):
        hidden1 =
tf.layers.dense(inputs=z,units=128,activation=tf.nn.leaky_relu)
        hidden2 =
tf.layers.dense(inputs=hidden1,units=128,activation=tf.nn.leaky_relu)
        output =
tf.layers.dense(inputs=hidden2,units=784,activation=tf.nn.tanh)
        return output
```

Defining the discriminator

We know that discriminator D returns the probability of the given image being real. We define the discriminator also as a feedforward network with three layers:

```
def discriminator(X,reuse=None):
    with tf.variable_scope('discriminator',reuse=reuse):
        hidden1 =
tf.layers.dense(inputs=X,units=128,activation=tf.nn.leaky_relu)
        hidden2 =
tf.layers.dense(inputs=hidden1,units=128,activation=tf.nn.leaky_relu)
        logits = tf.layers.dense(inputs=hidden2,units=1)
        output = tf.sigmoid(logits)
        return logits
```

Defining the input placeholders

Now we define the `placeholder` for the input x and the noise z:

```
x = tf.placeholder(tf.float32,shape=[None,784])
z = tf.placeholder(tf.float32,shape=[None,100])
```

Starting the GAN!

First, we feed the noise z to the generator and it will output the fake image, $fake\ x = G(z)$:

```
fake_x = generator(z)
```

Now we feed the real image to discriminator $D(x)$ and get the probability of the real image being real:

```
D_logits_real = discriminator(x)
```

Similarly, we feed the fake image to discriminator $D(x)$ and get the probability of the fake image being real:

```
D_logits_fake = discriminator(fake_x,reuse=True)
```

Computing the loss function

Now, we will see how to compute the loss function.

Discriminator loss

The discriminator loss is given as follows:

$$L^D = -\mathbb{E}_{x \sim p_r(x)}[\log D(x)] - \mathbb{E}_{z \sim p_z(z)}[\log(1 - D(G(z)))]$$

First, we will implement the first term, $-\mathbb{E}_{x \sim p_r(x)}[\log D(x)]$.

The first term, $-\mathbb{E}_{x \sim p_r(x)}[\log D(x)]$, implies the expectations of the log likelihood of images sampled from the real data distribution being real.

It is basically the binary cross-entropy loss. We can implement binary cross-entropy loss with the `tf.nn.sigmoid_cross_entropy_with_logits()` TensorFlow function. It takes two parameters as inputs, `logits` and `labels`, explained as follows:

- The `logits` input, as the name suggests, is the logits of the network so it is `D_logits_real`.
- The `labels` input, as the name suggests, is the true label. We learned that discriminator should return 1 for real images and 0 for fake images. Since we are calculating the loss for input images sampled from the real data distribution, the true label is 1.

We use `tf.ones_likes()` for setting the labels to 1 with the same shape as `D_logits_real`. That is, `labels = tf.ones_like(D_logits_real)`.

Then we compute the mean loss using `tf.reduce_mean()`. If you notice, there is a minus sign in our loss function, which we added for converting our loss to a minimization objective. But, in the following code, there is no minus sign, because TensorFlow optimizers will only minimize and not maximize. So we don't have to add minus sign in our implementation because in any case, it will be minimized by the TensorFlow optimizer:

```
D_loss_real =
tf.reduce_mean(tf.nn.sigmoid_cross_entropy_with_logits(logits=D_logits_real
,
    labels=tf.ones_like(D_logits_real)))
```

Now we will implement the second term, $-\mathbb{E}_{z \sim p_z(z)}[\log(1 - D(G(z)))]$.

The second term, $-\mathbb{E}_{z \sim p_z(z)}[\log(1 - D(G(z)))]$, implies the expectations of the log likelihood of images generated by the generator being fake.

Similar to the first term, we can use `tf.nn.sigmoid_cross_entropy_with_logits()` for calculating the binary cross-entropy loss. In this, the following holds true:

- Logits is `D_logits_fake`
- Since we are calculating the loss for the fake images generated by the generator, the `true` label is `0`

We use `tf.zeros_like()` for setting the labels to `0` with the same shape as `D_logits_fake`. That is, `labels = tf.zeros_like(D_logits_fake)`:

```
D_loss_fake =
tf.reduce_mean(tf.nn.sigmoid_cross_entropy_with_logits(logits=D_logits_fake
,
  labels=tf.zeros_like(D_logits_fake)))
```

Now we will implement the final loss.

So, combining the preceding two terms, the loss function of the discriminator is given as follows:

```
D_loss = D_loss_real + D_loss_fake
```

Generator loss

The generator loss is given as $L^G = -\mathbb{E}_{z \sim p_z(z)}[\log(D(G(z)))]$.

It implies the probability of the fake image being classified as a real image. As we calculated binary cross-entropy in the discriminator, we use `tf.nn.sigmoid_cross_entropy_with_logits()` for calculating the loss in the generator.

Here, the following should be borne in mind:

- Logits is `D_logits_fake`.

- Since our loss implies the probability of the fake input image being classified as real, the true label is 1. Because, as we learned, the goal of the generator is to generate the fake image and fool the discriminator to classify the fake image as a real image.

We use `tf.ones_like()` for setting the labels to 1 with the same shape as
`D_logits_fake`. That is, `labels = tf.ones_like(D_logits_fake)`:

```
G_loss =
tf.reduce_mean(tf.nn.sigmoid_cross_entropy_with_logits(logits=D_logits_fake
, labels=tf.ones_like(D_logits_fake)))
```

Optimizing the loss

Now we need to optimize our generator and discriminator. So, we collect the parameters of
the discriminator and generator as θ_D and θ_G respectively:

```
training_vars = tf.trainable_variables()
theta_D = [var for var in training_vars if 'dis' in var.name]
theta_G = [var for var in training_vars if 'gen' in var.name]
```

Optimize the loss using the Adam optimizer:

```
learning_rate = 0.001

D_optimizer =
tf.train.AdamOptimizer(learning_rate).minimize(D_loss,var_list = theta_D)
G_optimizer = tf.train.AdamOptimizer(learning_rate).minimize(G_loss,
var_list = theta_G)
```

Starting the training

Let's begin the training by defining the batch size and the number of epochs:

```
batch_size = 100
num_epochs = 1000
```

Initialize all the variables:

```
init = tf.global_variables_initializer()
```

Generating handwritten digits

Start the TensorFlow session and generate handwritten digits:

```
with tf.Session() as session:
```

Initialize all variables:

```
session.run(init)
```

To execute this for each epoch:

```
for epoch in range(num_epochs):
```

Select the number of batches:

```
num_batches = data.train.num_examples // batch_size
```

To execute this for each batch:

```
for i in range(num_batches):
```

Get the batch of data according to the batch size:

```
batch = data.train.next_batch(batch_size)
```

Reshape the data:

```
batch_images = batch[0].reshape((batch_size,784))
batch_images = batch_images * 2 - 1
```

Sample the batch noise:

```
batch_noise = np.random.uniform(-1,1,size=(batch_size,100))
```

Define the feed dictionaries with input x as `batch_images` and noise z as `batch_noise`:

```
feed_dict = {x: batch_images, z : batch_noise}
```

Train the discriminator and generator:

```
_ = session.run(D_optimizer,feed_dict = feed_dict)
_ = session.run(G_optimizer,feed_dict = feed_dict)
```

Compute loss of discriminator and generator:

```
discriminator_loss = D_loss.eval(feed_dict)
generator_loss = G_loss.eval(feed_dict)
```

Feed the noise to a generator on every 100th epoch and generate an image:

```
if epoch%100==0:
    print("Epoch: {}, iteration: {}, Discriminator Loss:{},
Generator Loss: {}".format(epoch,i,discriminator_loss,generator_loss))

    _fake_x = fake_x.eval(feed_dict)


    plt.imshow(_fake_x[0].reshape(28,28))
    plt.show()
```

During training, we notice how loss decreases and how GANs learn to generate images as shown follows:

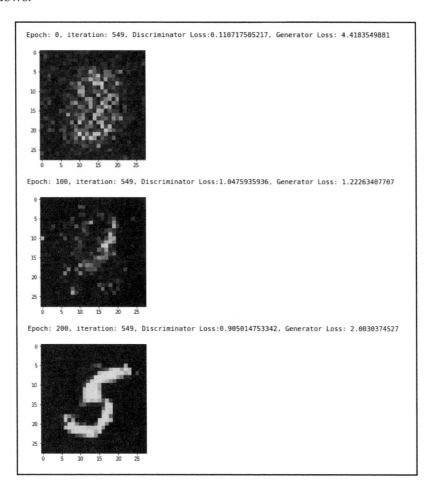

DCGAN – Adding convolution to a GAN

We just learned how effective GANs are and how can they be used to generate images. We know that a GAN has two components generators that generate the image and the discriminator, which acts as a critic to the generated image. As you can see, both of these generators and discriminators are basically feedforward neural networks. Instead of keeping them as feedforward networks, can we use convolutional networks?

In Chapter 6, *Demystifying Convolutional Networks*, we have seen the effectiveness of convolutional networks for image-based data and how they extract features from images in an unsupervised fashion. Since in GANs we are generating images, it is desirable to use convolutional networks instead of feedforward networks. So, we introduce a new type of GAN called **DCGAN**. It extends the design of GANs with convents. We basically replace the feedforward network in the generator and discriminator with a **Convolutional Neural Network** (**CNN**).

The discriminator uses convolutional layers for classifying the image as a fake or real image, while the generator uses convolutional transpose layers to generate a new image. Now we will go into detail and see how generators and discriminators differ in DCGAN compared to the vanilla GANs.

Deconvolutional generator

We know that the role of the generator is to generate a new image by learning the real data distribution. In DCGAN, a generator is composed of convolutional transpose and batch norm layers with ReLU activations.

 Note that convolutional transpose operation is also known as deconvolution operation or fractionally strided convolutions.

The input to the generator is the noise, z, which we draw from a standard normal distribution, and it outputs an image of the same size as the images in the training data, say, 64 x 64 x 3.

The architecture of the generator is shown in the following diagram:

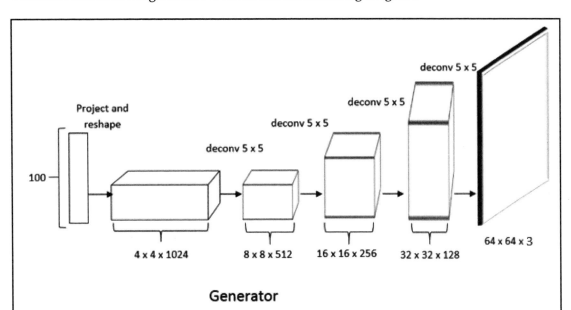

Generator

First, we convert the noise, z, of a 100 x 1 shape into 1024 x 4 x 4 to have the shape of width, height, and feature map and it is called the **project and reshape**. Following this, we perform a series of convolutional operations with fractionally strided convolutions. We apply batch normalization to every layer except at the last layer. Also, we apply ReLU activations to every layer but the last layer. We apply the tanh activation function to scale the generated image between -1 and +1.

Convolutional discriminator

Now we will see the architecture of a discriminator in DCGAN. As we know, the discriminator takes the image and it tells us whether the image is a real image or a fake image. Thus, it is basically a binary classifier. The discriminator is composed of a series of convolutional and batch norm layers with leaky ReLU activations.

The architecture of the discriminator is shown in the following diagram:

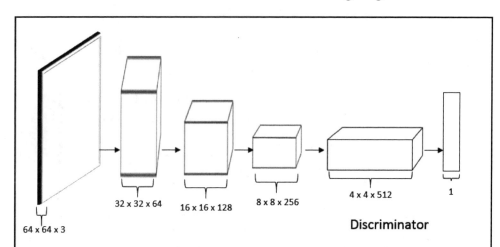

As you can see, it takes the input image of the 64 x 64 x 3 shape and performs a series of convolutional operations with a leaky ReLU activation function. We apply batch normalization at all layers except at the input layer.

Remember, we don't apply a max pooling operation in both the discriminator and the generator. Instead, we apply a strided convolution operation (that is convolution operation with strides).

In a nutshell, we enhance the vanilla GAN by replacing the feedforward network in the generator and the discriminator with the convolutional network.

Implementing DCGAN to generate CIFAR images

Now we will see how to implement DCGAN in TensorFlow. We will learn how to use DCGAN with images from the **Canadian Institute For Advanced Research (CIFAR)**-10 dataset. CIFAR-10 consists of 60,000 images from 10 different classes that include airplanes, cars, birds, cats, deer, dogs, frogs, horses, ships, and trucks. We will examine how we can use DCGAN to generate such images.

First, import the required libraries:

```
import warnings
warnings.filterwarnings('ignore')


import numpy as np
import tensorflow as tf
from tensorflow.examples.tutorials.mnist import input_data
tf.logging.set_verbosity(tf.logging.ERROR)

from keras.datasets import cifar10

import matplotlib.pyplot as plt
%matplotlib inline
from IPython import display

from scipy.misc import toimage
```

Exploring the dataset

Load the CIFAR dataset:

```
(x_train, y_train), _ = cifar10.load_data()

x_train = x_train.astype('float32')/255.0
```

Let's see what we have in our dataset. Define a helper function for plotting the image:

```
def plot_images(X):
    plt.figure(1)
    z = 0
    for i in range(0,4):
        for j in range(0,4):
            plt.subplot2grid((4,4),(i,j))
            plt.imshow(toimage(X[z]))
            z = z + 1

    plt.show()
```

Let's plot a few images:

```
plot_images(x_train[:17])
```

The plotted images are shown as follows:

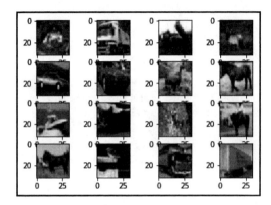

Defining the discriminator

We define a discriminator as a convolutional network with three convolutional layers followed by a fully connected layer. It is composed of a series of convolutional and batch norm layers with leaky ReLU activations. We apply batch normalization at all layers except at the input layer:

```
def discriminator(input_images, reuse=False, is_training=False, alpha=0.1):
    with tf.variable_scope('discriminator', reuse= reuse):
```

First convolutional layer with leaky ReLU activation:

```
layer1 = tf.layers.conv2d(input_images,
                          filters=64,
                          kernel_size=5,
                          strides=2,
                          padding='same',
                          kernel_initializer=kernel_init,
                          name='conv1')

layer1 = tf.nn.leaky_relu(layer1, alpha=0.2, name='leaky_relu1')
```

Second convolutional layer with batch normalization and leaky ReLU activation:

```
layer2 = tf.layers.conv2d(layer1,
                          filters=128,
                          kernel_size=5,
                          strides=2,
                          padding='same',
                          kernel_initializer=kernel_init,
```

```
                                name='conv2')
        layer2 = tf.layers.batch_normalization(layer2,
  training=is_training, name='batch_normalization2')
        layer2 = tf.nn.leaky_relu(layer2, alpha=0.2, name='leaky_relu2')
```

Third convolutional layer with batch normalization and leaky ReLU:

```
        layer3 = tf.layers.conv2d(layer2,
                                filters=256,
                                kernel_size=5,
                                strides=1,
                                padding='same',
                                name='conv3')
        layer3 = tf.layers.batch_normalization(layer3,
  training=is_training, name='batch_normalization3')
        layer3 = tf.nn.leaky_relu(layer3, alpha=0.1, name='leaky_relu3')
```

Flatten the output of the final convolutional layer:

```
        layer3 = tf.reshape(layer3, (-1,
  layer3.shape[1]*layer3.shape[2]*layer3.shape[3]))
```

Define the fully connected layer and return the `logits`:

```
        logits = tf.layers.dense(layer3, 1)
        output = tf.sigmoid(logits)
        return logits
```

Defining the generator

As we learned, the generator performs the transpose convolutional operation. The generator is composed of convolutional transpose and batch norm layers with ReLU activations. We apply batch normalization to every layer except for the last layer. Also, we apply ReLU activations to every layer, but for the last layer, we apply the `tanh` activation function to scale the generated image between -1 and +1:

```
def generator(z, z_dim, batch_size, is_training=False, reuse=False):
    with tf.variable_scope('generator', reuse=reuse):
```

First fully connected layer:

```
        input_to_conv = tf.layers.dense(z, 8*8*128)
```

Convert the shape of the input and apply batch normalization followed by ReLU activations:

```
        layer1 = tf.reshape(input_to_conv, (-1, 8, 8, 128))
        layer1 = tf.layers.batch_normalization(layer1,
  training=is_training, name='batch_normalization1')
        layer1 = tf.nn.relu(layer1, name='relu1')
```

The second layer, that is the transpose convolution layer, with batch normalization and the ReLU activation:

```
        layer2 = tf.layers.conv2d_transpose(layer1, filters=256,
  kernel_size=5, strides= 2, padding='same',
                                      kernel_initializer=kernel_init,
  name='deconvolution2')
        layer2 = tf.layers.batch_normalization(layer2,
  training=is_training, name='batch_normalization2')
        layer2 = tf.nn.relu(layer2, name='relu2')
```

Define the third layer:

```
        layer3 = tf.layers.conv2d_transpose(layer2, filters=256,
  kernel_size=5, strides= 2, padding='same',
                                      kernel_initializer=kernel_init,
  name='deconvolution3')
        layer3 = tf.layers.batch_normalization(layer3,training=is_training,
  name='batch_normalization3')
        layer3 = tf.nn.relu(layer3, name='relu3')
```

Define the fourth layer:

```
        layer4 = tf.layers.conv2d_transpose(layer3, filters=256,
  kernel_size=5, strides= 1, padding='same',
                                      kernel_initializer=kernel_init,
  name='deconvolution4')
        layer4 = tf.layers.batch_normalization(layer4,training=is_training,
  name='batch_normalization4')
        layer4 = tf.nn.relu(layer4, name='relu4')
```

In the final layer, we don't apply batch normalization and instead of ReLU, we use `tanh` activation:

```
        layer5 = tf.layers.conv2d_transpose(layer4, filters=3,
  kernel_size=7, strides=1, padding='same',
                                      kernel_initializer=kernel_init,
  name='deconvolution5')
        logits = tf.tanh(layer5, name='tanh')
        return logits
```

Defining the inputs

Define the `placeholder` for the input:

```
image_width = x_train.shape[1]
image_height = x_train.shape[2]
image_channels = x_train.shape[3]


x = tf.placeholder(tf.float32, shape= (None, image_width, image_height,
image_channels), name="d_input")
```

Define the `placeholder` for the learning rate and training boolean:

```
learning_rate = tf.placeholder(tf.float32, shape=(), name="learning_rate")
is_training = tf.placeholder(tf.bool, [], name='is_training')
```

Define the batch size and dimension of the noise:

```
batch_size = 100
z_dim = 100
```

Define the placeholder for the noise, z:

```
z = tf.random_normal([batch_size, z_dim], mean=0.0, stddev=1.0, name='z')
```

Starting the DCGAN

First, we feed the noise z to the generator and it will output the fake image, $fake\ x = G(z)$:

```
fake_x = generator(z, z_dim, batch_size, is_training=is_training)
```

Now we feed the real image to the discriminator $D(x)$ and get the probability of the real image being real:

```
D_logit_real = discriminator(x, reuse=False, is_training=is_training)
```

Similarly, we feed the fake image to the discriminator, $D(z)$, and get the probability of the fake image being real:

```
D_logit_fake = discriminator(fake_x, reuse=True,  is_training=is_training)
```

Computing the loss function

Now we will see how to compute the loss function.

Discriminator loss

The loss function is the same as for the vanilla GAN:

$$L^D = -\mathbb{E}_{x \sim p_r(x)}[\log D(x)] - \mathbb{E}_{z \sim p_z(z)}[\log(1 - D(G(z)))]$$

So, we can directly write the following:

```
D_loss_real =
tf.reduce_mean(tf.nn.sigmoid_cross_entropy_with_logits(logits=D_logits_real
,
  labels=tf.ones_like(D_logits_real)))

D_loss_fake =
tf.reduce_mean(tf.nn.sigmoid_cross_entropy_with_logits(logits=D_logits_fake
,
  labels=tf.zeros_like(D_logits_fake)))

D_loss = D_loss_real + D_loss_fake
```

Generator loss

The generator loss is also the same as for the vanilla GAN:

$$L^G = -\mathbb{E}_{z \sim p_z(z)}[\log(D(G(z)))]$$

We can compute it by using the following code:

```
G_loss =
tf.reduce_mean(tf.nn.sigmoid_cross_entropy_with_logits(logits=D_logits_fake
, labels=tf.ones_like(D_logits_fake)))
```

Optimizing the loss

As we saw in vanilla GANs, we collect the parameters of the discriminator and generator as θ_D and θ_G respectively:

```
training_vars = tf.trainable_variables()
theta_D = [var for var in training_vars if 'dis' in var.name]
theta_G = [var for var in training_vars if 'gen' in var.name]
```

Optimize the loss using the Adam optimizer:

```
d_optimizer = tf.train.AdamOptimizer(learning_rate).minimize(D_loss,
var_list=theta_D)
g_optimizer = tf.train.AdamOptimizer(learning_rate).minimize(G_loss,
var_list=theta_G)
```

Train the DCGAN

Let's begin the training. Define the number of batches, epochs, and learning rate:

```
num_batches = int(x_train.shape[0] / batch_size)
steps = 0
num_epcohs = 500
lr = 0.00002
```

Define a helper function for generating and plotting the generated images:

```
def generate_new_samples(session, n_images, z_dim):

    z = tf.random_normal([1, z_dim], mean=0.0, stddev=1.0)

    is_training = tf.placeholder(tf.bool, [], name='training_bool')

    samples = session.run(generator(z, z_dim, batch_size, is_training,
reuse=True),feed_dict={is_training: True})
    img = (samples[0] * 255).astype(np.uint8)
    plt.imshow(img)
    plt.show()
```

Start the training:

```
with tf.Session() as session:
```

Initialize all variables:

```
session.run(tf.global_variables_initializer())
```

To execute for each epoch:

```
for epoch in range(num_epcohs):
    #for each batch
    for i in range(num_batches):
```

Define the start and end of the batch:

```
start = i * batch_size
end = (i + 1) * batch_size
```

Sample batch of images:

```
batch_images = x_train[start:end]
```

Train the discriminator for every two steps:

```
if(steps % 2 == 0):
        _, discriminator_loss = session.run([d_optimizer,D_loss],
   feed_dict={x: batch_images, is_training:True, learning_rate:lr})
```

Train the generator and discriminator:

```
        _, generator_loss = session.run([g_optimizer,G_loss],
   feed_dict={x: batch_images, is_training:True, learning_rate:lr})
        _, discriminator_loss = session.run([d_optimizer,D_loss],
   feed_dict={x: batch_images, is_training:True, learning_rate:lr})
```

Generate a new image:

```
display.clear_output(wait=True)
generate_new_samples(session, 1, z_dim)
print("Epoch: {}, iteration: {}, Discriminator Loss:{},
   Generator Loss: {}".format(epoch,i,discriminator_loss,generator_loss))

steps += 1
```

In the first iteration, DCGAN will generate the raw pixels, but over the series of iterations, it will learn to generate real images with following parameters:

```
Epoch: 0, iteration: 0, Discriminator Loss:1.44706475735, Generator
Loss: 0.726667642593
```

The following image is generated by DCGAN in the first iteration:

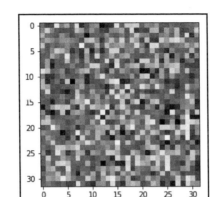

Least squares GAN

We just learned how GANs are used to generate images. **Least Squares GAN** (**LSGAN**) is another simple variant of a GAN. As the name suggests, here, we use the least square error as a loss function instead of sigmoid cross-entropy loss. With LSGAN, we can improve the quality of images being generated from the GAN. But how can we do that? Why do the vanilla GANs generate poor quality images?

If you can recollect the loss function of GAN, we used sigmoid cross-entropy as the loss function. The goal of the generator is to learn the distribution of the images in the training set, that is, real data distribution, map it to the fake distribution, and generate fake samples from the learned fake distribution. So, the GANs try to map the fake distribution as close to the true distribution as possible.

But once the fake samples are on the correct side of the decision surface, then gradients tend to vanish even though the fake samples are far away from the real distribution. This is due to the sigmoid cross-entropy loss.

Let's understand this with the following figure. A decision boundary of vanilla GANs with sigmoid cross-entropy as a loss function is shown in the following figure where fake samples are represented by a cross, and real samples are represented by a dot, and the fake samples for updating the generator are represented by a star.

As you can observe, once the fake samples (star) generated by the generator are on the correct side of the decision surface, that is, once the fake samples are on the side of real samples (dot) then the gradients tend to vanish even though the fakes samples are far away from real distribution. This is due to the sigmoid cross-entropy loss, because it does not care whether the fake samples are close to real samples; it only looks for whether the fake samples are on the correct side of the decision surface. This leads to a problem that when the gradients vanish even though the fake samples are far away from the real data distribution then the generator cannot learn the real distribution of the dataset:

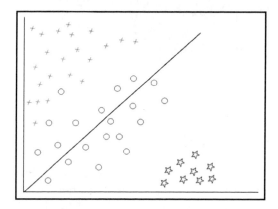

So, we can change this decision surface with sigmoid cross-entropy loss to a least squared loss. Now, as you can see in the following diagram, although the fake samples generated by the generator are on the correct side of the decision surface, gradients will not vanish until the fake samples match the true distribution. Least square loss forces the updates to match the fake samples to the true samples:

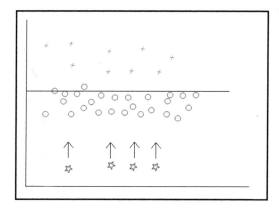

So, since we are matching a fake distribution to the real distribution, our image quality will be improved when we use the least square as a cost function.

In a nutshell gradient updates in the vanilla GANs will be stopped when the fake samples are correct side of the decision surface even though they are from the real samples, that is, the real distribution. This is due to the sigmoid cross-entropy loss and it does not care whether the fake samples are close to real samples, it only looks for whether the fake samples are on the correct side. This leads to the problem that we cannot learn the real data distribution perfectly. So, we use LSGAN, which uses least squared error as a loss function, where the gradient updates will not be stopped until the fake samples match the real sample, even though the fake samples are on the correct side of the decision boundary.

Loss function

Now that, we have learned that the least square loss function improves the generator's image quality, how can we rewrite our GANs loss function in terms of least squares?

Let's say a and b are the actual labels for the generated images and real images respectively, then we can write the loss function of the discriminator in terms of least square loss as follows:

$$L^D = \frac{1}{2}\mathbb{E}_{x \sim p_r(x)}\left[(D(x) - b)^2\right] + \frac{1}{2}\mathbb{E}_{z \sim p_z(z)}\left[(D(G(z)) - a)^2\right]$$

Similarly, let's say c is the actual label that the generator wants the discriminator to believe that the generated image is the real image, so label c represents the real image. Then we can write the loss function of a generator in terms of least square loss as follows:

$$L^G = \frac{1}{2}\mathbb{E}_{z \sim p_z(z)}\left[(D(G(z)) - c)^2\right]$$

We set labels for real images as 1 and for fake images 0, so b and c become 1 and a becomes 0. So, our final equation can be given as follows:

The loss function of the discriminator is given as follows:

$$\boxed{L^D = \frac{1}{2}\mathbb{E}_{x \sim p_r(x)}\left[(D(x) - 1)^2\right] + \frac{1}{2}\mathbb{E}_{z \sim p_z(z)}\left[(D(G(z)))^2\right]}$$

The loss function of the generator is given as follows:

$$L^G = \frac{1}{2}\mathbb{E}_{z \sim p_z(z)}\left[(D(G(z)) - 1)^2\right]$$

LSGAN in TensorFlow

Implementing LSGAN is the same as the vanilla GAN, except for the change in the loss function. So, instead of looking at the whole code, we will see only how to implement LSGAN's loss function in TensorFlow. The complete code for LSGAN is available in the GitHub repo at `http://bit.ly/2HMCrrx`.

Let's now see how the loss function of LSGAN is implemented.

Discriminator loss

The discriminator loss is given as follows:

$$L^D = \frac{1}{2}\mathbb{E}_{x \sim p_r(x)}\left[(D(x) - 1)^2\right] + \frac{1}{2}\mathbb{E}_{z \sim p_z(z)}\left[(D(G(z)))^2\right]$$

First, we will implement the first term, $\frac{1}{2}\mathbb{E}_{x \sim p_{data}(x)}\left[(D(x) - 1)^2\right]$:

```
D_loss_real = 0.5*tf.reduce_mean(tf.square(D_logits_real-1))
```

Now we will implement the second term, $\frac{1}{2}\mathbb{E}_{z \sim p_z(z)}\left[(D(G(z)))^2\right]$:

```
D_loss_fake = 0.5*tf.reduce_mean(tf.square(D_logits_fake))
```

The final discriminator loss can be written as follows:

```
D_loss = D_loss_real + D_loss_fake
```

Generator loss

The generator loss, $L^G = \frac{1}{2}\mathbb{E}_{z \sim p_z(z)}\left[(D(G(z)) - 1)^2\right]$, is given as follows:

```
G_loss = 0.5*tf.reduce_mean(tf.square(D_logits_fake-1))
```

GANs with Wasserstein distance

Now we will see another very interesting version of a GAN, called Wasserstein GAN (**WGAN**). It uses the Wasserstein distance in the GAN's loss function. First, let's understand why we need a Wasserstein distance measure and what's wrong with our current loss function.

Before going ahead, first, let's briefly explore two popular divergence measures that are used for measuring the similarity between two probability distributions.

The **Kullback-Leibler (KL)** divergence is one of the most popularly used measures for determining how one probability distribution diverges from the other. Let's say we have two discrete probability distributions, P and Q, then the KL divergence can be expressed as follows:

$$D_{\mathrm{KL}}(P\|Q) = \sum_x P(x)\log\left(\frac{P(x)}{Q(x)}\right)$$

When the two distributions are continuous, then the KL divergence can be expressed in the integral form as shown:

$$D_{KL}(P\|Q) = \int_x P(x)\log\frac{P(x)}{Q(x)}dx$$

The KL divergence is not symmetric, meaning the following:

$$D_{KL}(P\|Q) \neq D_{KL}(Q\|P)$$

The **Jensen-Shanon (JS)** divergence is another measure for measuring the similarity of two probability distributions. But unlike the KL divergence, the JS divergence is symmetric and it can be given as follows:

$$D_{JS}(P\|Q) = \frac{1}{2}D_{KL}\left(P\|\frac{P+Q}{2}\right) + \frac{1}{2}D_{KL}\left(Q\|\frac{P+Q}{2}\right)$$

Are we minimizing JS divergence in GANs?

We know that generators try to learn the real data distribution, p_r, so that it can generate new samples from the learned distribution, p_g, and the discriminator tells us whether the image is from a real or fake distribution.

We also learned that when $p_r = p_g$, then the discriminator cannot tell us whether the image is from a real or a fake distribution. It just outputs 0.5 because it cannot differentiate between p_r and p_g.

So, for a generator, the optimal discriminator can be given as follows:

$$D(x) = \frac{p_r(x)}{p_r(x) + p_g(x)} = \frac{1}{2} \tag{1}$$

Let's recall the loss function of the discriminator:

$$\max_d L(D, G) = \mathbb{E}_{x \sim p_r(x)} [\log D(x; \theta_d)] + \mathbb{E}_{z \sim p_z(z)} [\log(1 - D(G(z; \theta_g); \theta_d))]$$

It can be simply written as follows:

$$L = \mathbb{E}_{x \sim p_r} [logD(x)] + \mathbb{E}_{x \sim p_g} [log \, 1 - D(x)]$$

Substituting equation *(1)* in the preceding equation we get the following:

$$L = \mathbb{E}_{x \sim p_r} [log \frac{p_r(x)}{p_r(x) + p_g(x)}] + \mathbb{E}_{x \sim p_g} [log1 - \frac{p_r(x)}{p_r(x) + p_g(x)}]$$

It can be solved as follows:

$$
\begin{aligned}
L &= \mathbb{E}_{x \sim p_r} [log \frac{p_r(x)}{p_r(x) + p_g(x)}] + \mathbb{E}_{x \sim p_g} [log \, 1 - \frac{p_r(x)}{p_r(x) + p_g(x)}] \\
&= \mathbb{E}_{x \sim p_r} [log \frac{p_r(x)}{p_r(x) + p_g(x)}] + \mathbb{E}_{x \sim p_g} [log \frac{p_r(x) + p_g(x) - p_r(x)}{p_r(x) + p_g(x)}] \\
&= \mathbb{E}_{x \sim p_r} [log \frac{p_r(x)}{p_r(x) + p_g(x)}] + \mathbb{E}_{x \sim p_g} [log \frac{p_g(x)}{p_r(x) + p_g(x)}] \\
&= \mathbb{E}_{x \sim p_r} [log \frac{p_r(x)}{2 * \frac{1}{2}(p_r(x) + p_g(x))}] + \mathbb{E}_{x \sim p_g} [log \frac{p_g(x)}{2 * \frac{1}{2}(p_r(x) + p_g(x))}] \\
&= \mathbb{E}_{x \sim p_r} [log \frac{p_r(x)}{\frac{1}{2}(p_r(x) + p_g(x))}] - log \, 2 + \mathbb{E}_{x \sim p_g} [log \frac{p_g(x)}{\frac{1}{2}(p_r(x) + p_g(x))}] - log \, 2 \\
&= \mathbb{E}_{x \sim p_r} [log \frac{p_r(x)}{p_{average}(x)}] + \mathbb{E}_{x \sim p_g} [log \frac{p_g(x)}{p_{average}(x)}] - 2log \, 2 \\
&= 2JS(p_r | p_g) - 2log \, 2
\end{aligned}
$$

As you can see, we are basically minimizing the JS divergence in the loss function of GAN. So, minimizing the loss function of the GAN basically implies that minimizing the JS divergence between the real data distribution, P_r and the fake data distribution, P_g as shown:

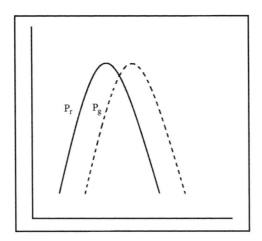

Minimizing the JS divergence between P_r and P_g denotes that the generator G makes their distribution P_g similar to the real data distribution P_r. But there is a problem with JS divergence. As you can see from the following figure, there is no overlap between the two distributions. When there is no overlap or when the two distributions do not share the same support, JS divergence will explode or return a constant value and the GANs cannot learn properly:

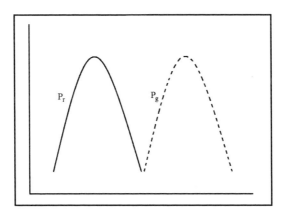

So, to avoid this, we need to change our loss function. Instead of minimizing the JS divergence, we use a new distance metric called the Wasserstein distance, which tells us how the two distributions are apart from each other even when they don't share the same support.

What is the Wasserstein distance?

The Wasserstein distance, also known as the **Earth Movers** (**EM**) distance, is one of the most popularly used distance measures in the optimal transport problems where we need to move things from one configuration to another.

So, when we have two distributions, p_r and p_g, $W(p_r, p_g)$ implies that how much amount of work is required for the probability distribution, p_r to match the probability distribution p_g.

Let's try to understand the intuition behind the EM distance. We can view the probability distribution as a collection of mass. Our goal is to convert one probability distribution to another. There are many possible ways to convert one distribution to another, but the Wasserstein metric seeks to find the optimal and minimum way that has the least cost in conversion.

The cost of conversion can be given as a distance multiplied by the mass.

The amount of information moved from point x to point y is given as $\gamma(x, y)$. It is called a **transport plan**. It tells us how much information we need to transport from x to y, and the distance between x and y is given as $\|x - y\|$.

So, the cost is given as follows:

$$Cost = \gamma(x, y)\|x - y\|$$

We have many *(x,y)* pairs so the expectations across all *(x,y)* pairs are given as follows:

$$\mathbb{E}_{x,y \sim \gamma}\|x - y\|$$

It implies the cost of moving from point x to y. There are many ways to move from x to y, but we are interested only in the optimal path, that is, minimum cost, so we rewrite our preceding equation as follows:

$$\inf_{\gamma \sim \Pi(p_r, p_g)} \mathbb{E}_{(x,y) \sim \gamma}[\|x - y\|]$$

Here, *inf* basically implies the minimum value. $\Pi(p_r, p_g)$ is the set of all possible joint distributions between p_r and p_g.

So, out of all the possible joint distributions between p_r and p_g we are finding the minimum cost required to make one distribution look like another.

Our final equation can be given as follows:

$$W(p_r, p_g) = \inf_{\gamma \sim \Pi(p_r, p_g)} \mathbb{E}_{(x,y) \sim \gamma}[\|x - y\|]$$

However, calculating the Wasserstein distance is not a simple task because it is difficult to exhaust all possible joint distributions, $\Pi(p_r, p_g)$, and it turns into another optimization problem.

In order to avoid that, we introduce **Kantorovich-Rubinstein duality**. It converts our equation into a simple maximization problem, as follows:

$$W(p_r, p_g) = \frac{1}{K} \sup_{\|f\|_L \leq K} \mathbb{E}_{x \sim p_r}[f(x)] - \mathbb{E}_{x \sim p_g}[f(x)]$$

Okay, but what does the above equation mean? We are basically applying the **supremum** over all **k-Lipschitz function**. Wait. What is the Lipschitz function and what is supremum? Let's discuss that in the next section.

Demystifying the k-Lipschitz function

A Lipschitz continuous function is a function that must be continuous and almost differentiable everywhere. So, for any function to be a Lipschitz continuous, the absolute value of a slope of the function's graph cannot be more than a constant K. This constant K is called the **Lipschitz constant**:

$$|f(x_1) - f(x_2)| \leq K|x_1 - x_2|$$

To put it in simple terms, we can say a function is Lipschitz continuous when the derivation of a function is bounded by some constant K and it never exceeds the constant.

Let's say $K = 1$, for instance, *sin x* is Lipschitz continuous since its derivative *cos x* is bounded by 1. Similarly, $|x|$ is Lipschitz continuous, since its slope is -1 or 1 everywhere. However, it is not differentiable at 0.

So, let's recall our equation:

$$W(p_r, p_g) = \frac{1}{K} \sup_{\|f\|_L \leq K} \mathbb{E}_{x \sim p_r}[f(x)] - \mathbb{E}_{x \sim p_g}[f(x)]$$

Here, supremum is basically an opposite to infimum. So, supremum over the Lipschitz function implies a maximum over k-Lipschitz functions. So, we can write the following:

$$W(p_r, p_g) = \max_{w \in \mathcal{W}} \mathbb{E}_{x \sim p_r}[f_w(x)] - \mathbb{E}_{x \sim p_g}[f_w(x)]$$

The preceding equation basically tells us that we are basically finding a maximum distance between the expected value over real samples and the expected value over generated samples.

The loss function of WGAN

Okay, why are we learning all this? We saw previously that there is a problem with JS divergence in the loss function, so we resorted to the Wasserstein distance. Now, our goal of the discriminator is no longer to say whether the image is from the real or fake distribution; instead, it tries to maximize the distance between real and generated sample. We train the discriminator to learn the Lipschitz continuous function for computing the Wasserstein distance between a real and fake data distribution.

So, the discriminator loss is given as follows:

$$L^D = W(p_r, p_g) = \max_{w \in \mathcal{W}} \mathbb{E}_{x \sim p_r}[D_w(x)] - \mathbb{E}_{x \sim p_g}[D_w(x)]$$

Now we need to ensure that our function is a k-Lipschitz function during training. So, for every gradient update, we clip the weights of our gradients between a lower bound and upper bound, say between -0.01 and +0.01.

We know that the discriminator loss is given as:

$$\max_{w \in \mathcal{W}} \mathbb{E}_{x \sim p_r}[D_w(x)] - \mathbb{E}_{x \sim p_g}[D_w(x)]$$

Instead of maximizing, we convert this into minimization objective by adding a negative sign:

$$\min_{w \in \mathcal{W}} -(\mathbb{E}_{x \sim p_r}[D_w(x)] - \mathbb{E}_{x \sim p_g}[D_w(x)])$$

$$\min_{w \in \mathcal{W}} -\mathbb{E}_{x \sim p_r}[D_w(x)] + \mathbb{E}_{x \sim p_g}[D_w(x)]$$

The generator loss is the same as we learned in vanilla GANs.

Thus, the loss function of the discriminator is given as:

$$L^D = -\mathbb{E}_{x \sim p_r} D_w(x) + \mathbb{E}_z D_w(G(z))$$

The loss function of the generator is given as:

$$L^G = -\mathbb{E}_z D_w(G(z))$$

WGAN in TensorFlow

Implementing the WGAN is the same as implementing a vanilla GAN except that the loss function of WGAN varies and we need to clip the gradients of the discriminator. Instead of looking at the whole, we will only see how to implement the loss function of the WGAN and how to clip the gradients of the discriminator.

We know that loss of the discriminator is given as:

$$L^D = -\mathbb{E}_{x \sim p_r} D_w(x) + \mathbb{E}_z D_w(G(z))$$

And it can be implemented as follows:

```
D_loss = - tf.reduce_mean(D_real) + tf.reduce_mean(D_fake)
```

We know that generator loss is given as:

$$L^G = -\mathbb{E}_z D_w(G(z))$$

And it can be implemented as follows:

```
G_loss = -tf.reduce_mean(D_fake)
```

We clip the gradients of the discriminator as follows:

```
clip_D = [p.assign(tf.clip_by_value(p, -0.01, 0.01)) for p in theta_D]
```

Summary

We started this chapter by understanding the difference between generative and discriminative models. We learned that the discriminative models learn to find the good decision boundary that separates the classes in an optimal way, while the generative models learn about the characteristics of each class.

Later, we understood how GANs work. They basically consist of two neural networks called generators and discriminators. The role of the generators is to generate a new image by learning the real data distribution, while the discriminator acts as a critic and its role is to tell us whether the generated image is from the true data distribution or the fake data distribution, basically whether it is a real image or a fake image.

Next, we learned about DCGAN where we basically replace the feedforward neural networks in the generator and discriminator with convolutional neural networks. The discriminator uses convolutional layers for classifying the image as a fake or a real image, while the generator uses convolutional transpose layers to generate a new image.

Then, we learned about the LSGAN, which replaces the loss function of both the generator and the discriminator with a least squared error loss. Because, when we use sigmoid cross-entropy as a loss function, our gradients tend to vanish once the fake samples on the correct side of the decision boundary even though they are not close to the real distribution. So, we replace the cross-entropy loss with the least squared error loss where the gradients will not vanish till the fake samples match the true distribution. It forces the gradient updates to match the fake samples to the real samples.

Finally, we learned another interesting type of GAN called the Wassetrtain GAN where we use the Wasserstein distance measure in the discriminator's loss function. Because in vanilla GANs we are basically minimizing JS divergence and it will be constant or results in 0 when the distributions of real data and fake does not overlap. To overcome this, we used the Wasserstein distance measure in the discriminator's loss function.

In the next chapter, we will learn about several other interesting types of GANs called CGAN, InfoGAN, CycleGAN, and StackGAN.

Questions

Let's evaluate our knowledge of GANs by answering the following questions:

1. What is the difference between generative and discriminative models?
2. Explain the role of a generator.
3. Explain the role of a discriminator.
4. What is the loss function of the generator and discriminator?
5. How does a DCGAN differ from a vanilla GAN?
6. What is KL divergence?
7. Define the Wasserstein distance.
8. What is the k-Lipschitz continuous function?

Further reading

Refer to the following papers for further information:

- *Generative Adversarial Nets* by Ian J Goodfellow, et al., `https://arxiv.org/pdf/1406.2661.pdf`
- *Unsupervised Representation Learning with Deep Convolutional Generative Adversarial Networks* by Alec Radford, Soumith Chintala, and Luke Metz, `https://arxiv.org/pdf/1511.06434.pdf`
- *Least Squares Generative Adversarial Networks* by Xudong Mao, et al., `https://arxiv.org/pdf/1611.04076.pdf`
- *Wasserstein GAN* by Martin Arjovsky, Soumith Chintala, and Léon Bottou, `https://arxiv.org/pdf/1701.07875.pdf`

9
Learning More about GANs

We learned what **Generative Adversarial Networks** (**GANs**) are and how different types of GANs are used to generate images in Chapter 8, *Generating Images Using GANs*.

In this chapter, we will uncover various interesting different types of GANs. We've learned that GANs can be used to generate new images but we do not have any control over the images that they generate. For instance, if we want our GAN to generate a human face with specific traits how do we tell this information to the GAN? We can't because we have no control over the images generated by the generator.

To resolve this, we use a new type of GAN called a **Conditional GAN (CGAN)** where we can condition the generator and discriminator by specifying what we want to generate. We will start off the chapter by comprehending how CGANs can be used to generate images of our interest and then we learn how to implement CGANs using **TensorFlow**.

We then understand about the **InfoGANs** which is an unsupervised version of a CGAN. We will understand what InfoGANs are and how they differ from CGANs, and how can we implement them using TensorFlow to generate new images.

Then, we shall learn about **CycleGANs**, which are a very intriguing type of GAN. They try to learn the mapping from the distribution of images in one domain to the distribution of images in another domain. For instance, to convert a grayscale image to a colored image, we train the CycleGAN to learn the mapping between grayscale and colored images, which means they learn to map from one domain, to another and the best part is, unlike other architectures, they even don't require a paired dataset. We will investigate how exactly they learn these mappings and their architecture in detail. We will explore how to implement CycleGAN to convert real pictures to paintings.

At the end of the chapter, we will explore, **StackGAN**, which can convert the text description to a photo-realistic image. We will perceive how StackGANs do this by gaining a deeper understanding of their architecture in detail.

In this chapter, we will learn about the following:

- Conditional GANs
- Generating specific digits using CGAN
- InfoGAN
- Architecture of InfoGAN
- Constructing InfoGAN using TensorFlow
- CycleGAN
- Converting pictures to paintings using CycleGAN
- StackGAN

Conditional GANs

We know that the generator generates new images by learning the real data distribution, while the discriminator examines whether the image generated by the generator is from the real data distribution or fake data distribution.

However, the generator has the capability to generate new and interesting images by learning the real data distribution. We have no control or influence over the images generated by the generator. For instance, let's say our generator is generating human faces; how can we tell the generator to generate a human face with certain features, say big eyes and a sharp nose?

We can't! Because we have no control over the images that are being generated by the generator.

To overcome this, we introduce a small variant of a GAN called a **CGAN**, which imposes a condition to both the generator and the discriminator. This condition tells the GAN that what image we want our generator to generate. So, both of our components—the discriminator and the generator—act upon this condition.

Let's consider a simple example. Say we are generating handwritten digits using CGAN with the MNIST dataset. Let's assume that we are more focused on generating digit 7 instead of other digits. Now, we need to impose this condition to both of our generators and discriminators. How do we do that?

The generator takes the noise z as an input and generates an image. But along with z, we also pass an additional input, c. This c is a one-hot encoded class label. As we are interested in generating digit 7, we set the seventh index to 1 and set all other indices to 0, that is, [0,0,0,0,0,0,0,1,0,0].

We concatenate the latent vector, z, and the one-hot encoded conditional variable, c, and pass that as an input to the generator. Then, the generator starts generating the digit 7.

What about the discriminator? We know that the discriminator takes image x as an input and tells us whether the image is a real or fake image. In CGAN, We want the discriminator to discriminate based on the condition, which means it has to identify whether the generated image is a real digit 7 or a fake digit 7. So, along with passing input x, we also pass the conditional variable c to the discriminator by concatenating x and c.

As you can see in the following figure, we are passing z and c to the generator:

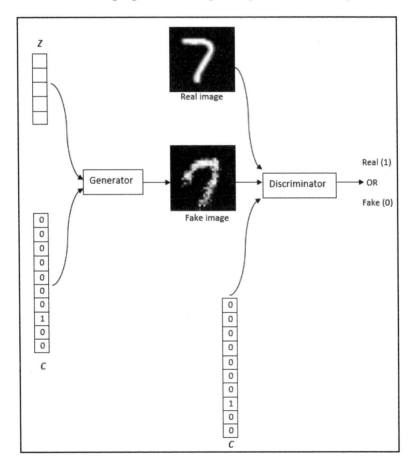

The generator is conditioned on information imposed on c. Similarly, along with passing real and fake images to the discriminator, we also pass c to the discriminator. So, the generator generates digit 7 and the discriminator learns to discriminate between the real 7 and the fake 7.

We've just learned how to generate a specific digit using CGAN, and yet the applications of CGAN do not end here. Assume we need to generate a digit with a specific width and height. We can also impose this condition on c and make the GAN to generate any desired image.

Loss function of CGAN

As you may have noticed, there is not much difference between our vanilla GAN and CGAN except that in CGANs, we are concatenating the additional input, which is the conditioning variable c with the inputs of the generator and the discriminator. So, the loss function for both generator and discriminator is the same as the vanilla GAN with the exception that it is conditioned on c.

Thus, the loss function of the discriminator is as follows:

$$\boxed{L^D = -\mathbb{E}_{x \sim p_r}\left[\log D(x|c)\right] - \mathbb{E}_{z \sim p(z)}\left[\log(1 - D(G(z|c)))\right]}$$

The loss function of the generator is given as:

$$\boxed{L^G = -\mathbb{E}_{z \sim p(z)}\left[\log D(G(z|c))\right]}$$

The CGAN learns by minimizing the loss function using gradient descent.

Generating specific handwritten digits using CGAN

We just learned how CGAN works and the architecture of CGAN. To strengthen our understanding, now we will learn how to implement CGAN in TensorFlow for generating an image of specific handwritten digit, say digit 7.

First, Load the required libraries:

```
import warnings
warnings.filterwarnings('ignore')

import numpy as np
import tensorflow as tf
from tensorflow.examples.tutorials.mnist import input_data
tf.logging.set_verbosity(tf.logging.ERROR)
tf.reset_default_graph()

import matplotlib.pyplot as plt
%matplotlib inline

from IPython import display
```

Load the MNIST dataset:

```
data = input_data.read_data_sets("data/mnist",one_hot=True)
```

Defining the generator

Generator, G, takes the noise, z, and also the conditional variable, c, as an input and returns an image. We define the generator as a simple two- layer feed-forward network:

```
def generator(z, c,reuse=False):
    with tf.variable_scope('generator', reuse=reuse):
```

Initialize the weights:

```
        w_init = tf.contrib.layers.xavier_initializer()
```

Concatenate the noise, z, and the conditional variable, c:

```
        inputs = tf.concat([z, c], 1)
```

Define the first layer:

```
        dense1 = tf.layers.dense(inputs, 128,
    kernel_initializer=w_init)
        relu1 = tf.nn.relu(dense1)
```

Define the second layer and compute the output with the `tanh` activation function:

```
        logits = tf.layers.dense(relu1, 784, kernel_initializer=w_init)
        output = tf.nn.tanh(logits)

        return output
```

Defining discriminator

We know that the discriminator, D, returns the probability; that is, it will tell us the probability of the given image being real. Along with the input image, x, it also takes the conditional variable, c, as an input. We define the discriminator also as a simple two-layer feed-forward network:

```
def discriminator(x, c, reuse=False):
    with tf.variable_scope('discriminator', reuse=reuse):
```

Initialize the weights:

```
        w_init = tf.contrib.layers.xavier_initializer()
```

Concatenate input, x and the conditional variable, c:

```
        inputs = tf.concat([x, c], 1)
```

Define the first layer:

```
        dense1 = tf.layers.dense(inputs, 128,
kernel_initializer=w_init)
        relu1 = tf.nn.relu(dense1)
```

Define the second layer and compute the output with the `sigmoid` activation function:

```
        logits = tf.layers.dense(relu1, 1, kernel_initializer=w_init)
        output = tf.nn.sigmoid(logits)

        return output
```

Define the placeholder for the input, x, conditional variable, c, and the noise, z:

```
x = tf.placeholder(tf.float32, shape=(None, 784))
c = tf.placeholder(tf.float32, shape=(None, 10))
z = tf.placeholder(tf.float32, shape=(None, 100))
```

Start the GAN!

First, we feed the noise, z and the conditional variable, c, to the generator, and it will output the fake image, that is, $fake\ x = G(z|c)$:

```
fake_x = generator(z, c)
```

Now we feed the real image x along with conditional variable, c, to the discriminator, $D(x|c)$, and get the probability of them being real:

```
D_logits_real = discriminator(x,c)
```

Similarly, we feed the fake image, `fake_x`, and the conditional variable, c, to the discriminator, $D(z|c)$, and get the probability of them being real:

```
D_logits_fake = discriminator(fake_x, c, reuse=True)
```

Computing the loss function

Now we will see how to compute the loss function. It is essentially the same as the Vanilla GAN except that we add a conditional variable.

Discriminator loss

The discriminator loss is given as follows:

$$L^D = -\mathbb{E}_{x\sim p_r(x)}[\log D(x|c)] - \mathbb{E}_{z\sim p_z(z)}[\log(1 - D(G(z|c)))]$$

First, we will implement the first term, that is, $\mathbb{E}_{x\sim p_r(x)}[\log D(x|c)]$:

```
D_loss_real =
tf.reduce_mean(tf.nn.sigmoid_cross_entropy_with_logits(logits=D_logits_real
,
            labels=tf.ones_like(D_logits_real)))
```

Now we will implement the second term, $\mathbb{E}_{z\sim p_z(z)}[\log(1 - D(G(z|c)))]$:

```
D_loss_fake =
tf.reduce_mean(tf.nn.sigmoid_cross_entropy_with_logits(logits=D_logits_fake
,
            labels=tf.zeros_like(D_logits_fake)))
```

The final loss can be written as follows:

```
D_loss = D_loss_real + D_loss_fake
```

Generator loss

The generator loss is given as follows:

$$L^G = -\mathbb{E}_{z \sim p_z(z)}[\log(D(G(z|c)))]$$

Generator loss can be implemented as:

```
G_loss =
tf.reduce_mean(tf.nn.sigmoid_cross_entropy_with_logits(logits=D_logits_fake
,
                labels=tf.ones_like(D_logits_fake)))
```

Optimizing the loss

We need to optimize our generator and discriminator. So, we collect the parameters of the discriminator and generator as `theta_D` and `theta_G` respectively:

```
training_vars = tf.trainable_variables()
theta_D = [var for var in training_vars if
var.name.startswith('discriminator')]
theta_G = [var for var in training_vars if
var.name.startswith('generator')]
```

Optimize the loss using the Adam optimizer:

```
learning_rate = 0.001

D_optimizer = tf.train.AdamOptimizer(learning_rate,
beta1=0.5).minimize(D_loss,
                var_list=theta_D)
G_optimizer = tf.train.AdamOptimizer(learning_rate,
beta1=0.5).minimize(G_loss,
                var_list=theta_G)
```

Start training the CGAN

Start the TensorFlow session and initialize the variables:

```
session = tf.InteractiveSession()
tf.global_variables_initializer().run()
```

Define the `batch_size`:

```
batch_size = 128
```

Define the number of epochs and the number of classes:

```
num_epochs = 500
num_classes = 10
```

Define the images and labels:

```
images = (data.train.images)
labels = data.train.labels
```

Generate the handwritten digit, 7

We set the digit (label) to generate as 7:

```
label_to_generate = 7
onehot = np.eye(10)
```

Set the number of iterations:

```
for epoch in range(num_epochs):

    for i in range(len(images) // batch_size):
```

Sample images based on the batch size:

```
        batch_image = images[i * batch_size:(i + 1) * batch_size]
```

Sample the condition that is, digit we want to generate:

```
        batch_c = labels[i * batch_size:(i + 1) * batch_size]
```

Sample noise:

```
        batch_noise = np.random.normal(0, 1, (batch_size, 100))
```

Train the generator and compute the generator loss:

```
        generator_loss, _ = session.run([D_loss, D_optimizer], {x:
    batch_image, c: batch_c, z: batch_noise})
```

Train the discriminator and compute the discriminator loss:

```
        discriminator_loss, _ = session.run([G_loss, G_optimizer], {x:
    batch_image, c: batch_c, z: batch_noise})
```

Randomly sample noise:

```
        noise = np.random.rand(1,100)
```

Select the digit we want to generate:

```
gen_label = np.array([[label_to_generate]]).reshape(-1)
```

Convert the selected digit into a one-hot encoded vector:

```
one_hot_targets = np.eye(num_classes)[gen_label]
```

Feed the noise and one hot encoded condition to the generator and generate the fake image:

```
_fake_x = session.run(fake_x, {z: noise, c: one_hot_targets})
_fake_x = _fake_x.reshape(28,28)
```

Print the loss of generator and discriminator and plot the generator image:

```
print("Epoch: {},Discriminator Loss:{}, Generator Loss:
{}".format(epoch,discriminator_loss,generator_loss))
#plot the generated image
display.clear_output(wait=True)
plt.imshow(_fake_x)
plt.show()
```

As you can see following plot, the generator has now learned to generate the digit 7 instead of generating other digits randomly:

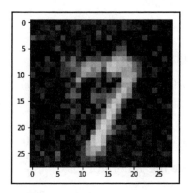

Understanding InfoGAN

InfoGAN is an unsupervised version of CGAN. In CGAN, we learned how to condition the generator and discriminator to generate the image we want. But how can we do that when we have no labels in the dataset? Assume we have an MNIST dataset with no labels; how can we tell the generator to generate the specific image that we are interested in? Since the dataset is unlabeled, we do not even know about the classes present in the dataset.

We know that generators use noise z as an input and generate the image. Generators encapsulate all the necessary information about the image in the z and it is called **entangled representation**. It is basically learning the semantic representation of the image in z. If we can disentangle this vector, then we can discover interesting features of our image.

So, we will split this z into two:

- Usual noise
- Code c

What is the code? The code c is basically interpretable disentangled information. Assuming we have MNIST data, then, code $c1$ implies the digit label, code $c2$ implies the width, $c3$ implies the stroke of the digit, and so on. We collectively represent them with the term c.

Now that we have z and c, how can we learn meaningful code c? Can we learn meaningful code with the image generated from the generator? Say a generator generates the image of 7. Now we can say code $c1$ *is 7* as we know c1 implies the digit label.

But since code can mean anything, say, a label, a width of the digit, stroke, rotation angle, and so on—how can we learn what we want? The code c will be learned based on the choice of the prior. For instance, if we chose a multinomial prior for c, then our InfoGAN might assign a digit label for c. Say, we assign a Gaussian prior, then it might assign a rotation angle, and so on. We can also have more than one prior.

The distribution for prior c can be anything. InfoGAN assigns different properties according to the distribution. In InfoGAN, the code c is inferred automatically based on the generator output, unlike CGAN, where we explicitly specify the c.

In a nutshell, we are inferring C based on the generator output, $G(z, c)$. But how exactly we are inferring C? We use a concept from information theory called **mutual information**.

Mutual information

Mutual information between two random variables tells us the amount of information we can obtain from one random variable through another. Mutual information between two random variables x and y can be given as follows:

$$I(x, y) = H(y) - H(y|x)$$

It is basically the difference between the entropy of y and the conditional entropy of y given x.

Mutual information between code c and the generator output $G(z|c)$ tells us how much information we can obtain about c through $G(z|c)$. If the mutual information c and $G(z|c)$ is high, then we can say knowing the generator output helps us to infer c. But if the mutual information is low, then we cannot infer c from the generator output. Our goal is to maximize the mutual information.

The mutual information between code c and the generator output, $G(z|c)$, can be given as follows:

$$I(c, G(z, c)) = H(c) - H(c|G(z, c))$$

Let's look at the elements of the formula:

- $H(c)$ is the entropy of the code
- $H(c|G(z, c))$ is the conditional entropy of the code c given the generator output $G(z|c)$

But the problem is, how do we compute $H(c|G(z, c))$? Because to compute this value, we need to know the posterior, $p(c|G(z, c))$, which we don't know yet. So, we estimate the posterior with the auxiliary distribution, $Q(c|x)$:

$$I(c, G(z, c)) = H(c) - H(c|G(z, c))$$

Let's say $x = G(z, c)$, then we can deduce mutual information as follows:

$$
\begin{aligned}
I(c, x) &= H(c) - \mathbb{E}_x H[c|x] \\
&= H(c) + \mathbb{E}_x \mathbb{E}_{c|x} \log p(c|x) \\
&= H(c) + \mathbb{E}_x \mathbb{E}_{c|x} \log \frac{p(c|x)q(c|x)}{q(c|x)} \\
&= H(c) + \mathbb{E}_x \mathbb{E}_{c|x} \log q(c|x) + \mathbb{E}_x \mathbb{E}_{c|x} \log \frac{p(c|x)}{q(c|x)} \\
&= H(c) + \mathbb{E}_x \mathbb{E}_{c|x} \log q(c|x) + \mathbb{E}_x KL[p(c|x)|q(c|x)] \\
&\geq H(c) + \mathbb{E}_x \mathbb{E}_{c|x} \log q(c|x)
\end{aligned}
$$

Thus, we can say:

$$I(c, G(z, c)) = \mathbb{E}_{c \sim P(c), x \sim G(z, c)} \ H(c) + \log Q(c|x)$$

Maximizing mutual information, $I(c, G(z, c))$ basically implies we are maximizing our knowledge about c given the generated output, that is, knowing about one variable through another.

Architecture of the InfoGAN

Okay. What is really going on here? why are we doing this? To put it in simple terms, we split the input to the generator into two: z and c. Since both z and c are used for generating the image, they capture the semantic meaning of the image. The code c gives us the interpretable disentangled information about the image. So we try to find c given the generator output. However, we can't do this easily since we don't know the posterior, $p(c|G(z,c))$, so we use an auxiliary distribution, $Q(c|x)$, to learn c.

This auxiliary distribution is basically another neural network; let's call this network as Q network. The role of the Q network is to predict the likelihood of c given a generator image x and is given by $Q(c|x)$.

First, we sample c from a prior, $p(c)$. Then we concatenate c and z and feed them to the generator. Next, we feed the generator result given by $G(z,c)$ to the discriminator. We know that the role of the discriminator is to output the probability of the given image being real. So, it takes the image generated by the generator and returns the probability. Also, the Q network takes the generated image and returns the estimates of c given the generated image.

Both the discriminator D and Q networks take the generator image and return the output so they both share some layers. Since they both share some layers, we attach the Q network to the discriminator, as shown in the following diagram:

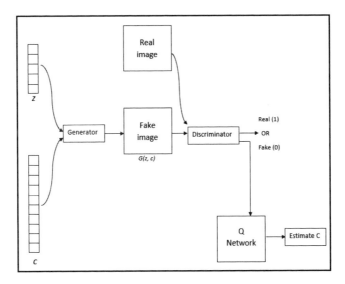

Thus, the discriminator returns two outputs:

- The probability of the image being real
- The estimates of c, which is the probability of c given the generator image

We add the mutual information term to our loss function.

Thus, the loss function of the discriminator is given as:

$$L^D = -\mathbb{E}_{x \sim p_r} \log D(\boldsymbol{x}) - \mathbb{E}_{z,c} \log(1 - D(G(z,c))) - \lambda I(c, G(z,c))$$

The loss function of the generator is given as:

$$L^G = -\mathbb{E}_{z,c} \log D(G(z,c)) - \lambda I(c, G(z,c))$$

Both of the preceding equations implies we are minimizing the loss of GAN along with maximizing the mutual information. Still confused about InfoGANs? Don't worry! We will learn about InfoGANs step by step better by implementing them in TensorFlow.

Constructing an InfoGAN in TensorFlow

We will better understand InfoGANs by implementing them in TensorFlow step by step. We will use the MNIST dataset and learn how the InfoGAN infers the code c automatically based on the generator output. We build an Info-DCGAN; that is, we use convolutional layers in the generator and discriminator instead of a vanilla neural network.

First, we will import all the necessary libraries:

```
import warnings
warnings.filterwarnings('ignore')

import numpy as np
import tensorflow as tf

from tensorflow.examples.tutorials.mnist import input_data
tf.logging.set_verbosity(tf.logging.ERROR)

import matplotlib.pyplot as plt
%matplotlib inline
```

Load the MNIST dataset:

```
data = input_data.read_data_sets("data/mnist",one_hot=True)
```

Define the leaky ReLU activation function:

```
def lrelu(X, leak=0.2):
    f1 = 0.5 * (1 + leak)
    f2 = 0.5 * (1 - leak)
    return f1 * X + f2 * tf.abs(X)
```

Defining generator

Generator G which takes the noise, z, and also a variable, c, as an input and returns an image. Instead of using a fully connected layer in the generator, we use a deconvolutional network, just like when we studied DCGANs:

```
def generator(c, z,reuse=None):
```

First, concatenate the noise, z, and the variable, c:

```
    input_combined = tf.concat([c, z], axis=1)
```

Define the first layer, which is a fully connected layer with batch normalization and ReLU activations:

```
    fuly_connected1 = tf.layers.dense(input_combined, 1024)
    batch_norm1 = tf.layers.batch_normalization(fuly_connected1,
training=is_train)
    relu1 = tf.nn.relu(batch_norm1)
```

Define the second layer, which is also fully connected with batch normalization and ReLU activations:

```
    fully_connected2 = tf.layers.dense(relu1, 7 * 7 * 128)
    batch_norm2 = tf.layers.batch_normalization(fully_connected2,
training=is_train)
    relu2 = tf.nn.relu(batch_norm2)
```

Flatten the result of the second layer:

```
    relu_flat = tf.reshape(relu2, [batch_size, 7, 7, 128])
```

The third layer consists of **deconvolution**; that is, a transpose convolution operation and it is followed by batch normalization and ReLU activations:

```
deconv1 = tf.layers.conv2d_transpose(relu_flat,
                                      filters=64,
                                      kernel_size=4,
                                      strides=2,
                                      padding='same',
                                      activation=None)
batch_norm3 = tf.layers.batch_normalization(deconv1, training=is_train)
relu3 = tf.nn.relu(batch_norm3)
```

The fourth layer is another transpose convolution operation:

```
deconv2 = tf.layers.conv2d_transpose(relu3,
                                      filters=1,
                                      kernel_size=4,
                                      strides=2,
                                      padding='same',
                                      activation=None)
```

Apply sigmoid function to the result of the fourth layer and get the output:

```
output = tf.nn.sigmoid(deconv2)

return output
```

Defining the discriminator

We learned that both the discriminator D and Q network take the generator image and return the output so they both share some layers. Since they both share some layers, we attach the Q network to the discriminator, as we learned in the architecture of InfoGAN. Instead of using fully connected layers in the discriminator, we use a convolutional network, as we learned in the discriminator of DCGAN:

```
def discriminator(x, reuse=None):
```

Define the first layer, which performs the convolution operation followed by a leaky ReLU activation:

```
conv1 = tf.layers.conv2d(x,
                         filters=64,
                         kernel_size=4,
                         strides=2,
                         padding='same',
         kernel_initializer=tf.contrib.layers.xavier_initializer(),
```

```
                                      activation=None)
        lrelu1 = lrelu(conv1, 0.2)
```

We also perform convolution operation in the second layer, which is followed by batch normalization and a leaky ReLU activation:

```
        conv2 = tf.layers.conv2d(lrelu1,
                                 filters=128,
                                 kernel_size=4,
                                 strides=2,
                                 padding='same',
        kernel_initializer=tf.contrib.layers.xavier_initializer(),
                                 activation=None)
        batch_norm2 = tf.layers.batch_normalization(conv2, training=is_train)
        lrelu2 = lrelu(batch_norm2, 0.2)
```

Flatten the result of the second layer:

```
        lrelu2_flat = tf.reshape(lrelu2, [batch_size, -1])
```

Feed the flattened result to a fully connected layer which is the third layer and it is followed by batch normalization and leaky ReLU activation:

```
        full_connected = tf.layers.dense(lrelu2_flat,
                                 units=1024,
                                 activation=None)
        batch_norm_3 = tf.layers.batch_normalization(full_connected,
        training=is_train)
        lrelu3 = lrelu(batch_norm_3, 0.2)
```

Compute the discriminator output:

```
        d_logits = tf.layers.dense(lrelu3, units=1, activation=None)
```

As we learned that we attach the Q network to the discriminator. Define the first layer of the Q network that takes the final layer of the discriminator as inputs:

```
        full_connected_2 = tf.layers.dense(lrelu3,
                                     units=128,
                                     activation=None)

        batch_norm_4 = tf.layers.batch_normalization(full_connected_2,
        training=is_train)
        lrelu4 = lrelu(batch_norm_4, 0.2)
```

Define the second layer of the Q network:

```
q_net_latent = tf.layers.dense(lrelu4,
                               units=74,
                               activation=None)
```

Estimate c:

```
q_latents_categoricals_raw = q_net_latent[:,0:10]

c_estimates = tf.nn.softmax(q_latents_categoricals_raw, dim=1)
```

Return the discriminator `logits` and the estimated c value as output:

```
return d_logits, c_estimates
```

Define the input placeholders

Now we define the placeholder for the input, x, the noise, z, and the code, c:

```
batch_size = 64
input_shape = [batch_size, 28,28,1]


x = tf.placeholder(tf.float32, input_shape)
z = tf.placeholder(tf.float32, [batch_size, 64])
c = tf.placeholder(tf.float32, [batch_size, 10])

is_train = tf.placeholder(tf.bool)
```

Start the GAN

First, we feed the noise, z and the code, c to the generator, and it will output the fake image according to the equation $fake\ x = G(z, c)$:

```
fake_x = generator(c, z)
```

Now we feed the real image, x, to the discriminator, $D(x)$, and get the probability that the image being real. Along with this, we also obtain the estimate of c for the real image:

```
D_logits_real, c_posterior_real = discriminator(x)
```

Similarly, we feed the fake image to the discriminator and get the probability of the image being real and also the estimate of c for the fake image:

```
D_logits_fake, c_posterior_fake = discriminator(fake_x,reuse=True)
```

Computing loss function

Now we will see how to compute the loss function.

Discriminator loss

The discriminator loss is given as follows:

$$L^D = -\mathbb{E}_{x \sim p_r(x)}[\log D(x)] - \mathbb{E}_{z \sim p_z(z)}[\log(1 - D(G(z)))]$$

As the discriminator loss of an InfoGAN is same as with a CGAN, implementing the discriminator loss is the same as what we learned in the CGAN section:

```
#real loss
D_loss_real =
tf.reduce_mean(tf.nn.sigmoid_cross_entropy_with_logits(logits=D_logits_real
,
 labels=tf.ones(dtype=tf.float32, shape=[batch_size, 1])))

#fake loss
D_loss_fake =
tf.reduce_mean(tf.nn.sigmoid_cross_entropy_with_logits(logits=D_logits_fake
,
 labels=tf.zeros(dtype=tf.float32, shape=[batch_size, 1])))

#final discriminator loss
D_loss = D_loss_real + D_loss_fake
```

Generator loss

The loss function of the generator is given as follows:

$$L^G = -\mathbb{E}_{z \sim p_z(z)}[\log(D(G(z)))]$$

Generator loss is implemented as:

```
G_loss =
tf.reduce_mean(tf.nn.sigmoid_cross_entropy_with_logits(logits=D_logits_fake
,
 labels=tf.ones(dtype=tf.float32, shape=[batch_size, 1]))))
```

Mutual information

We subtract **mutual information** from both the discriminator and the generator loss. So, the final loss function of discriminator and generator is given as follows:

$$L^D = L^D - \lambda I(c, G(z, c))$$

$$L^G = L^G - \lambda I(c, G(z, c))$$

The mutual information can be calculated as follows:

$$I(c, G(z, c)) = \mathbb{E}_{c \sim P(c), x \sim G(z, c)} \; H(c) + \log Q(c|X)$$

First, we define a prior for c:

```
c_prior = 0.10 * tf.ones(dtype=tf.float32, shape=[batch_size, 10])
```

The entropy of c is represented as $H(c)$. We know that the entropy is calculated as $H(c) = - \sum_i p(c) \log p(c)$:

```
entropy_of_c = tf.reduce_mean(-tf.reduce_sum(c *
tf.log(tf.clip_by_value(c_prior, 1e-12, 1.0)),axis=-1))
```

The conditional entropy of c when X is given is $\log Q(c|X)$. The code for the conditional entropy is as follows:

```
log_q_c_given_x = tf.reduce_mean(tf.reduce_sum(c *
tf.log(tf.clip_by_value(c_posterior_fake, 1e-12, 1.0)), axis=-1))
```

The mutual information is given as $I(c, G(z, c)) = H(c) + \log Q(c|X)$:

```
mutual_information = entropy_of_c + log_q_c_given_x
```

The final loss of the discriminator and the generator is given as:

```
D_loss = D_loss - mutual_information
G_loss = G_loss - mutual_information
```

Optimizing the loss

Now we need to optimize our generator and discriminator. So, we collect the parameters of the discriminator and generator as θ_D and θ_G respectively:

```
training_vars = tf.trainable_variables()

theta_D = [var for var in training_vars if 'discriminator' in var.name]
theta_G = [var for var in training_vars if 'generator' in var.name]
```

Optimize the loss using the Adam optimizer:

```
learning_rate = 0.001

D_optimizer =
tf.train.AdamOptimizer(learning_rate).minimize(D_loss,var_list = theta_D)
G_optimizer = tf.train.AdamOptimizer(learning_rate).minimize(G_loss,
var_list = theta_G)
```

Beginning training

Define the batch size and the number of epochs and initialize all the TensorFlow variables:

```
num_epochs = 100
session = tf.InteractiveSession()
session.run(tf.global_variables_initializer())
```

Define a helper function for visualizing results:

```
def plot(c, x):
    c_ = np.argmax(c, 1)

    sort_indices = np.argsort(c_, 0)
    x_reshape = np.reshape(x[sort_indices], [batch_size, 28, 28])
    x_reshape = np.reshape( np.expand_dims(x_reshape, axis=0), [4,
(batch_size // 4), 28, 28])

    values = []
    for i in range(0,4):
        row = np.concatenate( [x_reshape[i,j,:,:] for j in
range(0,(batch_size // 4))], axis=1)
        values.append(row)
    return np.concatenate(values, axis=0)
```

Generating handwritten digits

Start training and generate the image. For every 100 iterations, we print the image generated by the generator:

```
onehot = np.eye(10)

for epoch in range(num_epochs):

    for i in range(0, data.train.num_examples // batch_size):
```

Sample the images:

```
x_batch, _ = data.train.next_batch(batch_size)
x_batch = np.reshape(x_batch, (batch_size, 28, 28, 1))
```

Sample the value of c:

```
c_ = np.random.randint(low=0, high=10, size=(batch_size,))
c_one_hot = onehot[c_]
```

Sample noise z:

```
z_batch = np.random.uniform(low=-1.0, high=1.0,
size=(batch_size,64))
```

Optimize the loss of the generator and the discriminator:

```
feed_dict={x: x_batch, c: c_one_hot, z: z_batch, is_train: True}

_ = session.run(D_optimizer, feed_dict=feed_dict)
_ = session.run(G_optimizer, feed_dict=feed_dict)
```

Print the generator image for every 100th iteration:

```
if i % 100 == 0:
    discriminator_loss = D_loss.eval(feed_dict)
    generator_loss = G_loss.eval(feed_dict)
    _fake_x = fake_x.eval(feed_dict)

    print("Epoch: {}, iteration: {}, Discriminator Loss:{},
Generator Loss: {}".format(epoch,i,discriminator_loss,generator_loss))
    plt.imshow(plot(c_one_hot, _fake_x))
    plt.show()
```

We can see how the generator is evolving on each iteration and generating better digits:

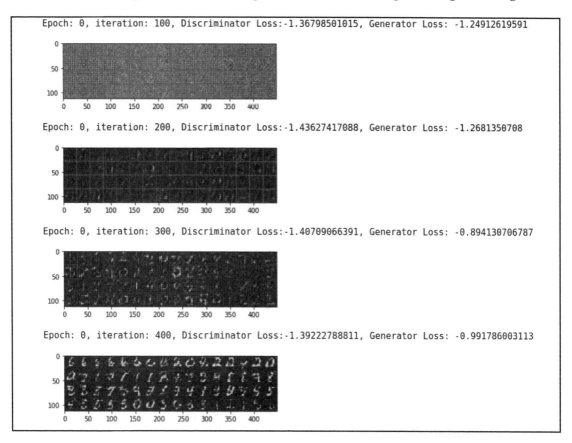

Translating images using a CycleGAN

We have learned several types of GANs, and the applications of them are endless. We have seen how the generator learns the distribution of real data and generates new realistic samples. We will now see a really different and very innovative type of GAN called the **CycleGAN**.

Unlike other GANs, the CycleGAN maps the data from one domain to another domain, which implies that here we try to learn the mapping from the distribution of images from one domain to the distribution of images in another domain. To put it simply, we translate images from one domain to another.

What does this mean? Assume we want to convert a grayscale image to a colored image. The grayscale image is one domain and the colored image is another domain. A CycleGAN learns the mapping between these two domains and translates between them. This means that given a grayscale image, a CycleGAN converts the image into a colored one.

Applications of CycleGANs are numerous, such as converting real photos to artistic pictures, season transfer, photo enhancement, and many more. As shown in the following figure, you can see how a CycleGAN converts images between different domains:

But what is so special about CycleGANs? It's their ability to convert images from one domain to another without any paired examples. Let's say we are converting photos (source) to paintings (target). In a normal image-to-image translation, how do we that? We prepare the training data by collecting some photos and also their corresponding paintings in pairs, as shown in the following image:

Collecting these paired data points for every use case is an expensive task, and we might not have many records or pairs. Here is where the biggest advantage of a CycleGAN lies. It doesn't require data in aligned pairs. To convert from photos to paintings, we just need a bunch of photos and a bunch of paintings. They don't have to map or align with each other.

As shown in the following figure, we have some photos in one column and some paintings in another column; as you can see, they are not paired with each other. They are completely different images:

Thus, to convert images from any source domain to the target domain, we just need a bunch of images in both of the domains and it does not have to be paired. Now let's see how they work and how they learn the mapping between the source and the target domain.

Unlike other GANs, the CycleGAN consists of two generators and two discriminators. Let's represent an image in the source domain by x and target domain by y. We need to learn the mapping between x and y.

Let's say we are learning to convert a real picture, x, to a painting, y, as shown in the following figure:

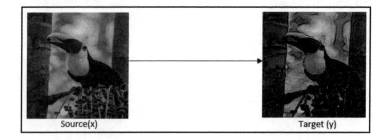

Source(x)   Target (y)

Role of generators

We have two generators, G and F. The role of G is to learn the mapping from x to y. As mentioned above, the role of G is to learn to translate photos to paintings, as shown in the following figure:

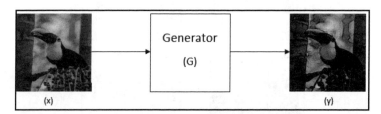

It tries to generate a fake target image, which implies it takes a source image, x, as input and generates a fake target image, y:

$$y = G(x)$$

The role of the generator, F, is to learn the mapping from y to x and learn to translate from the painting to a real picture, as shown in following figure:

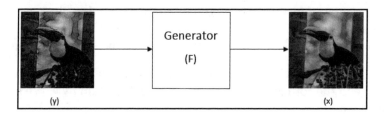

It tries to generate a fake source image, which implies it takes a target image, y, as input and generates a fake source image, x:

$$x = F(y)$$

Role of discriminators

Similar to the generators, we have two discriminators, D_x and D_y. The role of the discriminator D_x is to discriminate between the real source image, x, and the fake source image $F(y)$. We know that the fake source image is generated by the generator F.

Given an image to discriminator D_x, it returns the probability of the image being a real source image:

$$log\ D_x(x) + log\ (1 - D_x(F(y)))$$

The following figure shows the discriminator D_x, as you can observe it takes the real source image x and the fake source image generated by the generator F as inputs and returns the probability of the image being a real source image:

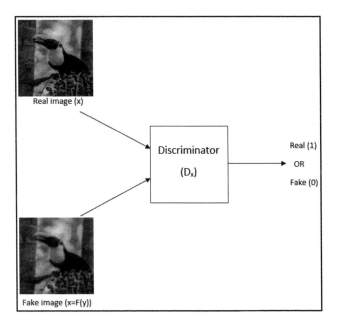

The role of discriminator D_y is to discriminate between the real target image, y, and the fake target image, $G(x)$. We know that fake target image is generated by the generator, G. Given an image to discriminator D_y, it returns the probability of the image being a real target image:

$$log\ D_y(y) + log\ (1 - D_y(G(x)))$$

The following figure shows the discriminator D_y, as you can observe it takes the real target image, y and the fake target image generated by the generator, G as inputs and returns the probability of the image being a real target image:

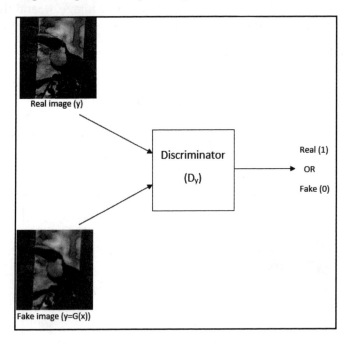

Loss function

In CycleGANs, we have two generators and two discriminators. Generators learn to translate images from one domain to another and a discriminator tries to discriminate between the translated images.

So, we can say the loss function of discriminator D_x can be represented as follows:

$$L^{D_x} = -\mathbb{E}_{x \sim p_r(x)}[log\ D_x(x)] - \mathbb{E}_{y \sim p_r(y)}log\ (1 - D_x(F(y)))$$

Similarly, the loss function of discriminator D_y can be represented as follows:

$$L^{D_y} = -\mathbb{E}_{y \sim p_r(y)}[log \, D_y(y)] - \mathbb{E}_{x \sim p_r(x)}log \, (1 - D_y(G(x)))$$

The loss function of generator G can be represented as follows:

$$L^G = -\mathbb{E}_{x \sim p_r(x)}[log \, D_y(G(x))]$$

The loss function of generator F can be given as follows:

$$L^F = -\mathbb{E}_{y \sim p_r(y)}[log \, D_x(F(y))]$$

Altogether, the final loss can be written as follows:

$$L^{Gan} = L^{D_x} + L^{D_y} + L^G + L^F$$

Cycle consistency loss

The adversarial loss alone does not ensure the proper mapping of the images. For instance, a generator can map the images from the source domain to a random permutation of images in the target domain which can match the target distribution.

So, to avoid this, we introduce an additional loss called **cycle consistent loss**. It enforces both generators G and F to be cycle-consistent.

Let's recollect the function of the generators:

- **Generator G**: Converts x to y
- **Generator F**: Converts y to x

We know that generator G takes the source image x and converts it to a fake target image y. Now if we feed this generated fake target image y to generator F, it has to return the original source image x. Confusing, right?

Look at the following figure; we have a source image, x. First, we feed this image to generator G, and it returns the fake target image. Now we take this fake target image, y, and feed it to generator F, which has to return the original source image:

$$x \rightarrow G(x) \rightarrow F(G(x)) \rightarrow x$$

The above equation can be represented as follows:

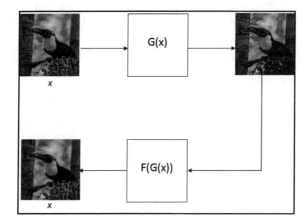

This is called the **forward consistency loss** and can be represented as follows:

$$L^{forward} = \mathbb{E}_{x \sim p_r(x)}[\|F(G(x)) - x\|]$$

Similarly, we can specify backward consistent loss, as shown in the following figure. Let's say we have an original target image, y. We take this y and feed it to discriminator F, and it returns the fake source image x. Now we feed this fake source image x to the generator G, and it has to return the original target image y:

$$y \rightarrow F(y) \rightarrow F(G(y)) \rightarrow y$$

The preceding equation can be represented as:

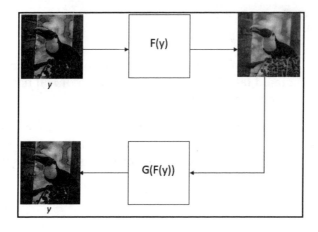

The backward consistency loss can be represented as follows:

$$L^{backward} = \mathbb{E}_{y \sim p_r(y)}[\|G(F(y)) - y\|]$$

So, together with a backward and forward consistent loss, we can write the cycle consistency loss as:

$$L^{cycle} = L^{forward} + L^{backward}$$

$$L^{cycle} = \mathbb{E}_{x \sim p_r(x)}[\|F(G(x)) - x\|] + \mathbb{E}_{y \sim p_r(y)}[\|G(F(y)) - y\|]$$

We want our generators to cycle consistent so, we multiply their loss with the cycle consistent loss. So, the final loss function can be given as:

$$L = L^{D_x} + L^{D_y} + \lambda L^{cycle}(L^G + L^F)$$

Converting photos to paintings using a CycleGAN

Now we will learn how to implement a CycleGAN in TensorFlow. We will see how to convert pictures to paintings using a CycleGAN:

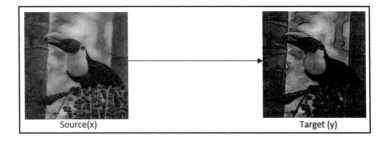

The dataset used in this section can be downloaded from `https://people.eecs.berkeley. edu/~taesung_park/CycleGAN/datasets/monet2photo.zip`. Once you have downloaded the dataset, unzip the archive; it will consist of four folders, `trainA`, `trainB`, `testA`, and `testB`, with training and testing images.

The `trainA` folder consists of paintings (Monet) and the `trainB` folder consists of photos. Since we are mapping photos (x) to the paintings (y), the `trainB` folder, which consists of photos, will be our source image, x, and the `trainA`, which consists of paintings, will be our target image, y.

The complete code for the CycleGAN with a step-by-step explanation is available as a Jupyter Notebook at `https://github.com/PacktPublishing/Hands-On-Deep-Learning-Algorithms-with-Python`.

Instead of looking at the whole code, we will see only how the CycleGAN is implemented in TensorFlow and maps the source image to the target domain. You can also check the complete code at `https://github.com/PacktPublishing/Hands-On-Deep-Learning-Algorithms-with-Python`.

Define the `CycleGAN` class:

```
class CycleGAN:
        def __init__(self):
```

Define the placeholders for the input, X, and the output, Y:

```
        self.X = tf.placeholder("float", shape=[batchsize, image_height,
image_width, 3])
        self.Y = tf.placeholder("float", shape=[batchsize, image_height,
image_width, 3])
```

Define the generator, G, that maps x to y:

```
        G = generator("G")
```

Define the generator, F, that maps y to x:

```
        F = generator("F")
```

Define the discriminator, D_x, that discriminates between the real source image and the fake source image:

```
        self.Dx = discriminator("Dx")
```

Define the discriminator, D_y, that discriminates between the real target image and the fake target image:

```
        self.Dy = discriminator("Dy")
```

Generate the fake source image:

```
        self.fake_X = F(self.Y)
```

Generate the fake target image:

```
        self.fake_Y = G(self.X)
```

Get the `logits`:

```
#real source image logits
self.Dx_logits_real = self.Dx(self.X)

#fake source image logits
self.Dx_logits_fake = self.Dx(self.fake_X, True)

#real target image logits
self.Dy_logits_fake = self.Dy(self.fake_Y, True)

#fake target image logits
self.Dy_logits_real = self.Dy(self.Y)
```

We know cycle consistency loss is given as follows:

$$L^{cycle} = \mathbb{E}_{x \sim p_r(x)}[\|F(G(x)) - x\|] + \mathbb{E}_{y \sim p_r(y)}[\|G(F(y)) - y\|]$$

We can implement the cycle consistency loss as follows:

```
    self.cycle_loss = tf.reduce_mean(tf.abs(F(self.fake_Y, True) -
self.X)) + \
                    tf.reduce_mean(tf.abs(G(self.fake_X, True) -
self.Y))
```

Define the loss for both of our discriminators, D_x and D_y.

We can rewrite our loss function of discriminator with Wasserstein distance as:

$$L^{D_x} = -\mathbb{E}_{x \sim p_r} D_x(x) + \mathbb{E}_z D_x(F(y))$$

$$L^{D_y} = -\mathbb{E}_{x \sim p_r} D_y(y) + \mathbb{E}_z D_y(G(x))$$

Thus, the loss of both the discriminator is implemented as follows:

```
    self.Dx_loss = -tf.reduce_mean(self.Dx_logits_real) +
tf.reduce_mean(self.Dx_logits_fake)
    self.Dy_loss = -tf.reduce_mean(self.Dy_logits_real) +
tf.reduce_mean(self.Dy_logits_fake)
```

Define the loss for both of the generators, G and F. We can rewrite our loss function of generators with Wasserstein distance as:

$$L^G = -\mathbb{E}_{x \sim p_r(x)} D_y(G(x))$$

$$L^F = -\mathbb{E}_{y \sim p_r(y)} D_x(F(y))$$

Thus, the loss of both the generators multiplied with the cycle consistency loss, `cycle_loss` is implemented as:

```
        self.G_loss = -tf.reduce_mean(self.Dy_logits_fake) + 10. *
    self.cycle_loss

        self.F_loss = -tf.reduce_mean(self.Dx_logits_fake) + 10. *
    self.cycle_loss
```

Optimize the discriminators and generators using the Adam optimizer:

```
        #optimize the discriminator
        self.Dx_optimizer = tf.train.AdamOptimizer(2e-4, beta1=0.,
    beta2=0.9).minimize(self.Dx_loss, var_list=[self.Dx.var])


        self.Dy_optimizer = tf.train.AdamOptimizer(2e-4, beta1=0.,
    beta2=0.9).minimize(self.Dy_loss, var_list=[self.Dy.var])

        #optimize the generator
        self.G_optimizer = tf.train.AdamOptimizer(2e-4, beta1=0.,
    beta2=0.9).minimize(self.G_loss, var_list=[G.var])


        self.F_optimizer = tf.train.AdamOptimizer(2e-4, beta1=0.,
    beta2=0.9).minimize(self.F_loss, var_list=[F.var])
```

Once we start training the model, we can see how the loss of discriminators and generators decreases over the iterations:

```
Epoch: 0, iteration: 0, Dx Loss: -0.6229429245, Dy Loss: -2.42867970467, G
Loss: 1385.33557129, F Loss: 1383.81530762, Cycle Loss: 138.448059082


Epoch: 0, iteration: 50, Dx Loss: -6.46077537537, Dy Loss: -7.29514217377,
G Loss: 629.768066406, F Loss: 615.080932617, Cycle Loss: 62.6807098389


Epoch: 1, iteration: 100, Dx Loss: -16.5891685486, Dy Loss: -16.0576553345,
G Loss: 645.53137207, F Loss: 649.854919434, Cycle Loss: 63.9096908569
```

StackGAN

Now we will see one of the most intriguing and fascinating types of GAN, which is called a **StackGAN**. Can you believe if I say StackGANs can generate photo-realistic images just based on the textual descriptions? Well, yes. They can do that. Given a text description, they can generate a realistic image.

Let's first understand how an artist draws an image. In the first stage, artists draw primitive shapes and create a basic outline that forms an initial version of the image. In the next stage, they enhance the image by making it more realistic and appealing.

StackGANs works in a similar manner. They divide the process of generating images into two stages. Just like artists draw pictures, in the first stage, they generate a basic outline, primitive shapes, and create a low-resolution version of the image, and in the second stage, they enhance the picture generated in the first stage by making it more realistic, and then convert them into a high-resolution image.

But how do StackGANs do this?

They use two GANs, one for each stage. The GAN in the first stage generates a basic image and sends it to the GAN in the next stage, which converts basic low-resolution image into a proper high-resolution image. The following figure shows how StackGANs generate images in each of the stages based on the text description:

Source: https://arxiv.org/pdf/1612.03242.pdf

As you can see , in the first stage, we have a low-resolution version of the image, but in the second stage, we have good clarity high-resolution image. But, still, how StackGAN are doing this? Remember, when we learned with conditional GANs that we can make our GAN generate images that we want by conditioning them?

We just use them in both of the stages. In stage one, our network is conditioned based on the text description. With this text description, they generate a basic version of an image. In stage II, our network is conditioned based on the image generated from stage I and also on the text description.

But why do we have to have to condition on the text description again in stage II? Because in stage I, we miss some details specified in the text description to create a basic version of an image. So, in stage II, we again condition on the text description to fix the missing information and also to make our image more realistic.

With this ability to generate pictures just based on the text, it is used for numerous applications. It is heavily used in the entertainment industry, for instance, for creating frames just based on descriptions, and it can also be used for generating comics and many more.

The architecture of StackGANs

Now that we have a basic understanding of how StackGANs work, we will take a closer look into their architecture and see how exactly they generate a picture from the text.

The complete architecture of a StackGAN is shown in the following diagram:

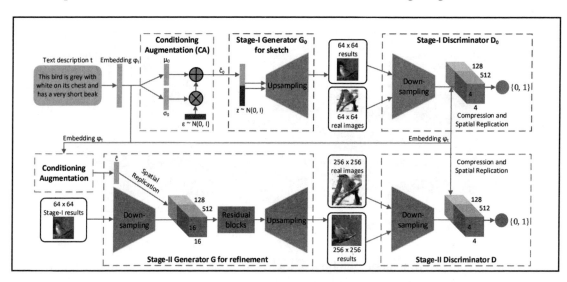

Source: https://arxiv.org/pdf/1612.03242.pdf

We will look at each component one by one.

Conditioning augmentation

We have a text description as an input to the GAN. Based on these descriptions, it has to generate the images. But how do they understand the meaning of the text to generate a picture?

First, we convert the text into an embedding using an encoder. We represent this text embedding by φ_t. Can we create variations of φ_t? By creating variations of text embeddings, φ_t, we can have additional training pairs, and we can also increase the robustness to small perturbations.

Let $\mu(\varphi_t)$ be mean and $\Sigma(\varphi_t)$ be the diagonal covariance matrix of our text embedding, φ_t. Now we randomly sample an additional conditioning variable, $\hat{c}$, from the independent Gaussian distribution, $\mathcal{N}(\mu(\varphi_t), \Sigma(\varphi_t))$. It helps us create variations of text descriptions with their meanings. We know that same text can be written in various ways, so with the conditioning variable, $\hat{c}$, we can have various versions of the text mapping to the image.

Thus, once we have the text description, we will extract their embeddings using the encoder, and then we compute their mean and covariance. Then, we sample $\hat{c}$ from the Gaussian distribution of the text embedding, φ_t.

Stage I

OK, now we have a text embedding, φ_t, and also a conditioning variable, $\hat{c}$. We will see how it is being used to generate the basic version of the image.

Generator

We know that the goal of the generator is to generate a fake image by learning the real data distribution. First, we sample noise from a Gaussian distribution and create z. Then we concatenate z with our conditional variable, $\hat{c}$, and feed this as an input to the generator which outputs a basic version of the image.

The loss function of the generator is given as follows:

$$\mathcal{L}_{G_0} = \mathbb{E}_{z \sim p_z, t \sim p_r} \left[\log(1 - D_0\left(G_0\left(z, \hat{c}_0\right), \varphi_t\right)) \right]$$

Let's examine this formula:

- $z \sim p_z$ implies we sample z from the fake data distribution, that is, the noise prior.
- $t \sim p_r$ implies we sample the text description, t, from the real data distribution.
- $G_0(z, \hat{c}_0)$ implies that the generator takes the noise and the conditioning variable returns the image. We feed this generated image to the discriminator.
- $(1 - D_0(G_0(z, \hat{c}_0), \varphi_t))$ implies the log probability of the generated image being fake.

Along with this loss, we also add a regularizer term, $\lambda D_{KL} (\mathcal{N} (\mu_0 (\varphi_t), \Sigma_0 (\varphi_t)) \| \mathcal{N}(0, I))$, to our loss function, which implies the KL divergence between the standard Gaussian distribution and the conditioning Gaussian distribution. It helps us to avoid overfitting.

So, our final loss function for the generator becomes:

$$\mathcal{L}_{G_0} = \mathbb{E}_{z \sim p_z, t \sim p_r} \left[\log(1 - D_0 (G_0 (z, \hat{c}_0), \varphi_t)) \right] + \lambda D_{KL} (\mathcal{N} (\mu_0 (\varphi_t), \Sigma_0 (\varphi_t)) \| \mathcal{N}(0, I))$$

Discriminator

Now we feed this generated image to the discriminator, which returns the probability of the image being real. The discriminator loss is given as follows:

$$\mathcal{L}_{D_0} = \mathbb{E}_{(I_0, t) \sim p_r} \left[\log D_0 (I_0, \varphi_t) \right] + \mathbb{E}_{z \sim p_z, t \sim p_r} \left[\log(1 - D_0 (G_0 (z, \hat{c}_0), \varphi_t)) \right]$$

Here:

- $D_0 (I_0, \varphi_t)$ implies the real image, I_0, conditional on the text description, φ_t
- $G_0(z, \hat{c}_0)$ implies the generated fake image

Stage II

We have learned how the basic version of the image is generated in stage I. Now, in stage II, we fix the defects of the image produced in stage I and generate a more realistic version of the image. We condition our network with the image generated from the previous stage and also on the text embeddings.

Generator

Instead of taking noise as an input, the generator in stage II takes the image generated from the previous stage as an input and it is conditioned on the text description.

Here, $s_0 \sim p_{G_0}$ implies that we are sampling s_0 from the p_{G_0}. It basically means that we are sampling the image generated from stage I.

$t \sim p_r$ implies that we are sampling text from the given real data distribution, p_r.

Then the generator loss can be given as follows:

$$\mathcal{L}_G = \mathbb{E}_{s_0 \sim p_{G_0}, t \sim p_r} \left[\log(1 - D\left(G\left(s_0, \hat{c}\right), \varphi_t\right))\right]$$

Along with the regularizer, our loss function of the generator becomes:

$$\mathcal{L}_G = \mathbb{E}_{s_0 \sim p_{G_0}, t \sim p_r} \left[\log(1 - D\left(G\left(s_0, \hat{c}\right), \varphi_t\right))\right] + \lambda D_{KL}\left(\mathcal{N}\left(\mu\left(\varphi_t\right), \Sigma\left(\varphi_t\right)\right) \| \mathcal{N}(0, I)\right)$$

Discriminator

The goal of the discriminator is to tell us whether the image is from the real distribution or the generator distribution. Thus, the loss function of the discriminator is given as follows:

$$\mathcal{L}_D = \mathbb{E}_{(I,t) \sim p_r} \left[\log D\left(I, \varphi_t\right)\right] + \mathbb{E}_{s_0 \sim p_{G_0}, t \sim p_r} \left[\log(1 - D\left(G\left(s_0, \hat{c}\right), \varphi_t\right))\right]$$

Summary

We started the chapter by learning about conditional GANs and how they can be used to generate our image of interest.

Later, we learned about InfoGANs, where the code c is inferred automatically based on the generated output, unlike CGAN, where we explicitly specify c. To infer c, we need to find the posterior, $p(c|G(z,c))$, which we don't have access to. So, we use an auxiliary distribution. We used mutual information to maximize the mutual information, $I(c, G(z, c))$, to maximize our knowledge about c given the generator output.

Then, we learned about CycleGANs, which map the data from one domain to another domain. We tried to learn the mapping from the distribution of images from photos domain to the distribution of images in paintings domain. Finally, we understood how StackGANs generate photorealistic images from a text description.

In the next chapter, we will learn about **autoencoders** and their types.

Questions

Answer the following questions to gauge how much you have learned from this chapter:

1. How do conditional GANs differ from vanilla GANs?
2. What is the code called in InfoGAN?
3. What is mutual information?
4. Why do we need auxiliary distribution in InfoGANs?
5. What is cycle consistency loss?
6. Explain the role of generators in a CycleGAN.
7. How do StackGANs convert text descriptions into pictures?

Further reading

Refer to the following links for more information:

- *Conditional Generative Adversarial Nets* by Mehdi Mirza and Simon Osindero, `https://arxiv.org/pdf/1411.1784.pdf`
- *InfoGAN: Interpretable Representation Learning by Information Maximizing Generative Adversarial Nets* by Xi Chen et al., `https://arxiv.org/pdf/1606.03657.pdf`
- *Unpaired Image-to-Image Translation using Cycle-Consistent Adversarial Networks* by Jun-Yan Zhu et al., `https://arxiv.org/pdf/1703.10593.pdf`

10
Reconstructing Inputs Using Autoencoders

Autoencoders are unsupervised learning algorithm. Unlike other algorithms, autoencoders learn to reconstruct the input, that is, an autoencoder takes the input and learns to reproduce the input as an output. We start the chapter by understanding what are autoencoders and how exactly they reconstruct the input. Then, we will learn how autoencoders reconstruct MNIST images.

Going ahead, we will learn about the different variants of autoencoders; first, we will learn about **convolutional autoencoders** (**CAEs**), which use convolutional layers; then, we will learn about how **denoising autoencoders** (**DAEs**) which learn to remove noise in the input. After this, we will understand sparse autoencoders and how they learn from sparse inputs. At the end of the chapter, we will learn about an interesting generative type of autoencoders called **variational autoencoders**. We will understand how variational autoencoders learn to generate new inputs and how they differ from other autoencoders.

In this chapter, we will cover the following topics:

- Autoencoders and their architecture
- Reconstructing MNIST images using autoencoders
- Convolutional autoencoders
- Building convolutional autoencoders
- Denoising autoencoders

- Removing noise in the image using denoising autoencoders
- Sparse autoencoders
- Contractive autoencoders
- Variational autoencoders

What is an autoencoder?

An **autoencoder** is an interesting unsupervised learning algorithm. Unlike other neural networks, the objective of the autoencoder is to reconstruct the given input; that is, the output of the autoencoders is the same as the input. It consists of two important components called the **encoder** and the **decoder**.

The role of the encoder is to encode the input by learning the latent representation of the input, and the role of the decoder is to reconstruct the input from the latent representation produced by the encoder. The latent representation is also called **bottleneck** or **code**. As shown in the following diagram, an image is passed as an input to the autoencoder. An encoder takes the image and learns the latent representation of the image. The decoder takes the latent representation and tries to reconstruct the image:

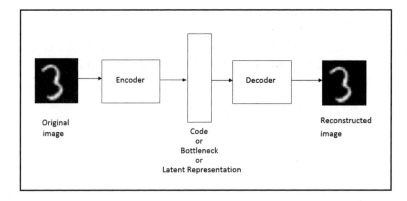

A simple vanilla autoencoder with two layers is shown in the following diagram; as you may notice, it consists of an input layer, a hidden layer, and an output layer. First, we feed the input to the input layer, and then the encoder learns the representation of the input and maps it to the bottleneck. From the bottleneck, the decoder reconstructs the input:

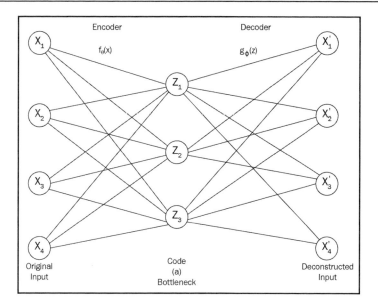

We might wonder what the use of this is. Why do we need to encode and decode the inputs? Why do we just have to reconstruct the input? Well, there are various applications such as dimensionality reduction, data compression, image denoising, and more.

Since the autoencoder reconstructs the inputs, the number of nodes in the input and output layers are always the same. Let's assume we have a dataset that contains 100 input features, and we have a neural network with an input layer of 100 units, a hidden layer of 50 units, and an output layer with 100 units. When we feed the dataset to the autoencoder, the encoder tries to learn the important features in the dataset and reduces the number of features to 50 and forms the bottleneck. The bottleneck holds the representations of the data, that is, the embeddings of the data, and encompasses only the necessary information. Then, the bottleneck is fed to the decoder to reconstruct the original input. If the decoder reconstructs the original input successfully, then it means that the encoder has successfully learned the encodings or representations of the given input. That is, the encoder has successfully encoded or compressed the dataset of 100 features into a representation with only 50 features by capturing the necessary information.

So, essentially the encoder tries to learn to reduce the dimensionality of the data without losing the useful information. We can think of autoencoders as similar to dimensionality reduction techniques such as **Principal Component Analysis (PCA)**. In PCA, we project the data into a low dimension using linear transformation and remove the features that are not required. The difference between PCA and an autoencoder is that PCA uses linear transformation for dimensionality reduction while the autoencoder uses a nonlinear transformation.

Apart from dimensionality reduction, autoencoders are also widely used for denoising noise in the images, audio, and so on. We know that the encoder in the autoencoder reduces the dimensionality of the dataset by learning only the necessary information and forms the bottleneck or code. Thus, when the noisy image is fed as an input to the autoencoder, the encoder learns only the necessary information of the image and forms the bottleneck. Since the encoder learns only the important and necessary information to represent the image, it learns that noise is unwanted information and removes the representations of noise from the bottleneck.

Thus, now we will have a bottleneck, that is, a representation of the image without any noise information. When this learned representation of the encoder, that is, the bottleneck, is fed to the decoder, the decoder reconstructs the input image from the encodings produced by the encoder. Since the encodings have no noise, the reconstructed image will not contain any noise.

In a nutshell, autoencoders map our data of a high dimension data to a low-level representation. This low-level data representation of data is called as **latent representations** or **bottleneck** which have only meaningful and important features that represent the input.

Since the role of our autoencoder is to reconstruct its input, we use a reconstruction error as our loss function, which implies we try to understand how much of the input is properly reconstructed by the decoder. So, we can use mean squared error loss as our loss function to quantify the performance of autoencoders.

Now that we have understood what autoencoders are, we will explore the architecture of autoencoders in the next section.

Understanding the architecture of autoencoders

As we have just learned, autoencoders consist of two important components: an encoder $f_\theta(\cdot)$ and a decoder $g_\phi(\cdot)$. Let's look at each one of them closely:

- **Encoder**: The encoder $f(\cdot)$ learns the input and returns the latent representation of the input. Let's assume we have an input, x. When we feed the input to the encoder, it returns a low-dimensional latent representation of the input called code or a bottleneck, z. We represent the parameter of the encoder by θ:

$$z = f_\theta(x)$$

$$z = \sigma(\theta x + b)$$

- **Decoder**: The decoder $g(\cdot)$ tries to reconstruct the original input x using the output of the encoder that is code z as an input. The reconstructed image is represented by x'. We represent the parameters of the decoder by ϕ:

$$x' = g_\phi(z)$$

$$x' = \sigma(\phi z + b)$$

We need to learn the optimal parameters, θ and ϕ, of our encoder and decoder respectively so that we can minimize the reconstruction loss. We can define our loss function as the mean squared error between the actual input and reconstructed input:

$$L(\theta, \phi) = \frac{1}{n} \sum_{i=1}^{n} (\mathbf{x}_i - g_\phi(f_\theta(\mathbf{x}_i)))^2$$

Here, n is the number of training samples.

When the latent representation has a lesser dimension than the input, then it is called an **undercomplete autoencoder.** Since the dimensions are less, undercomplete autoencoders try to learn and retain the only useful distinguishing and important features of the input and remove the rest. When the latent representation has a dimension greater than or the same as the input, the autoencoders will just copy the input without learning any useful features, and such a type of autoencoder is called **overcomplete autoencoders.**

Undercomplete and overcomplete autoencoders are shown in the following diagram. Undercomplete autoencoders have fewer neurons in the hidden layer (code) than the number of neurons in the input layer; while in the overcomplete autoencoders, the number of neurons in the hidden layer (code) is greater than the number of units in the input layer:

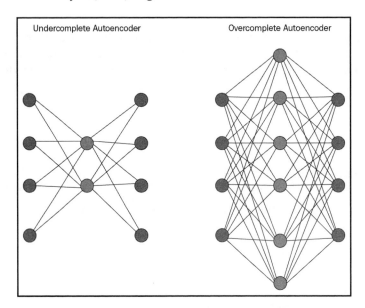

Thus, by limiting the neurons in the hidden layer (code), we can learn the useful representations of the input. Autoencoders can also have any number of hidden layers. Autoencoders with multiple hidden layers are called **multilayer autoencoders** or **deep autoencoders**. What we have learned so far is just the **vanilla** or the **shallow autoencoder**.

Reconstructing the MNIST images using an autoencoder

Now we will learn how autoencoders reconstruct handwritten digits using the MNIST dataset. First, let's import the necessary libraries:

```
import warnings
warnings.filterwarnings('ignore')

import numpy as np
import tensorflow as tf

from tensorflow.keras.models import Model
```

```
from tensorflow.keras.layers import Input, Dense
tf.logging.set_verbosity(tf.logging.ERROR)

#plotting
import matplotlib.pyplot as plt
%matplotlib inline

#dataset
from tensorflow.keras.datasets import mnist
```

Preparing the dataset

Let's load the MNIST dataset. Since we are reconstructing the given input, we don't need the labels. So, we just load x_train for training and x_test for testing:

```
(x_train, _), (x_test, _) = mnist.load_data()
```

Normalize the data by dividing by the max pixel value, which is 255:

```
x_train = x_train.astype('float32') / 255
x_test = x_test.astype('float32') / 255
```

Print the shape of our dataset:

```
print(x_train.shape, x_test.shape)

((60000, 28, 28), (10000, 28, 28))
```

Reshape the images as a 2D array:

```
x_train = x_train.reshape((len(x_train), np.prod(x_train.shape[1:])))
x_test = x_test.reshape((len(x_test), np.prod(x_test.shape[1:])))
```

Now, the shape of the data would become as follows:

```
print(x_train.shape, x_test.shape)

((60000, 784), (10000, 784))
```

Defining the encoder

Now we define the encoder layer, which takes the images as an input and returns the encodings.

Define the size of the encodings:

```
encoding_dim = 32
```

Define the placeholders for the input:

```
input_image = Input(shape=(784,))
```

Define the encoder which takes `input_image` and returns the encodings:

```
encoder   = Dense(encoding_dim, activation='relu')(input_image)
```

Defining the decoder

Let's define the decoder which takes the encoded values from the encoder and returns the reconstructed image:

```
decoder = Dense(784, activation='sigmoid')(encoder)
```

Building the model

Now that we defined encoder and decoder, we define the model which takes images as input and returns the output of the decoder which is the reconstructed image:

```
model = Model(inputs=input_image, outputs=decoder)
```

Let's look at a summary of the model:

```
model.summary()
```

Layer (type)	Output Shape	Param #
input_1 (InputLayer)	(None, 784)	0
dense (Dense)	(None, 32)	25120
dense_1 (Dense)	(None, 784)	25872

```
Total params: 50,992
Trainable params: 50,992
Non-trainable params: 0
```

Compile the model with `loss` as a binary cross-entropy and minimize the loss using the `adadelta` optimizer:

```
model.compile(optimizer='adadelta', loss='binary_crossentropy')
```

Now let's train the model.

Generally, we train the model as `model.fit(x,y)` where x is the input and y is the label. But since autoencoders reconstruct their inputs, the input and output to the model should be the same. So, here, we train the model as `model.fit(x_train, x_train)`:

```
model.fit(x_train, x_train, epochs=50, batch_size=256, shuffle=True,
validation_data=(x_test, x_test))
```

Reconstructing images

Now that we have trained the model, we see how the model is reconstructing the images of the test set. Feed the test images to the model and get the reconstructed images:

```
reconstructed_images = model.predict(x_test)
```

Plotting reconstructed images

First, let us plot the actual images, that is, input images:

```
n = 7
plt.figure(figsize=(20, 4))
for i in range(n):

    ax = plt.subplot(1, n, i+1)
    plt.imshow(x_test[i].reshape(28, 28))
    plt.gray()
    ax.get_xaxis().set_visible(False)
    ax.get_yaxis().set_visible(False)
plt.show()
```

The plot of the actual images is as follows:

Plot the reconstructed image as follows:

```
n = 7
plt.figure(figsize=(20, 4))
for i in range(n):
    ax = plt.subplot(2, n, i + n + 1)
    plt.imshow(reconstructed_images[i].reshape(28, 28))
    plt.gray()
    ax.get_xaxis().set_visible(False)
    ax.get_yaxis().set_visible(False)

plt.show()
```

The following shows the reconstructed images:

As you can see, the autoencoder has learned better representations of the input images and reconstructed them.

Autoencoders with convolutions

We just learned what autoencoders are in the previous section. We learned about a vanilla autoencoder, which is basically the feedforward shallow network with one hidden layer. Instead of keeping them as a feedforward network, can we use them as a convolutional network? Since we know that a convolutional network is good at classifying and recognizing images (provided that we use convolutional layers instead of feedforward layers in the autoencoders), it will learn to reconstruct the inputs better when the inputs are images.

Thus, we introduce a new type of autoencoders called CAEs that use a convolutional network instead of a vanilla neural network. In the vanilla autoencoders, encoders and decoders are basically a feedforward network. But in CAEs, they are basically convolutional networks. This means the encoder consists of convolutional layers and the decoder consists of transposed convolutional layers, instead of a feedforward network. A CAE is shown in the following diagram:

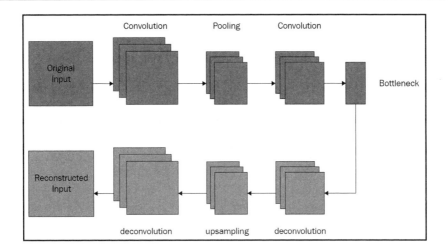

As shown, we feed the input image to the encoder that consists of a convolutional layer, and the convolutional layer performs the convolution operation and extracts important features from the image. We then perform max pooling to keep only the important features of the image. In a similar fashion, we perform several convolutional and max-pooling operations and obtain a latent representation of the image, called a **bottleneck**.

Next, we feed the bottleneck to the decoder that consists of deconvolutional layers, and the deconvolutional layer performs the deconvolution operation and tries to reconstruct the image from the bottleneck. It consists of several deconvolutional and upsampling operations to reconstruct the original image.

Thus, this is how CAE uses convolutional layers in the encoder and transpose convolutional layers in the decoders to reconstruct the image.

Building a convolutional autoencoder

Just as we learned how to implement an autoencoder in the previous section, implementing a CAE is also the same, but the only difference is here we use convolutional layers in the encoder and decoder instead of a feedforward network. We will use the same MNIST dataset to reconstruct the images using CAE.

Import the libraries:

```
import warnings
warnings.filterwarnings('ignore')

#modelling
from tensorflow.keras.models import Model
from tensorflow.keras.layers import Input, Dense, Conv2D, MaxPooling2D,
UpSampling2D
from tensorflow.keras import backend as K

#plotting
import matplotlib.pyplot as plt
%matplotlib inline

#dataset
from keras.datasets import mnist
import numpy as np
```

Read and reshape the dataset:

```
(x_train, _), (x_test, _) = mnist.load_data()

# Normalize the dataset

x_train = x_train.astype('float32') / 255.
x_test = x_test.astype('float32') / 255.

# reshape

x_train = np.reshape(x_train, (len(x_train), 28, 28, 1))
x_test = np.reshape(x_test, (len(x_test), 28, 28, 1))
```

Let's define the shape of our input image:

```
input_image = Input(shape=(28, 28, 1))
```

Defining the encoder

Now, let's define our encoder. Unlike vanilla autoencoders, where we use feedforward networks, here we use a convolutional network. Hence, our encoder comprises three convolutional layers, followed by a max pooling layer with `relu` activations.

Define the first convolutional layer, followed by a max pooling operation:

```
x = Conv2D(16, (3, 3), activation='relu', padding='same')(input_image)
x = MaxPooling2D((2, 2), padding='same')(x)
```

Define the second convolutional and max pooling layer:

```
x = Conv2D(8, (3, 3), activation='relu', padding='same')(x)
x = MaxPooling2D((2, 2), padding='same')(x)
```

Define the final convolutional and max pooling layer:

```
x = Conv2D(8, (3, 3), activation='relu', padding='same')(x)
encoder = MaxPooling2D((2, 2), padding='same')(x)
```

Defining the decoder

Now, we define our decoder; in the decoder, we perform the deconvolution operation with three layers, that is, we upsample the encodings created by the encoder and reconstruct the original image.

Define the first convolutional layer followed by upsampling:

```
x = Conv2D(8, (3, 3), activation='relu', padding='same')(encoder)
x = UpSampling2D((2, 2))(x)
```

Define the second convolutional layer with upsampling:

```
x = Conv2D(8, (3, 3), activation='relu', padding='same')(x)
x = UpSampling2D((2, 2))(x)
```

Define the final convolutional layer with upsampling:

```
x = Conv2D(16, (3, 3), activation='relu')(x)
x = UpSampling2D((2, 2))(x)
decoded = Conv2D(1, (3, 3), activation='sigmoid', padding='same')(x)
```

Building the model

Define the model that takes an input image and returns the images generated by the decoder, which is reconstructed images:

```
model = Model(input_image, decoder)
```

Let's compile the model with loss as binary cross-entropy and we use adadelta as our optimizer:

```
model.compile(optimizer='adadelta', loss='binary_crossentropy')
```

Then, train the model as follows:

```
model.fit(x_train, x_train, epochs=50,batch_size=128, shuffle=True,
validation_data=(x_test, x_test))
```

Reconstructing the images

Reconstruct the images using our trained model:

```
reconstructed_images = model.predict(x_test)
```

First, let us plot the input images:

```
n = 7
plt.figure(figsize=(20, 4))
for i in range(n):

    ax = plt.subplot(1, n, i+1)
    plt.imshow(x_test[i].reshape(28, 28))
    plt.gray()
    ax.get_xaxis().set_visible(False)
    ax.get_yaxis().set_visible(False)
plt.show()
```

The plot of the input images is as follows:

Now, we plot the reconstructed images:

```
n = 7
plt.figure(figsize=(20, 4))
for i in range(n):
    ax = plt.subplot(2, n, i + n + 1)
    plt.imshow(reconstructed_images[i].reshape(28, 28))
    plt.gray()
    ax.get_xaxis().set_visible(False)
    ax.get_yaxis().set_visible(False)

plt.show()
```

The plot of the reconstructed images is as follows:

Exploring denoising autoencoders

DAE are another small variant of the autoencoder. They are mainly used to remove noise from the image, audio, and other inputs. So, when we feed the corrupted input to the DAE, it learns to reconstruct the original uncorrupted input. Now we inspect how DAEs remove the noise.

With a DAE, instead of feeding the raw input to the autoencoder, we corrupt the input by adding some stochastic noise and feed the corrupted input. We know that the encoder learns the representation of the input by keeping only important information and maps the compressed representation to the bottleneck. When the corrupted input is sent to the encoder, while learning the representation of the input encoder will learn that noise is unwanted information and removes its representation. Thus, encoders learn the compact representation of the input without noise by keeping only necessary information, and map the learned representation to the bottleneck.

Now the decoder tries to reconstruct the image using the representation learned by the encoder, that is, the bottleneck. Since the representation, does not contain any noise, decoders reconstruct the input without noise. This is how a DAE removes noise from the input.

A typical DAE is shown in the following diagram. First, we corrupt the input by adding some noise and feed the corrupted input to the encoder, which learns the representation of the input without the noise, while the decoder reconstructs the uncorrupted input using the representation learned by the encoder:

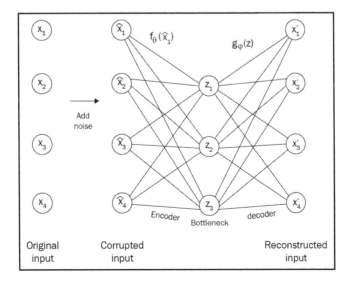

Mathematically, this can be expressed as follows.

Say we have an image, x, and we add noise to the image and get $\hat{x}$ which is the corrupted image:

$$\hat{x} = x + noise$$

Now feed this corrupted image to the encoder:

$$z = f_\theta(\hat{x})$$

$$z = \sigma(\hat{x}\theta + b)$$

The decoder tries to reconstruct the actual image:

$$x' = g_\phi(z)$$

$$x' = \sigma(\phi z + b)$$

Denoising images using DAE

In this section, we will learn how to denoise the images using DAE. We use CAE for denoising the images. The code for DAE is just the same as CAE, except that here, we use noisy images in the input. Instead of looking at the whole code, we will see only the respective changes. The complete code is available on GitHub at https://github.com/PacktPublishing/Hands-On-Deep-Learning-Algorithms-with-Python.

Set the noise factor:

```
noise_factor = 1
```

Add noise to the train and test images:

```
x_train_noisy = x_train + noise_factor * np.random.normal(loc=0.0,
scale=1.0, size=x_train.shape)
x_test_noisy = x_test + noise_factor * np.random.normal(loc=0.0, scale=1.0,
size=x_test.shape)
```

Clip the train and test set by 0 and 1:

```
x_train_noisy = np.clip(x_train_noisy, 0., 1.)
x_test_noisy = np.clip(x_test_noisy, 0., 1.)
```

Let's train the model. Since, we want the model to learn to remove noise in the image, input to the model is the noisy images, that is, x_train_noisy and output is the denoised images, that is, x_train:

```
model.fit(x_train_noisy, x_train, epochs=50,batch_size=128, shuffle=True,
validation_data=(x_test_noisy, x_test))
```

Reconstruct images using our trained model:

```
reconstructed_images = model.predict(x_test_noisy)
```

First, let us plot the input image which is the corrupted image:

```
n = 7
plt.figure(figsize=(20, 4))
for i in range(n):

    ax = plt.subplot(1, n, i+1)
    plt.imshow(x_test_noisy[i].reshape(28, 28))
    plt.gray()
    ax.get_xaxis().set_visible(False)
    ax.get_yaxis().set_visible(False)
plt.show()
```

The plot of the input noisy image is shown as follows:

Now, let us plot the reconstructed images by the model:

```
n = 7
plt.figure(figsize=(20, 4))
for i in range(n):
    ax = plt.subplot(2, n, i + n + 1)
    plt.imshow(reconstructed_images[i].reshape(28, 28))
    plt.gray()
    ax.get_xaxis().set_visible(False)
    ax.get_yaxis().set_visible(False)

plt.show()
```

As you can see, our model has learned to remove noise from the image:

Understanding sparse autoencoders

We know that autoencoders learn to reconstruct the input. But when we set the number of nodes in the hidden layer greater than the number of nodes in the input layer, then it will learn an identity function which is not favorable, because it just completely copies the input.

Having more nodes in the hidden layer helps us to learn robust latent representation. But when there are more nodes in the hidden layer, the autoencoder tries to completely mimic the input and thus it overfits the training data. To resolve the problem of overfitting, we introduce a new constraint to our loss function called the **sparsity constraint** or **sparsity penalty**. The loss function with sparsity penalty can be represented as follows:

$$L = \|X - X'\|_2^2 + \beta \text{ Sparse penalty}$$

The first term $\|X - X'\|_2^2$ represents the reconstruction error between the original input X and reconstructed input, x'. The second term implies the sparsity constraint. Now we will explore how this sparse constraint mitigates the problem of overfitting.

Using the sparsity constraint, we activate only specific neurons on the hidden layer instead of activating all the neurons. Based on the input, we activate and deactivate specific neurons so the neurons, when they are activated, will learn to extract important features from the input. By having the sparse penalty, autoencoders will not copy the input exactly to the output and it can also learn the robust latent representation.

As shown in the following diagram, sparse autoencoders have more units in the hidden layer than the input layer; however, only a few neurons in the hidden layer are activated. The unshaded neurons represent the neurons that are currently activated:

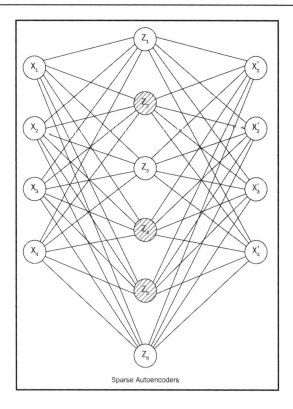

Sparse Autoencoders

A neuron returns 1 if it's active and 0 if it is inactive. In sparse autoencoders, we set most of the neurons in the hidden layer to inactive. We know that the sigmoid activation function squashes the value to between 0 and 1. So, when we use the sigmoid activation function, we try to keep the values of neurons close to 0.

We typically try to keep the average activation value of each neuron in the hidden layer close to zero, say 0.05, but not equal to zero, and this value is called *ρ*, which is our sparsity parameter. We typically set the value of *ρ* to 0.05.

First, we calculate the average activation of a neuron.

The average activation of the jth neuron in the hidden layer h over the whole training set can be calculated as follows:

$$\hat{\rho}_j = \frac{1}{n} \sum_{i=1}^{n} [a_j^{(h)} (x^{(i)})]$$

Here, the following holds true:

- $\hat{\rho}_j$ denotes the average activation of the jth neuron in the hidden layer h
- n is the number of the training sample
- a_j is the activation of jth neuron in the hidden layer h
- $x^{(i)}$ is the ith training sample
- $a_j x^{(i)}$ implies the activation of the jth neuron in the hidden layer h for the ith training sample

We try to keep the average activation value, $\hat{\rho}$, of the neurons close to ρ. That is, we try to keep the average activation values of the neurons to 0.05:

$$\hat{\rho} \approx \rho$$

So, we penalize the value of $\hat{\rho}_j$, which is varying from ρ. We know that the **Kullback–Leibler** (**KL**) divergence is widely used for measuring the difference between the two probability distributions. So, here, we use the KL divergence to measure the difference between two **Bernoulli distributions**, that is, mean ρ and mean $\hat{\rho}_j$ and it can be given as follows:

$$KL = \sum_{j=1}^{l^{(h)}} log\frac{\rho}{\hat{\rho}_j} + (1-\rho)log\frac{1-\rho}{1-\hat{\rho}_j} \tag{1}$$

In the earlier equation, $l^{(h)}$ denotes the hidden layer h, and j denotes the jth neurons in the hidden layer $l^{(h)}$. The earlier equation is basically the sparse penalty or sparsity constraint. Thus, with the sparsity constraint, all the neurons will never be active at the same time, and on average, they are set to 0.05.

Now we can rewrite the loss function with a sparse penalty as follows:

$$L = \|X - \hat{X}\|_2^2 + \beta(\sum_{j=1}^{l^{(h)}} log\frac{\rho}{\hat{\rho}_j} + (1-\rho)log\frac{1-\rho}{1-\hat{\rho}_j})$$

Thus, sparse autoencoders allow us to have a greater number of nodes in the hidden layer than the input layer, yet reduce the problem of overfitting with the help of the sparsity constraint in the loss function.

Building the sparse autoencoder

Building a sparse autoencoder is just as same as building a regular autoencoder, except that we use a sparse regularizer in the encoder and decoder, so instead of looking at the whole code in the following sections, we will only look at the parts related to implementing the sparse regularizer; the complete code with an explanation is available on GitHub.

Defining the sparse regularizer

The following is the code to define the sparse regularizer:

```
def sparse_regularizer(activation_matrix):
```

Set our ρ value to 0.05:

```
rho = 0.05
```

Calculate the $\hat{\rho}j$, which is the mean activation value:

```
rho_hat = K.mean(activation_matrix)
```

Compute the KL divergence between the mean ρ and the mean $\hat{\rho}j$ according to equation *(1)*:

```
KL_divergence = K.sum(rho*(K.log(rho/rho_hat)) + (1-rho)*(K.log(1-
rho/1-rho_hat)))
```

Sum the KL divergence values:

```
sum = K.sum(KL_divergence)
```

Multiply the `sum` by `beta` and return the results:

```
return beta * sum
```

The whole function for the sparse regularizer is given as follows:

```
def sparse_regularizer(activation_matrix):
    p = 0.01
    beta = 3
    p_hat = K.mean(activation_matrix)
    KL_divergence = p*(K.log(p/p_hat)) + (1-p)*(K.log(1-p/1-p_hat))
    sum = K.sum(KL_divergence)
    return beta * sum
```

Learning to use contractive autoencoders

Like sparse autoencoders, **contractive autoencoders** add a new regularization term to the loss function of the autoencoders. They try to make our encodings less sensitive to the small variations in the training data. So, with contractive autoencoders, our encodings become more robust to small perturbations such as noise present in our training dataset. We now introduce a new term called the **regularizer** or **penalty term** to our loss function. It helps to penalize the representations that are too sensitive to the input.

Our loss function can be mathematically represented as follows:

$$L = \|X - \hat{X}\|_2^2 + \lambda \|J_f(X)\|_F^2$$

The first term represents the reconstruction error and the second term represents the penalty term or the regularizer and it is basically the **Frobenius norm** of the **Jacobian matrix**. Wait! What does that mean?

The Frobenius norm, also called the **Hilbert-Schmidt norm**, of a matrix is defined as the square root of the sum of the absolute square of its elements. A matrix comprising a partial derivative of the vector-valued function is called the **Jacobian matrix**.

Thus, calculating the Frobenius norm of the Jacobian matrix implies our penalty term is the sum of squares of all partial derivatives of the hidden layer with respect to the input. It is given as follows:

$$\|J_f(X)\|_F^2 = \sum_{ij} \left(\frac{\partial h_j(X)}{\partial X_i} \right)^2$$

Calculating the partial derivative of the hidden layer with respect to the input is similar to calculating gradients of loss. Assuming we are using the sigmoid activation function, then the partial derivative of the hidden layer with respect to the input is given as follows:

$$\|J_f(x)\|_F^2 = \sum_j [h_j(1 - h_j)]^2 \sum_i (W_{ji}^T)^2$$

Adding the penalty term to our loss function helps in reducing the sensitivity of the model to the variations in the input and makes our model more robust to the outliers. Thus, contractive autoencoders reduce the sensitivity of the model to the small variations in the training data.

Implementing the contractive autoencoder

Building the contractive autoencoder is just as same as building the autoencoder, except that we use the contractive loss regularizer in the model, so instead of looking at the whole code, we will only look at the parts related to implementing the contractive loss.

Defining the contractive loss

Now let's see how to define the loss function in Python.

Define the mean squared loss as follows:

```
MSE = K.mean(K.square(actual - predicted), axis=1)
```

Obtain the weights from our encoder layer and transpose the weights:

```
weights =
K.variable(value=model.get_layer('encoder_layer').get_weights()[0])
weights = K.transpose(weights)
```

Get the output of our encoder layer:

```
h = model.get_layer('encoder_layer').output
```

Define the penalty term:

```
penalty_term =  K.sum(((h * (1 - h))**2) * K.sum(weights**2,
axis=1), axis=1)
```

The final loss is the sum of mean squared error and the penalty term multiplied by `lambda`:

```
Loss = MSE + (lambda * penalty_term)
```

The complete code for contractive loss is given as follows:

```
def contractive_loss(y_pred, y_true):

    lamda = 1e-4

    MSE = K.mean(K.square(y_true - y_pred), axis=1)

    weights =
K.variable(value=model.get_layer('encoder_layer').get_weights()[0])
    weights = K.transpose(weights)

    h = model.get_layer('encoder_layer').output
```

```
    penalty_term = K.sum(((h * (1 - h))**2) * K.sum(weights**2, axis=1),
axis=1)

    Loss = MSE + (lambda * penalty_term)


    return Loss
```

Dissecting variational autoencoders

Now we will see another very interesting type of autoencoders called **variational autoencoders** (**VAE**). Unlike other autoencoders, VAEs are generative models that imply they learn to generate new data just like GANs.

Let's say we have a dataset containing facial images of many individuals. When we train our variational autoencoder with this dataset, it learns to generate new realistic faces that are not seen in the dataset. VAEs have various applications because of their generative nature and some of them include generating images, songs, and so on. But what makes VAE generative and how is it different than other autoencoders? Let's learn that in the coming section.

Just as we learned when discussing GANs, for a model to be generative, it has to learn the distribution of the inputs. For instance, let's say we have a dataset that consists of handwritten digits, such as the MNIST dataset. Now, in order to generate new handwritten digits, our model has to learn the distribution of the digits in the given dataset. Learning the distribution of the digits present in the dataset helps VAE to learn useful properties such as digit width, stroke, height, and so on. Once the model encodes this property in its distribution, then it can generate new handwritten digits by sampling from the learned distribution.

Say we have a dataset of human faces, then learning the distribution of the faces in the dataset helps us to learn various properties such as gender, facial expression, hair color, and so on. Once the model learns and encode these properties in its distribution, then it can generate a new face just by sampling from the learned distribution.

Thus, in VAE, instead of mapping the encoder's encodings directly to the latent vector (bottleneck), we map the encodings to a distribution; usually, it is a Gaussian distribution. We sample a latent vector from this distribution and feed it to a decoder then the decoder learns to reconstruct the image. As shown in the following diagram, an encoder maps its encodings to a distribution and we sample a latent vector from this distribution and feed it to a decoder to reconstruct an image:

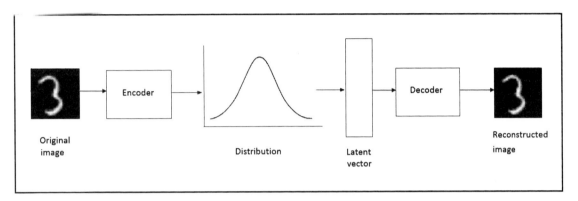

A **gaussian distribution** can be parameterized by its mean and covariance matrix. Thus, we can make our encoder generate its encoding and maps it to a mean vector and standard deviation vector that approximately follows the Gaussian distribution. Now, from this distribution, we sample a latent vector and feed it to our decoder which then reconstructs an image:

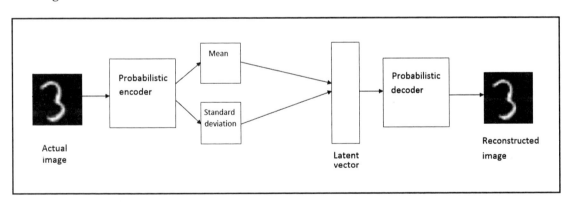

In a nutshell, the encoder learns the desirable properties of the given input and encodes them into distribution. We sample a latent vector from this distribution and feed the latent vector as input to the decoder which then generates images learned from the encoder's distribution.

In VAE, the encoder is also called as **recognition model** and the decoder is also called as **generative model**. Now that we have an intuitive understanding of VAE, in the next section, we will go into detail and learn how VAE works.

Variational inference

Before going ahead, let's get familiar with the notations:

- Let's represent the distribution of the input dataset by $p_\theta(x)$, where θ represents the parameter of the network that will be learned during training
- We represent the latent variable by z, which encodes all the properties of the input by sampling from the distribution
- $p(x, z)$ denotes the joint distribution of the input x with their properties, z
- $p(z)$ represents the distribution of the latent variable

Using the Bayesian theorem, we can write the following:

$$p_\theta(x) = \int p_\theta(z|x)p(x)dz$$

The preceding equation helps us to compute the probability distribution of the input dataset. But the problem lies in computing $p_\theta(z|x)$, because computing it is intractable. Thus, we need to find a tractable way to estimate the $p_\theta(z|x)$. Here, we introduce a concept called **variational inference**.

Instead of inferring the distribution of $p_\theta(z|x)$ directly, we approximate them using another distribution, say a Gaussian distribution $q_\phi(z|x)$. That is, we use $q_\phi(z|x)$ which is basically a neural network parameterized by ϕ parameter to estimate the value of $p_\theta(z|x)$:

- $q(z|x)$ is basically our probabilistic encoder; that is, they to create a latent vector z given x
- $p(x|z)$ is the probabilistic decoder; that is, it tries to construct the input x given the latent vector z

The following diagram helps you attain good clarity on the notations and what we have seen so far:

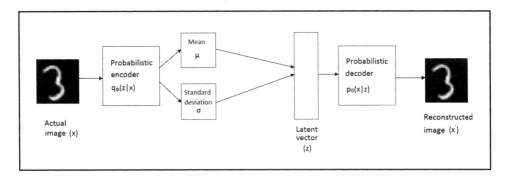

The loss function

We just learned that we use $q_\phi(z|x)$ to approximate $p_\theta(z|x)$. Thus, the estimated value of $q_\phi(z|x)$ should be close to $p_\theta(z|x)$. Since these both are distributions, we use KL divergence to measure how $q_\phi(z|x)$ diverges from $p_\theta(z|x)$ and we need to minimize the divergence.

A KL divergence between $q_\phi(z|x)$ and $p_\theta(z|x)$ is given as follows:

$$D_{KL}\left(q_\phi(z|x)\|p_\theta(z|x)\right) = E_{z\sim q}\left[\log q_\phi(z|x) - \log p_\theta(z|x)\right]$$

Since we know $p_\theta(z|x) = \dfrac{p_\theta(x|z)p_\theta(z)}{p_\theta(x)}$, substituting this in the preceding equation, we can write the following:

$$D_{KL}\left(q_\phi(z|x)\|p_\theta(z|x)\right) = E_{z\sim q}\left[\log q_\phi(z|x) - \log\left(\frac{p_\theta(x|z)p_\theta(z)}{p_\theta(x)}\right)\right]$$

Since we know $log\ (a/b) = log(a) - log(b)$, we can rewrite the preceding equation as follows:

$$\begin{aligned}D_{KL}\left(q_\phi(z|x)\|p_\theta(z|x)\right) &= E_{z\sim q}\left[\log q_\phi(z|x) - (log\left((p_\theta(x|z)p_\theta(z)) - log p_\theta(x)\right))\right]\\ &= E_{z\sim q}\left[\log q_\phi(z|x) - log((p_\theta(x|z)p_\theta(z)) + log p_\theta(x)\right]\end{aligned}$$

We can take the $p_\theta(x)$ outside the expectations since it has no dependency on $z\sim q$:

$$D_{KL}\left(q_\phi(z|x)\|p_\theta(z|x)\right) = E_{z\sim q}\left[\log q_\phi(z|x) - log((p_\theta(x|z)p_\theta(z))\right] + log p_\theta(x)$$

Since we know $log(ab) = log(a) + log(b)$, we can rewrite the preceding equation as follows:

$$D_{KL}\left(q_\phi(z|x)\|p_\theta(z|x)\right) = E_{z\sim q}\left[\log q_\phi(z|x) - (log\, p_\theta(x|z) + log\, p_\theta(z))\right] + log\, p_\theta(x)$$

$$D_{KL}\left(q_\phi(z|x)\|p_\theta(z|x)\right) = E_{z\sim q}\left[\log q_\phi(z|x) - log\, p_\theta(x|z) - log\, p_\theta(z)\right] + log\, p_\theta(x) \qquad (1)$$

We know that KL divergence between $q_\phi(z|x)$ and $p_\theta(z)$ can be given as:

$$D_{KL}\left(q_\phi(z|x)\|p_\theta(z)\right) = E_{z\sim q}\left[\log q_\phi(z|x) - \log p_\theta(z)\right] \qquad (2)$$

Substituting equation *(2)* in equation *(1)* we can write:

$$D_{KL}\left(q_\phi(z|x)\|p_\theta(z|x)\right) = \mathbb{E}_{z\sim q}\left[\log p_\theta(x|z)\right] - D_{KL}\left(q_\phi(z|x)\|p_\theta(z)\right) + \log p_\theta(x)$$

Rearranging the left and right-hand sides of the equation, we can write the following:

$$\mathbb{E}_{z\sim q}\left[\log p_\theta(x|z)\right] - D_{KL}\left(q_\phi(z|x)\|p_\theta(z)\right) = \log p_\theta(x) - D_{KL}\left(q_\phi(z|x)\|p_\theta(z|x)\right)$$

Rearranging the terms, our final equation can be given as follows:

$$\boxed{\log p_\theta(x) - D_{KL}\left(q_\phi(z|x)\|p_\theta(z|x)\right) = \mathbb{E}_{z\sim q}\left[\log p_\theta(x|z)\right] - D_{KL}\left(q_\phi(z|x)\|p_\theta(z)\right)}$$

What does the above equation imply?

The left-hand side of the equation is also known as the **variational lower bound** or the **evidence lower bound (ELBO)**. The first term in the left-hand side $p_\theta(x)$ implies the distribution of the input x, which we want to maximize and $-D_{KL}\left(q_\phi(z|x)\|p_\theta(z|x)\right)$ implies the KL divergence between the estimated and the real distribution.

The loss function can be written as follows:

$$L = log(p_\theta(x)) - D_{KL}\left(q_\phi(z|x)\|p_\theta(z|x)\right)$$

In this equation, you will notice the following:

- $log(p_\theta(x))$ implies we are maximizing the distribution of the input; we can convert the maximization problem into minimization by simply adding a negative sign; thus, we can write $-log(p_\theta(x))$
- $-D_{KL}\left(q_\phi(z|x)\|p_\theta(z|x)\right)$ implies we are maximizing the KL divergence between the estimated and real distribution, but we want to minimize them, so we can write $+D_{KL}\left(q_\phi(z|x)\|p_\theta(z|x)\right)$ to minimize the KL divergence

Thus, our loss function becomes the following:

$$L = -log(p_\theta(x)) + D_{KL}\left(q_\phi(z|x)\|p_\theta(z|x)\right)$$

$$L = -\mathbb{E}_{z\sim Q}\left[\log p_\theta(x|z)\right] + D_{KL}\left(q_\phi(z|x)\|p_\theta(z|x)\right)$$

If you look at this equation, $\mathbb{E}_{z\sim Q}\left[\log p_\theta(x|z)\right]$ basically implies the reconstruction of the input, that is, the decoder which takes the latent vector z and reconstructs the input x.

Thus, our final loss function is the sum of the reconstruction loss and the KL divergence:

$$\boxed{L = -\mathbb{E}_{z\sim Q}\left[\log p_\theta(x|z)\right] + D_{KL}\left(q_\phi(z|x)\|p_\theta(z|x)\right)}$$

The value for KL divergence is simplified as follows:

$$D_{KL}\left(q_\phi(z|x)\|p_\theta(z|x)\right) = -\frac{1}{2}\sum_{k=1}^{K}\{1 + \log\sigma_k^2 - \mu_k^2 - \sigma_k^2\}$$

Thus, minimizing the preceding loss function implies we are minimizing the reconstruction loss and also minimizing the KL divergence between the estimated and real distribution.

Reparameterization trick

We face a problem while training VAE a through gradient descent. Remember, we are performing a sampling operation to generate a latent vector. Since a sampling operation is not differentiable, we cannot calculate gradients. That is, while backpropagating the network to minimize the error, we cannot calculate the gradients of the sampling operation as shown in the following diagram:

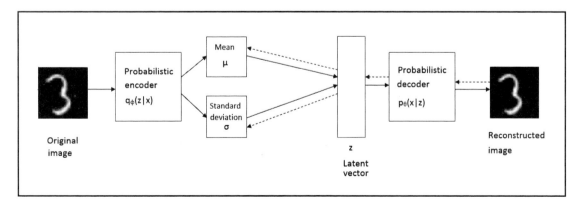

So, to combat this, we introduce a new trick called the **reparameterization trick**. We introduce a new parameter called **epsilon**, which we randomly sample from a unit Gaussian, which is given as follows:

$$\epsilon \in N(0,1)$$

And now we can rewrite our latent vector z as:

$$z = \mu + \sigma \odot \epsilon$$

The reparameterization trick is shown in the following diagram:

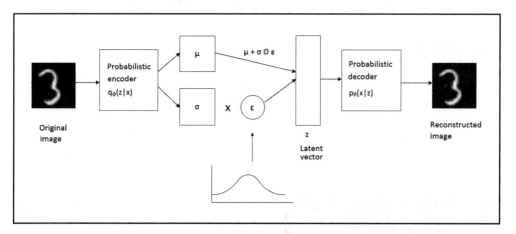

Thus, with the reparameterization trick, we can train the VAE with the gradient descent algorithm.

Generating images using VAE

Now that we have understood how the VAE model works, in this section, we will learn how to use VAE to generate images.

Import the required libraries:

```
import warnings
warnings.filterwarnings('ignore')

import numpy as np
import matplotlib.pyplot as plt
from scipy.stats import norm
```

```
from tensorflow.keras.layers import Input, Dense, Lambda
from tensorflow.keras.models import Model
from tensorflow.keras import backend as K
from tensorflow.keras import metrics
from tensorflow.keras.datasets import mnist

import tensorflow as tf
tf.logging.set_verbosity(tf.logging.ERROR)
```

Preparing the dataset

Load the MNIST dataset:

```
(x_train, _), (x_test, _) = mnist.load_data()
```

Normalize the dataset:

```
x_train = x_train.astype('float32') / 255.
x_test = x_test.astype('float32') / 255.
```

Reshape the dataset:

```
x_train = x_train.reshape((len(x_train), np.prod(x_train.shape[1:])))
x_test = x_test.reshape((len(x_test), np.prod(x_test.shape[1:])))
```

Now let's define some important parameters:

```
batch_size = 100
original_dim = 784
latent_dim = 2
intermediate_dim = 256
epochs = 50
epsilon_std = 1.0
```

Defining the encoder

Define the input:

```
x = Input(shape=(original_dim,))
```

Encoder hidden layer:

```
h = Dense(intermediate_dim, activation='relu')(x)
```

Compute the mean and the variance:

```
z_mean = Dense(latent_dim)(h)
z_log_var = Dense(latent_dim)(h)
```

Defining the sampling operation

Define the sampling operation with a reparameterization trick that samples the latent vector from the encoder's distribution:

```
def sampling(args):
    z_mean, z_log_var = args
    epsilon = K.random_normal(shape=(K.shape(z_mean)[0], latent_dim),
mean=0., stddev=epsilon_std)
    return z_mean + K.exp(z_log_var / 2) * epsilon
```

Sample the latent vector z from the mean and variance:

```
z = Lambda(sampling, output_shape=(latent_dim,))([z_mean, z_log_var])
```

Defining the decoder

Define the decoder with two layers:

```
decoder_hidden = Dense(intermediate_dim, activation='relu')
decoder_reconstruct = Dense(original_dim, activation='sigmoid')
```

Reconstruct the images using decoder which takes the latent vector z as input and returns the reconstructed image:

```
decoded = decoder_hidden(z)
reconstructed = decoder_reconstruct(decoded)
```

Building the model

We build the model as follows:

```
vae = Model(x, reconstructed)
```

Define the reconstruction loss:

```
Reconstruction_loss = original_dim * metrics.binary_crossentropy(x,
reconstructed)
```

Define KL divergence:

```
kl_divergence_loss = - 0.5 * K.sum(1 + z_log_var - K.square(z_mean) -
K.exp(z_log_var), axis=-1)
```

Thus, the total loss can be defined as:

```
total_loss = K.mean(Reconstruction_loss + kl_divergence_loss)
```

Add loss and compile the model:

```
vae.add_loss(total_loss)
vae.compile(optimizer='rmsprop')
vae.summary()
```

Train the model:

```
vae.fit(x_train,
        shuffle=True,
        epochs=epochs,
        batch_size=batch_size,
        verbose=2,
        validation_data=(x_test, None))
```

Defining the generator

Define the generator samples from the learned distribution and generates an image:

```
decoder_input = Input(shape=(latent_dim,))
_decoded = decoder_hidden(decoder_input)

_reconstructed = decoder_reconstruct(_decoded)
generator = Model(decoder_input, _reconstructed)
```

Plotting generated images

Now we plot the image generated by the generator:

```
n = 7
digit_size = 28
figure = np.zeros((digit_size * n, digit_size * n))

grid_x = norm.ppf(np.linspace(0.05, 0.95, n))
grid_y = norm.ppf(np.linspace(0.05, 0.95, n))

for i, yi in enumerate(grid_x):
```

```
    for j, xi in enumerate(grid_y):
        z_sample = np.array([[xi, yi]])
        x_decoded = generator.predict(z_sample)
        digit = x_decoded[0].reshape(digit_size, digit_size)
        figure[i * digit_size: (i + 1) * digit_size,
               j * digit_size: (j + 1) * digit_size] = digit

plt.figure(figsize=(4, 4), dpi=100)
plt.imshow(figure, cmap='Greys_r')
plt.show()
```

The following is the plot of the image generated by a generator:

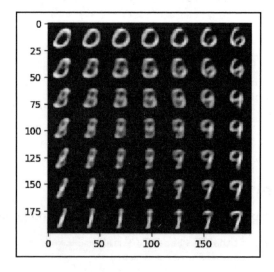

Summary

We started this chapter by learning what autoencoders are and how autoencoders are used to reconstruct their own input. We explored convolutional autoencoders, where instead of using feedforward networks, we used convolutional and deconvolutional layers for encoding and decoding, respectively. Following this, we learned about sparse which activate only certain neurons. Then, we learned about another type of regularizing autoencoder, called a contractive autoencoder, and at the end of the chapter, we learned about VAE which is a generative autoencoder model.

In the next chapter, we will learn about how to learn from a less data points using few-shot learning algorithms.

Questions

Let's examine our knowledge of autoencoders by answering the following questions:

1. What are autoencoders?
2. What is the objective function of autoencoders?
3. How do convolutional autoencoders differ from vanilla autoencoders?
4. What are denoising autoencoders?
5. How is the average activation of the neuron computed?
6. Define the loss function of contractive autoencoders.
7. What are the Frobenius norm and Jacobian matrix?

Further reading

You can also check the following links for more information:

- *Sparse autoencoder* notes by Andrew Ng, `https://web.stanford.edu/class/cs294a/sparseAutoencoder_2011new.pdf`
- *Contractive Auto-Encoders: Explicit Invariance During Feature Extraction* by Salah Rifai, et al., `http://www.icml-2011.org/papers/455_icmlpaper.pdf`
- *Variational Autoencoder for Deep Learning of Images, Labels and Captions* by Yunchen Pu, et al., `https://papers.nips.cc/paper/6528-variational-autoencoder-for-deep-learning-of-images-labels-and-captions.pdf`

11
Exploring Few-Shot Learning
Algorithms

Congratulations! We have made it to the final chapter. We have come a long way. We started off by learning what neural networks are and how they are used to recognize handwritten digits. Then we explored how to train neural networks with gradient descent algorithms. We also learned how recurrent neural networks i used for sequential tasks and how convolutional neural networks are used for image recognition. Following this, we investigated how the semantics of a text can be understood using word embedding algorithms. Then we got familiar with several different types of generative adversarial networks and autoencoders.

So far, we have learned that deep learning algorithms perform exceptionally well when we have a substantially large dataset. But how can we handle the situation when we don't have a large number of data points to learn from? For most use cases, we might not get a large dataset. In such cases, we can use few-shot learning algorithms, which do not require huge datasets to learn from. In this chapter, we will understand how exactly few-shot learning algorithms learn from a smaller number of data points and we explore different types of few-shot learning algorithms. First, we will study a popular few-shot learning algorithm called a **siamese network**. Following this, we will learn some other few-shot learning algorithms such as the prototypical, relation, and matching networks intuitively.

In this chapter, we will study the following topics:

- What is few-shot learning?
- Siamese networks
- Architecture of siamese networks
- Prototypical networks
- Relation networks
- Matching networks

What is few-shot learning?

Learning from a few data points is called **few-shot learning** or **k-shot learning**, where k specifies the number of data points in each of the class in the dataset.

Consider we are performing an image classification task. Say we have two classes – apple and orange – and we try to classify the given image as an apple or orange. When we have exactly one apple and one orange image in our training set, it is called one-shot learning; that is, we are learning from just one data point per each of the class. If we have, say, 11 images of an apple and 11 images of an orange, then that is called 11-shot learning. So, k in k-shot learning implies the number of data points we have per class.

There is also **zero-shot learning**, where we don't have any data points per class. Wait. What? How can we learn when there are no data points at all? In this case, we will not have data points, but we will have meta information about each of the class and we will learn from the meta information.

Since we have two classes in our dataset, that is, apple and orange, we can call it **two-way k-shot learning**. So, in n-way k-shot learning, n-way implies the number of classes we have in our dataset and k-shot implies a number of data points we have in each class.

We need our models to learn from just a few data points. In order to attain this, we train them in the same way; that is, we train the model on a very few data points. Say we have a dataset, D. We sample a few data points from each of the classes present in our dataset and we call it **support set**. Similarly, we sample some different data points from each of the classes and call it **query set**.

We train the model with a support set and test it with a query set. We train the model in an episodic fashion—that is, in each episode, we sample a few data points from our dataset, D, prepare our support set and query set, and train on the support set and test on the query set.

Siamese networks

Siamese networks are special types of neural networks and are among the simplest and most popularly used one-shot learning algorithms. As we have learned in the previous section, one-shot learning is a technique where we learn from only one training example per each class. So, siamese networks are predominantly used in applications where we don't have many data points for each of the class.

For instance, let's say we want to build a face recognition model for our organization and say about 500 people are working in our organization. If we want to build our face recognition model using a **convolutional neural network** (**CNN**) from scratch then we need many images of all these 500 people, to train the network and attain good accuracy. But, apparently, we will not have many images for all these 500 people and therefore it is not feasible to build a model using a CNN or any deep learning algorithm unless we have sufficient data points. So, in these kinds of scenarios, we can resort to a sophisticated one-shot learning algorithm such as a siamese network, which can learn from fewer data points.

But how do siamese networks work? Siamese networks basically consist of two symmetrical neural networks both sharing the same weights and architecture and both joined together at the end using an energy function, E. The objective of our siamese network is to learn whether the two inputs are similar or dissimilar.

Let's say we have two images, X_1 and X_2, and we want to learn whether the two images are similar or dissimilar. As shown in the following diagram, we feed **Image** X_1 to **Network** A and **Image** X_2 to **Network** B. The role of both of these networks is to generate embeddings (feature vectors) for the input image. So, we can use any network that will give us embeddings. Since our input is an image, we can use a convolutional network to generate the embeddings: that is, for extracting features. Remember that the role of the CNN here is only to extract features and not to classify.

As we know that these networks should have same weights and architecture, if **Network** *A* is a three-layer CNN then **Network** *B* should also be a three-layer CNN, and we have to use the same set of weights for both of these networks. So, **Network** *A* and **Network** *B* will give us the embeddings for input images X_1 and X_2 respectively. Then, we will feed these embeddings to the energy function, which tells us how similar the two input images are. Energy functions are basically any similarity measure, such as Euclidean distance and cosine similarity:

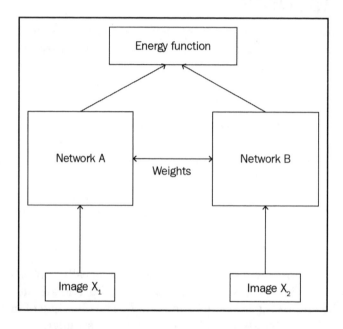

Siamese networks are not only used for face recognition, but are is also used extensively in applications where we don't have many data points and tasks where we need to learn the similarity between two inputs. The applications of siamese networks include signature verification, similar question retrieval, and object tracking. We will study siamese networks in detail in the next section.

Architecture of siamese networks

Now that we have a basic understanding of siamese networks, we will explore them in detail. The architecture of a siamese network is shown in the following figure:

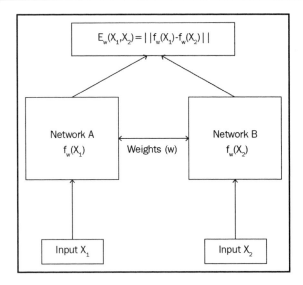

As you can see in the preceding figure, a siamese network consists of two identical networks, both sharing the same weights and architecture. Let's say we have two inputs, X_1 and X_2. We feed **Input** X_1 to **Network** A, that is, $f_W(X_1)$, and we feed **Input** X_2 to **Network** B, that is, $f_W(X_2)$.

As you can see, both of these networks have the same weights, W, and they will generate embeddings for our input, X_1 and X_2. Then, we feed these embeddings to the energy function, E, which will give us similarity between the two inputs. It can be expressed as follows:

$$E_W(X_1, X_2) = \|f_W(X_1) - f_W(X_2)\|$$

Let's say we use Euclidean distance as our energy function; then the value of E will be low if X_1 and X_2 are similar. The value of E will be large if the input values are dissimilar.

Assume that you have two sentences, sentence 1 and sentence 2. We feed sentence 1 to Network A and sentence 2 to Network B. Let's say both our Network A and Network B are **long short-term memory (LSTM)** networks and they share the same weights. So, Network A and Network B will generate the embeddings for sentence 1 and sentence 2 respectively.

Then, we feed these embeddings to the energy function, which gives us the similarity score between the two sentences. But how can we train our siamese networks? How should the data be? What are the features and labels? What is our objective function?

The input to the siamese networks should be in pairs, (X_1, X_2), along with their binary label, $Y \in (0, 1)$, stating whether the input pairs are a genuine pair the(same) or an imposite pair (different). As you can see in the following table, we have sentences as pairs and the label implies whether the sentence pairs are genuine (1) or imposite (0):

Sentence pairs		Label
She is a beautiful girl	She is a gorgeous girl	1
Birds fly in the sky	What are you doing ?	0
I Love Paris	I adore Paris	1
He just arrived	I am watching a movie	0

So, what is the loss function of our siamese network?

Since the goal of the siamese network is not to perform a classification task but to understand the similarity between the two input values, we use the contrastive loss function. It can be expressed as follows:

$$Contrastive\ loss = Y(E)^2 + (1 - Y)\max(\text{margin} - E, 0)^2$$

In the preceding equation, the value of Y is the true label, which will be 1 if the two input values are similar and 0 if the two input values are dissimilar, and E is our energy function, which can be any distance measure. The term **margin** is used to hold the constraint, that is, when two input values are dissimilar, and if their distance is greater than a margin, then they do not incur a loss.

Prototypical networks

Prototypical networks are yet another simple, efficient, and popular learning algorithm. Like siamese networks, they try to learn the metric space to perform classification.

The basic idea of the prototypical network is to create a prototypical representation of each class and classify a query point (new point) based on the distance between the class prototype and the query point.

Let's say we have a support set comprising images of lions, elephants, and dogs, as shown in the following diagram:

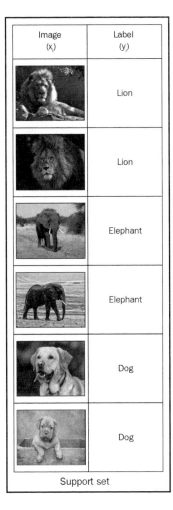

Support set

We have three classes (lion, elephant, and dog). Now we need to create a prototypical representation for each of these three classes. How can we build the prototype of these three classes? First, we will learn the embeddings of each data point using some embedding function. The embedding function, $f_\phi()$,can be any function that can be used to extract features. Since our input is an image, we can use the convolutional network as our embedding function, which will extract features from the input images, shown as follows:

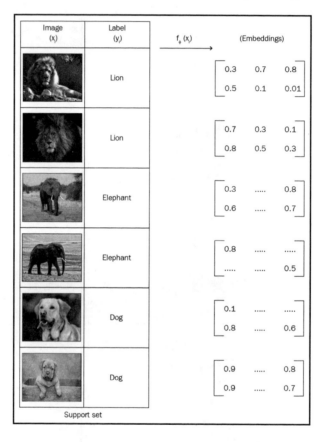

Once we learn the embeddings of each data point, we take the mean embeddings of data points in each class and form the class prototype, shown as follows. So, a class prototype is basically the mean embeddings of data points in a class:

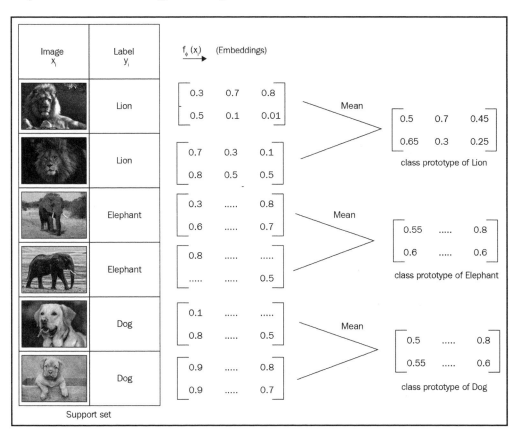

Similarly, when a new data point comes in, that is, a query point for which we want to predict the label, we will generate the embeddings for this new data point using the same embedding function that we used to create the class prototype: that is, we generate the embeddings for our query point using the convolutional network:

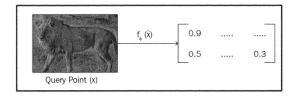

Once we have the embedding for our query point, we compare the distance between class prototypes and query point embeddings to find which class the query point belongs to. We can use Euclidean distance as a distance measure for finding the distance between the class prototypes and query points embeddings, as shown below:

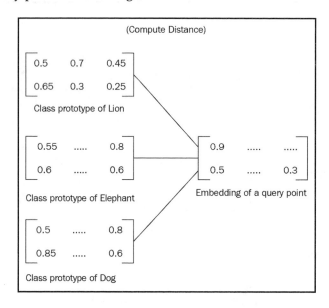

After finding the distance between the class prototype and query point embeddings, we apply softmax to this distance and get the probabilities. Since we have three classes, that is, lion, elephant, and dog, we will get three probabilities. The class that has high probability will be the class of our query point.

Since we want our network to learn from just a few data points, that is, since we want to perform few-shot learning, we train our network in the same way. We use **episodic training**; for each episode, we randomly sample a few data points from each of the classes in our dataset, and we call that a support set, and we train the network using only the support set, instead of the whole dataset. Similarly, we randomly sample a point from the dataset as a query point and try to predict its class. In this way, our network learns how to learn from data points.

The overall flow of the prototypical network is shown in the following figure. As you can see, first, we will generate the embeddings for all the data points in our support set and build the class prototype by taking the mean embeddings of data points in a class. We also generate the embeddings for our query point. Then we compute the distance between the class prototype and the query point embeddings. We use Euclidean distance as the distance measure. Then we apply softmax to this distance and get the probabilities.

As you can see in the following diagram, since our query point is a lion, the probability for lion is the highest, which is 0.9:

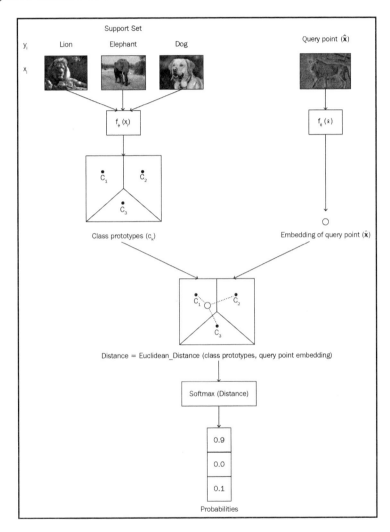

Prototypical networks are not only used for one-shot/few-shot learning, but are also used in zero-shot learning. Consider a case where we have no data points for each class but we have the meta-information containing a high-level description of each class.

In those cases, we learn the embeddings of meta- information of each class to form the class prototype and then perform classification with the class prototype.

Relation networks

Relation networks consist of two important functions: an embedding function, denoted by f_φ and the relation function, denoted by g_ϕ. The embedding function is used for extracting the features from the input. If our input is an image, then we can use a convolutional network as our embedding function, which will give us the feature vectors/embeddings of an image. If our input is text, then we can use LSTM networks to get the embeddings of the text. Let us say, we have a support set containing three classes, {lion, elephant, dog} as shown below:

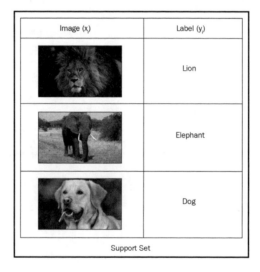

Image (x)	Label (y)
	Lion
	Elephant
	Dog

Support Set

And let's say we have a query image x_j, as shown in the following diagram, and we want to predict the class of this query image:

Query image (x$_j$)

First, we take each image, x_i, from the support set and pass it to the embedding function $f_\varphi(x_i)$ for extract the features. Since our support set has images, we can use a convolutional network as our embedding function for learning the embeddings. The embedding function will give us the feature vector of each of the data points in the support set. Similarly, we will learn the embeddings of our query image x_j by passing it to the embedding function $f_\varphi(x_j)$.

Once we have the feature vectors of the support set $f_\varphi(x_i)$ and query set $f_\varphi(x_j)$, we combine them using some operator Z. Here, Z can be any combination operator. We use concatenation as an operator to combining the feature vectors of the support and query set:

$$Z(f_\varphi(x_i), f_\varphi(x_j))$$

As shown in the following diagram, we will combine the feature vectors of the support set, $f_\varphi(x_i)$, and, query set, $f_\varphi(x_j)$. But what is the use of combining like this? Well, it will help us to understand how the feature vector of an image in the support set is related to the feature vector of a query image.

In our example, it will help us to understand how the feature vector of a lion is related to the feature vector of a query image, how the feature vector of an elephant is related to the feature vector of a query image, and how the feature vector of dog is related to the feature vector of a query image:

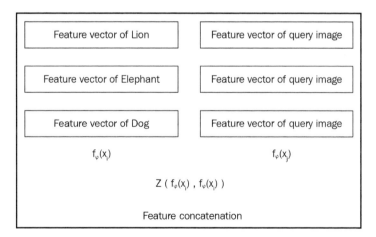

But how can we measure this relatedness? Well, that is why we use a relation function g_ϕ. We pass these combined feature vectors to the relation function, which will generate the relation score ranging from 0 to 1, representing the similarity between samples in the support set x_i and samples in the query set x_j.

The following equation shows how we compute relation score r_{ij} in the relation network:

$$r_{ij} = g_\phi(Z(f_\varphi(x_i), f_\varphi(x_j)))$$

Here, r_{ij} denotes the relation score representing the similarity between each of the classes in the support set and the query image. Since we have three classes in the support set and one image in the query set, we will have three scores indicating how all the three classes in the support set are similar to the query image.

The overall representation of the relation network in a one-shot learning setting is shown in the following diagram:

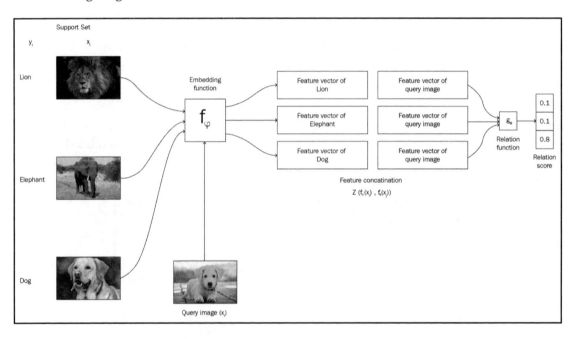

Matching networks

Matching networks are yet another simple and efficient one-shot learning algorithm published by Google's DeepMind. It can even produce labels for the unobserved class in the dataset. Let's say we have a support set, S, containing K examples as $(x_1, y_1), (x_2, y_2)...(x_k, y_k)$. When given a query point (new unseen example), $\hat{x}$, the matching network predicts the class of $\hat{x}$ by comparing it with the support set.

We can define this as $p(\hat{y}|\hat{x}, S)$, where p is the parameterized neural network, $\hat{y}$ is the predicted class for query point $\hat{x}$, and S is the support set. $p(\hat{y}|\hat{x}, S)$ will return the probability of $\hat{x}$ belonging to each class in the support set. Then we select the class of $\hat{x}$ as the one that has the highest probability. But how does this work, exactly? How is this probability computed? Let's see that now. The class, $\hat{y}$, of the query point, $\hat{x}$, can be predicted as follows:

$$\hat{y} = \sum_{i=1}^{k} a(\hat{x}, x_i) y_i$$

Let's decipher this equation. Here x_i and y_i are the input and labels of the support set. $\hat{x}$ is the query input, that is, the input to which we want to predict the label. Also a is the attention mechanism between $\hat{x}$ and x_i. But how do we perform attention? Here, we use a simple attention mechanism, which is softmax over the cosine distance between $\hat{x}$ and x_i:

$$a(\hat{x}, x_i) = softmax(cosine(\hat{x}, x_i))$$

We can't calculate cosine distance between the raw inputs $\hat{x}$ and x_i directly. So, first, we will learn their embeddings and calculate the cosine distance between the embeddings. We use two different embeddings, f and g, for learning the embeddings of $\hat{x}$ and x_i respectively. We will learn how exactly these two embedding functions f and g learn the embeddings in the upcoming section. So, we can rewrite our attention equation as follows:

$$a(\hat{x}, x_i) = softmax(cosine(f(\hat{x}), g(x_i))$$

We can rewrite the preceding equation as follows:

$$a(\hat{x}, x_i) = \frac{e^{cosine(f(\hat{x}), g(x_i))}}{\sum_{j=1}^{k} e^{cosine(f(\hat{x}), g(x_j))}}$$

After calculating the attention matrix, $a(\hat{x}, x_i)$, we multiply our attention matrix with support set labels y_i. But how can we multiply support set labels with our attention matrix? First, we convert our support set labels to the one hot encoded values and then multiply them with our attention matrix and, as a result, we get the probability of our query point $\hat{x}$ belonging to each of the classes in the support set. Then we apply *argmax* and select $\hat{y}$ as the one that has a maximum probability value.

Still not clear about matching networks? Look at the following diagram; you can see we have three classes in our support set (lion, elephant, and dog) and we have a new query image $\hat{x}$.

First, we feed the support set to embedding function g and the query image to the embedding function f and learn their embeddings and calculate the cosine distance between them, and then we apply softmax attention over this cosine distance. Then we multiply our attention matrix with the one-hot encoded support set labels and get the probabilities. Next, we select $\hat{y}$ as the one that has the highest probability. As you can see in the following diagram, the query set image is an elephant, and we have a high probability at the index 1, so we predict the class of $\hat{y}$ as 1 (elephant):

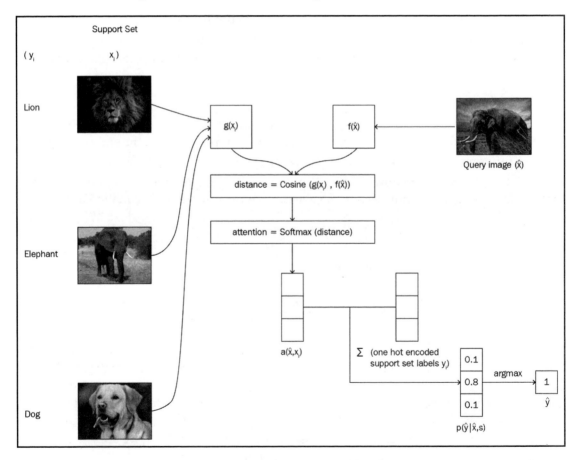

We have learned that we use two embedding functions, f and g, for learning the embeddings of $\hat{x}$ and x_i respectively. Now we will see exactly how these two functions learn the embeddings.

Support set embedding function

We use the embedding function g for learning the embeddings of the support set. We use bidirectional LSTM as our embedding function g. We can define our embedding function g as follows:

```
def g(self, x_i):

    forward_cell = rnn.BasicLSTMCell(32)
    backward_cell = rnn.BasicLSTMCell(32)
    outputs, state_forward, state_backward =
rnn.static_bidirectional_rnn(forward_cell, backward_cell, x_i,
dtype=tf.float32)

    return tf.add(tf.stack(x_i), tf.stack(outputs))
```

Query set embedding function

We use the embedding function f for learning the embedding of our query point $\hat{x}$. We use LSTM as our encoding function. Along with $\hat{x}$ as the input, we will also pass the embedding of our support set embeddings, which is $g(x)$, and we will pass one more parameter called K, which defines the number of processing steps. Let's see how we compute query set embeddings step-by-step. First, we will initialize our LSTM cell:

```
cell = rnn.BasicLSTMCell(64)
prev_state = cell.zero_state(self.batch_size, tf.float32)
```

Then, for the number of processing steps, we do the following:

```
for step in xrange(self.processing_steps):
```

We calculate embeddings of the query set, $\hat{x}$, by feeding it to the LSTM cell:

```
output, state = cell(XHat, prev_state)
h_k = tf.add(output, XHat)
```

Now, we perform softmax attention over the support set embeddings: that is, g_embedings. It helps us to avoid elements that are not required:

```
    content_based_attention = tf.nn.softmax(tf.multiply(prev_state[1],
g_embedding))
        r_k = tf.reduce_sum(tf.multiply(content_based_attention, g_embedding),
axis=0)
```

We update previous_state and repeat these steps for a number of processing steps, K:

```
    prev_state = rnn.LSTMStateTuple(state[0], tf.add(h_k, r_k))
```

The complete code for computing f_embeddings is as follows:

```
    def f(self, XHat, g_embedding):
        cell = rnn.BasicLSTMCell(64)
        prev_state = cell.zero_state(self.batch_size, tf.float32)

        for step in xrange(self.processing_steps):
            output, state = cell(XHat, prev_state)
            h_k = tf.add(output, XHat)

            content_based_attention =
tf.nn.softmax(tf.multiply(prev_state[1], g_embedding))
            r_k = tf.reduce_sum(tf.multiply(content_based_attention,
g_embedding), axis=0)

            prev_state = rnn.LSTMStateTuple(state[0], tf.add(h_k, r_k))

        return output
```

The architecture of matching networks

The overall flow of matching network is shown in the following diagram and it is different from the image we saw already. You can see how the support set x_i and query set $\hat{x}$ are calculated through the embedding functions g and f respectively.

As you can see, the embedding function f takes the query set along with the support set embeddings as input:

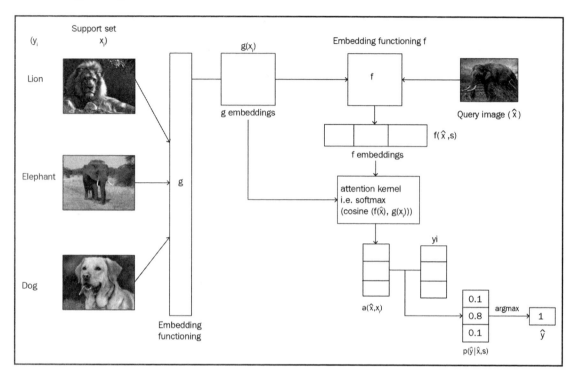

Congratulations again for learning all of the important and popular deep learning algorithms! Deep learning is an interesting and very popular field of AI that has revolutionized the world. Now that you've finished reading the book, you can start exploring various advancements in deep learning and start experimenting with various projects. Learn and deep learn!

Summary

We started off the chapter by understanding what k-shot learning is. We learned that in n-way k-shot learning, n-way implies the number of classes we have in our dataset and k-shot implies the number of data points we have in each class; and support set and the query set are equivalent to the train and test sets. Then we explored siamese networks. We learned how siamese networks use an identical network to learn the similarity of two inputs.

Followed by this, we learned about prototypical networks, which create a prototypical representation of each class and classify a query point (a new point) based on the distance between the class prototype and the query point. We also learned how relation networks use two different functions embedding and relation function to classify an image.

At the end of the chapter, we learned about matching networks and how it uses different embedding functions for the support set and the query set to classify an image.

Deep learning is one of the most interesting branches in the field of AI. Now that you've understood various deep learning algorithms, you can start building deep learning models and create interesting applications and also contribute to deep learning research.

Questions

Let's assess the knowledge acquired from this chapter by answering the following questions:

1. What is few-shot learning?
2. What are the support and query sets?
3. Define siamese networks.
4. Define energy functions.
5. What is the loss function of siamese networks?
6. How does the prototypical network work?
7. What are the different types of functions used in relation networks?

Further reading

To learn more about how to learn from a small number of data points, check out *Hands-On Meta Learning with Python* by Sudharsan Ravichandiran, published by Packt publishing available at, https://www.packtpub.com/big-data-and-business-intelligence/hands-meta-learning-python.

Assessments

The following are the answers to the questions mentioned at the end of each chapter.

Chapter 1 - Introduction to Deep Learning

1. The success of machine learning lies in the right set of features. Feature engineering plays a crucial role in machine learning. If we handcraft the right set of features to predict a certain outcome, then the machine learning algorithms can perform well, but finding and engineering the right set of features is not an easy task. With deep learning, we don't have to handcraft such features. Since deep **artificial neural networks (ANNs)** employ several layers, they learn the complex intrinsic features and multi-level abstract representation of the data by itself.

2. It is basically due to the structure of An ANN. ANNs consist of some n number of layers to perform any computation. We can build an ANN with several layers, where each layer is responsible for learning the intricate patterns in the data. Due to the computational advancements, we can build a network even with hundreds or thousands of layers deep. Since the ANN uses deep layers to perform learning, we call it deep learning, and when an ANN uses deep layers to learn, we call it a deep network.

3. The activation function is used to introduce non-linearity to the neural networks.

4. When we feed any negative input to the ReLU function, it converts them into zero. The snag for being zero for all negative values is a problem called **dying ReLU.**

5. The whole process of moving from the input layer to the output layer to predict output is known as **forward propagation**. During this propagation, the inputs are multiplied by their respective weights on each layer and an activation function is applied on top of them.

6. The whole process of backpropagating the network from the output layer to the input layer and updating the weights of the network using gradient descent to minimize the loss is called **backpropagation**.

7. Gradient checking is basically used for debugging the gradient descent algorithm and to validate that we have a correct implementation.

Chapter 2 - Getting to Know TensorFlow

1. Every computation in TensorFlow is represented by a computational graph. It consists of several nodes and edges, where nodes are the mathematical operations, such as addition, multiplication, and so on, and edges are the tensors. A computational graph is very efficient in optimizing resources and it also promotes distributed computing.

2. A computational graph with the operations on the node and tensors to its edges will only be created, and in order to execute the graph, we use a TensorFlow session.

3. A TensorFlow session can be created using `tf.Session()`, and it will allocate the memory for storing the current value of the variable.

4. Variables are the containers used to store values. Variables will be used as input to several other operations in the computational graph. We can think of placeholders as variables, where we only define the type and dimension, but will not assign the value. Values for the placeholders will be fed at runtime. We feed the data to the computational graphs using placeholders. Placeholders are defined with no values.

5. TensorBoard is TensorFlow's visualization tool that can be used to visualize the computational graph. It can also be used to plot various quantitative metrics and the results of several intermediate calculations. When we are training a really deep neural network, it would become confusing when we have to debug the model. As we can visualize the computational graph in TensorBoard, we can easily understand, debug and optimize such complex models. It also supports sharing.

6. Scoping is used to reduce complexity and helps us to better understand the model by grouping the related nodes together. Having a name scope helps us in grouping similar operations in a graph. It comes in handy when we are building a complex architecture. Scoping can be created using `tf.name_scope()`.

7. Eager execution in TensorFlow is more Pythonic and allows for rapid prototyping. Unlike the graph mode, where we need to construct a graph every time to perform any operations, eager execution follows the imperative programming paradigm, where any operations can be performed immediately without having to create a graph, just like we do in Python.

Chapter 3 - Gradient Descent and Its Variants

1. Unlike gradient descent, in SGD, in order to update the parameter, we don't have to iterate through all the data points in our training set. Instead, we just iterate through a single data point. That is, unlike gradient descent, we don't have to wait to update the parameter of the model after iterating all the data points in our training set. We just update the parameters of the model after iterating through every single data point in our training set.

2. In mini-batch gradient descent, instead of updating the parameters after iterating each training sample, we update the parameters after iterating some batches of data points. Let's say the batch size is 50, which means that we update the parameter of the model after iterating through 50 data points, instead of updating the parameter after iterating through each individual data point.

3. Performing mini-batch gradient descent with momentum helps us to reduce oscillations in the gradient steps and attain convergence faster.

4. The fundamental motivation behind the Nesterov momentum is that instead of calculating the gradient at the current position, we calculate gradients at the position where the momentum would take us to, and we call this position the lookahead position.

5. In Adagrad, we set the learning rate to a small value when the past gradients value is high and to a high value when the past gradient value is less. So, our learning rate value changes according to the past gradients updates of the parameter.

6. The update equation of Adadelta is given as follows:

$$\theta_t^i = \theta_{t-1}^i - \frac{RMS[\Delta\theta]_{t-1}}{RMS[g_t]} \cdot g_t^i$$

$$\theta_t^i = \theta_{t-1}^i + \nabla\theta_t$$

7. RMSProp is introduced to combat the decaying learning rate problem of Adagrad. So, in RMSProp, we compute the exponentially decaying running average of gradients as follows:

$$E[g^2]_t = \gamma E[g^2]_{t-1} + (1 - \gamma)g_t^2$$

Instead of a taking the sum of the square of all the past gradients, we use this running average of gradients. So, our update equation becomes the following:

$$\theta_t^i = \theta_{t-1}^i - \frac{\eta}{\sqrt{E[g^2]_t + \epsilon}} \cdot g_t^i$$

8. The update equation of Adam is given as follows:

$$\theta_t = \theta_{t-1} - \frac{\eta}{\sqrt{\hat{v}_t} + \epsilon} \hat{m}_t$$

Chapter 4 - Generating Song Lyrics Using an RNN

1. A normal feedforward neural network predicts output only based on the current input, but an RNN predicts output based on the current input and also the previous hidden state, which acts as a memory and stores the contextual information (input) that the network has seen so far.

2. The hidden state, h, at a time step, t, can be computed as follows:

$$h_t = \tanh(Ux_t + Wh_{t-1})$$

In other words, this is *hidden state at a time step, t = tanh([input to hidden layer weight x input] + [hidden to hidden layer weight x previous hidden state]).*

3. RNNs are widely applied for use cases that involve sequential data, such as time series, text, audio, speech, video, weather, and much more. They have been greatly used in various natural language processing (NLP) tasks, such as language translation, sentiment analysis, text generation, and so on.

4. While backpropagating the RNN, we multiply the weights and derivative of the *tanh* function at every time step. When we multiply smaller numbers at every step while moving backward, our gradient becomes infinitesimally small and leads to a number that the computer can't handle; this is called the **vanishing gradient problem.**

5. When we initialize the weights of the network to a very large number, the gradients will become very large at every step. While backpropagating, we multiply a large number together at every time step, and it leads to infinity. This is called the exploding gradient problem.

6. We use gradient clipping to bypass the exploding gradient problem. In this method, we normalize the gradients according to a vector norm (say, *L2*) and clip the gradient value to a certain range. For instance, if we set the threshold as 0.7, then we keep the gradients in the -0.7 to +0.7 range. If the gradient value exceeds -0.7, then we change it to -0.7, and similarly, if it exceeds 0.7, then we change it to +0.7.

7. Different types of RNN architectures include one-to-one, one-to-many, many-to-one, and many-to-many, and they are used for various applications.

Chapter 5 - Improvements to the RNN

1. **A Long Short-Term Memory (LSTM)** cell is a variant of an RNN that resolves the vanishing gradient problem by using a special structure called **gates**. Gates keep the information in the memory as long as it is required. They learn what information to keep and what information to discard from the memory.

2. LSTM consists of three types of gates, namely, the forget gate, the input gate, and the output gate. The forget gate is responsible for deciding what information should be removed from the cell state (memory). The input gate is responsible for deciding what information should be stored in the cell state. The output gate is responsible for deciding what information should be taken from the cell state to give as an output.

3. The cell state is also called internal memory where all the information will be stored.

4. While backpropagating the LSTM network, we need to update too many parameters on every iteration. This increases our training time. So, we introduce the **Gated Recurrent Units (GRU)** cell, which acts as a simplified version of the LSTM cell. Unlike LSTM, the GRU cell has only two gates and one hidden state.

5. In a bidirectional RNN, we have two different layers of hidden units. Both of these layers connect from the input to the output layer. In one layer, the hidden states are shared from left to right and in another layer, it is shared from right to left.

6. A Deep RNN computes the hidden state by taking the previous hidden state and also the previous layer's output as input.

7. The encoder learns the representation (embeddings) of the given input sentence. Once the encoder learns the embedding, it sends the embedding to the decoder. The decoder takes this embedding (a thought vector) as input and tries to construct a target sentence.

8. When the input sentence is long, the context vector does not capture the whole meaning of the sentence, since it is just the hidden state from the final time step. So, instead of taking the last hidden state as a context vector and using it for the decoder with an attention mechanism, we take the sum of all the hidden states from the encoder and use it as a context vector.

Chapter 6 - Demystifying Convolutional Networks

1. The different layers of CNN include convolution, pooling, and fully connected layers.

2. We slide over the input matrix with the filter matrix by one pixel and perform the convolution operation. But we can not only slide over the input matrix by one pixel-we can also slide over the input matrix by any number of pixels. The number of pixels we slide over the input matrix by the filter matrix is called **stride**.

3. With the convolution operation, we slide over the input matrix with a filter matrix. But in some cases, the filter does not perfectly fit the input matrix. That is, there exists a situation that when we move our filter matrix by two pixels, it reaches the border and the filter does not fit the input matrix, that is, some part of our filter matrix is outside the input matrix. In this case, we perform padding.

4. The pooling layer reduces spatial dimensions by keeping only the important features. The different types of pooling operation include max pooling, average pooling, and sum pooling.

5. VGGNet is one of the most popularly used CNN architectures. It was invented by the **Visual Geometry Group** (**VGG**) at the University of Oxford. The architecture of the VGG network consists of convolutional layers followed by a pooling layer. It uses 3 x 3 convolution and 2 x 2 pooling throughout the network.

6. With factorized convolution, we break down a convolutional layer with a larger filter size into a stack of convolutional layers, with a smaller filter size. So, in the inception block, a convolutional layer with a 5 x 5 filter can be broken down into two convolutional layers with 3 x 3 filters.

7. Like the CNN, the Capsule network checks the presence of certain features to classify the image, but apart from detecting the features, it will also check the spatial relationship between them- that is, it learns the hierarchy of the features.

8. In the Capsule networks, apart from calculating probabilities, we also need to preserve the direction of the vectors, so we use a different activation function, called the squash function. It is given as follows:

$$\vec{v}_j = \frac{\|\vec{s}_j\|^2}{1 + \|\vec{s}_j\|^2} \frac{\vec{s}_j}{\|\vec{s}_j\|}$$

Chapter 7 - Learning Text Representations

1. In the **continuous bag-of-words** (CBOW) model, we try to predict the target word given the context word, and in the skip-gram model, we try to predict the context word given the target word.

2. The loss function of the CBOW model is given as follows:

$$L = -u_{j^*} + log(\sum_{j'=1}^{V} \exp u'_j)$$

3. When we have millions of words in the vocabulary, we need to perform numerous weight updates until we predict the correct target word. It is time-consuming and also not an efficient method. So, instead of doing this, we mark the correct target word as a positive class and sample a few words from the vocabulary and mark it as a negative class, and this is called negative sampling

4. PV-DM is similar to a continuous bag of words model, where we try to predict the target word given a context word. In PV-DM, along with word vectors, we introduce one more vector, called the paragraph vector. As the name suggests, the paragraph vector learns the vector representation of the whole paragraph and it captures the subject of the paragraph.

5. The role of an encoder is to map the sentence to a vector and the role of the decoder is to generate the surrounding sentences; that is the previous and following sentences.

6. In QuickThoughts is an interesting algorithm for learning the sentence embeddings. In quick-thoughts, we try to learn whether a given sentence is related to the candidate sentence. So, instead of using a decoder, we use a classifier to learn whether a given input sentence is related to the candidate sentence.

Chapter 8 - Generating Images Using GANs

1. Discriminative models learn to find the decision boundary that separates the classes in an optimal way, while generative models learn about the characteristics of each class. That is, discriminative models predict the labels conditioned on the input, $p(y|x)$, whereas generative models learn the joint probability distribution, $p(x,y)$.

2. The generator learns the distribution of images in our dataset. It learns the distribution of handwritten digits in our training set. We feed random noise to the generator and it will convert the random noise into a new handwritten digit similar to the one in our training set.

3. The goal of the discriminator is to perform a classification task. Given an image, it classifies it as real or fake; that is, whether the image is from the training set or the one generated by the generator.

4. The loss function for the discriminator is given as follows:

$$L^D = -\mathbb{E}_{x \sim p_r(x)}[\log D(x)] - \mathbb{E}_{z \sim p_z(z)}[\log(1 - D(G(z)))]$$

The generator loss function is given as follows:

$$L^G = -\mathbb{E}_{z \sim p_z(z)}[\log(D(G(z)))]$$

5. DCGAN extends the design of GANs with convolutional networks. That is, we replace with feedforward network in the generator and discriminator with the **Convolutional Neural Network (CNN)**.

6. The **Kullback-Leibler** (**KL**) divergence is one of the most popularly used measures for determining how one probability distribution diverges from the other. Let's say we have two discrete probability distributions, P and Q, then the KL divergence can be expressed as follows:

$$D_{KL}(P\|Q) - \sum_x P(x) \log\left(\frac{P(x)}{Q(x)}\right)$$

7. The Wasserstein distance, also known as the **Earth Movers** (**EM**) distance, is one of the most popularly used distance measures in the optimal transport problems where we need to move things from one configuration to another.

8. A Lipschitz continuous function is a function that must be continuous and almost differentiable everywhere. So, for any function to be a Lipschitz continuous, the absolute value of a slope of the function's graph cannot be more than a constant, K. This constant, K, is called the **Lipschitz constant**.

Chapter 9 - Learning More about GANs

1. Unlike vanilla GANs, **CGAN**, is a condition to both the generator and the discriminator. This condition tells the GAN what image we are expecting our generator to generate. So, both of our components—the discriminator and the generator—act upon this condition.

2. The code, c, is basically interpretable disentangled information. Assuming we have some MNIST data, then, code, $c1$, implies the digit label, code, $c2$, implies the width, $c3$, implies the stroke of the digit, and so on. We collectively represent them by the term c.

3. Mutual information between two random variables tells us the amount of information we can obtain from one random variable through another. Mutual information between two random variables x and y can be given as follows:

$$I(x,y) = H(y) - H(y|x)$$

It is basically the difference between the entropy of y and the conditional entropy of y given x.

4. The code, c, gives us the interpretable disentangled information about the image. So, we try to find c given the image? However, we can't do this easily since we don't know the posterior $p(c|G(z,c))$, so, we use an auxiliary distribution $Q(c|x)$ to learn c.

5. The adversarial loss alone does not ensure the proper mapping of the images. For instance, a generator can map the images from the source domain to a random permutation of images in the target domain that matches the target distribution. So, to avoid this, we introduce an additional loss called **cycle consistent loss**. It enforces both of the generators G and F to be cycle consistent.

6. We have two generators: G and F. The role of G is to learn the mapping from to y and the role of the generator, F, is to learn the mapping from y to x.

7. Stack GANs convert text descriptions into pictures in two stages. In the first stage, artists draw primitive shapes and create a basic outline that forms an initial version of the image. In the next stage, they enhance the image by making it more realistic and appealing.

Chapter 10 - Reconstructing Inputs Using Autoencoders

1. Autoencoders are unsupervised learning algorithms. Unlike other algorithms, autoencoders learn to reconstruct the input, that is, an autoencoder takes the input and learns to reproduce the input as an output.

2. We can define our loss function as a difference between the actual input and reconstructed input as follows:

$$L(\theta, \phi) = \frac{1}{n} \sum_{i=1}^{n} (\mathbf{x}^{(i)} - g_\phi(f_\theta(\mathbf{x}^{(i)})))^2$$

Here, n is the number of training samples.

3. **Convolutional Autoencoder (CAE)** that uses a convolutional network instead of a vanilla neural network. In the vanilla autoencoders, encoders and decoders are basically a feedforward network. But in CAEs, they are basically convolutional networks. This means the encoder consists of convolutional layers and the decoder consists of transposed convolutional layers, instead of a raw feedforward network.

4. **Denoising Autoencoders (DAE)** are another small variant of the autoencoder. They are mainly used to remove noise from the image, audio, and other inputs. So, we feed the corrupted input to the autoencoder and it learns to reconstruct the original uncorrupted input.

5. The average activation of the jth neuron in the hidden layer, h, over the whole training set can be calculated as follows:

$$\hat{\rho}_j = \frac{1}{n} \sum_{i=1}^{n} [a_j^{(h)}(x^{(i)})]$$

6. The loss function of contractive autoencoders can be mathematically represented as follows:

$$L = \|X - \hat{X}\|_2^2 + \lambda \|J_f(X)\|_F^2$$

The first term represents the reconstruction error and the second term represents the penalty term or the regularizer, and it is basically the **Frobenius norm** of the **Jacobian matrix**.

7. The Frobenius norm, also called the **Hilbert-Schmidt norm**, of a matrix is defined as the square root of the sum of the absolute square of its elements. A matrix comprising a partial derivative of the vector-valued function is called the **Jacobian matrix**.

Chapter 11 - Exploring Few-Shot Learning Algorithms

1. Learning from a few data points is called **few-shot learning** or **k-shot learning**, where k specifies the number of data points in each of the classes in the dataset.
2. We need our models to learn from just a few data points. In order to attain this, we train them in the same way; that is, we train the model on very few data points. Say we have a dataset, D: we sample a few data points from each of the classes present in our dataset and we call it **support set**. Similarly, we sample some different data points from each of the classes and call it **query set**.
3. Siamese networks basically consist of two symmetrical neural networks both sharing the same weights and architecture and both joined together at the end using some energy function, E. The objective of our Siamese network is to learn whether the two inputs are similar or dissimilar.

4. The energy function, E, which will give us a similarity between the two inputs. It can be expressed as follows:

$$E_W(X_1, X_2) = \|f_W(X_1) - f_W(X_2)\|$$

5. Since the goal of the Siamese network is not to perform a classification task but to understand the similarity between the two input values, we use the contrastive loss function. It can be expressed as follows:

$$Contrastive\ loss = Y(E)^2 + (1 - Y)\max(\text{margin} - E, 0)^2$$

6. Prototypical networks are yet another simple, efficient, and popularly used few-shot learning algorithm. The basic idea of the prototypical network is to create a prototypical representation of each class and classify a query point (new point) based on the distance between the class prototype and the query point.

7. Relation networks consist of two important functions: an embedding function denoted by f_φ and the relation function denoted by g_ϕ.

Other Books You May Enjoy

If you enjoyed this book, you may be interested in these other books by Packt:

Deep Learning with PyTorch
Vishnu Subramanian

ISBN: 9781788624336

- Use PyTorch for GPU-accelerated tensor computations
- Build custom datasets and data loaders for images and test the models using torchvision and torchtext
- Build an image classifier by implementing CNN architectures using PyTorch
- Build systems that do text classification and language modeling using RNN, LSTM, and GRU
- Learn advanced CNN architectures such as ResNet, Inception, Densenet, and learn how to use them for transfer learning
- Learn how to mix multiple models for a powerful ensemble model
- Generate new images using GAN's and generate artistic images using style transfer

Python Deep Learning
Gianmario Spacagna, Daniel Slater, Et al
ISBN: 9781786464453

- Get a practical deep dive into deep learning algorithms
- Explore deep learning further with Theano, Caffe, Keras, and TensorFlow
- Learn about two of the most powerful techniques at the core of many practical deep learning implementations: Auto-Encoders and Restricted Boltzmann Machines
- Dive into Deep Belief Nets and Deep Neural Networks
- Discover more deep learning algorithms with Dropout and Convolutional Neural Networks
- Get to know device strategies so you can use deep learning algorithms and libraries in the real world

Leave a review - let other readers know what you think

Please share your thoughts on this book with others by leaving a review on the site that you bought it from. If you purchased the book from Amazon, please leave us an honest review on this book's Amazon page. This is vital so that other potential readers can see and use your unbiased opinion to make purchasing decisions, we can understand what our customers think about our products, and our authors can see your feedback on the title that they have worked with Packt to create. It will only take a few minutes of your time, but is valuable to other potential customers, our authors, and Packt. Thank you!

Index

T

target word 283
tensor 44
TensorBoard
  about 51, 52, 53
  name scope, creating 53, 54, 55
  word embeddings, visualizing 312, 313, 314
TensorFlow 2.0
  about 73
  used, for classifying MNIST digit 76, 77
TensorFlow implementation, song lyrics with RNNs
  BPTT, defining 155
  data, preparing 150, 151, 152, 153
  forward propagation, defining 154
  network parameters, defining 153
  placeholders, defining 154
  songs, generating 156, 157, 158, 159, 160
TensorFlow session 46, 47
TensorFlow, used for classifying handwritten digit
  about 55
  accuracy, calculating 60
  backpropagate 59
  dataset, loading 55, 56
  forward propagation 58
  graph, visualizing in TensorBoard 63, 64, 65, 66, 67
  model, training 61, 62
  neurons, defining in each layer 57
  placeholders, defining 57, 58
  required libraries, importing 55
  softmax cross-entropy, using 59
  summary, creating 61
TensorFlow
  about 43, 44
  images, generating GANs used 340
  math operations, exploring in 69, 70, 71, 72
  using 77, 78, 373
  Wasserstein GAN (WGAN) 369
term frequency-inverse document frequency (tf-idf) 281
thought vector 212
total loss 338
training strategies, word2vec
  hierarchical softmax 303, 304, 305
  negative sampling 305, 306

subsampling frequent words 306
transfer function 13, 15
transport plan 366
two-way k-shot learning 450

U

undercomplete autoencoder 417
units 14

V

VAE, using for generating images
  about 442
  dataset, preparing 443
  decoder, defining 444
  encoder, defining 443
  generated images, plotting 445
  generator, defining 445
  model, building 444
  sampling operation, defining 444
valid padding 231
vanilla 418
vanishing gradient problem 143, 144, 145, 146
vanishing gradients 168
variables 47, 48, 49
variational autoencoders (VAE)
  about 413
  dissecting 436, 437, 438
  loss function 439, 440, 441
  variational inference 438, 439
variational inference 438
variational lower bound 440
VGGNet
  architecture of 252, 254
visual geometry group (VGG) 252, 253
vocabulary 286

W

Wasserstein distance
  about 366, 367
  Generative Adversarial Networks (GANs), used 363
Wasserstein GAN (WGAN)
  about 328, 363
  in TensorFlow 369
  loss function 368, 369